The Pamphlets of
Lewis Carroll

Rev. Charles Lutwidge Dodgson

The Mathematical Pamphlets of Charles Lutwidge Dodgson and Related Pieces

COMPILED, WITH INTRODUCTORY ESSAYS, NOTES, AND ANNOTATIONS, BY

Francine F. Abeles

Published by the Lewis Carroll Society of North America, New York, and distributed by the University Press of Virginia, Charlottesville and London

1994

This volume and series is funded in part by the
Morton N. Cohen Publishing Trust for the Benefit of
the Lewis Carroll Society of North America

Copyright © 1994 by The Lewis Carroll Society of North America
First published 1994

Printed in the United States of America

LIBRARY OF CONGRESS CATALOGING-IN-PUBLICATION DATA
Carroll, Lewis, 1832–1898.
    The mathematical pamphlets of Charles Lutwidge Dodgson
and related pieces / compiled, with introductory essays, notes, and
annotations, by Francine F. Abeles.
        p.   cm. — (The pamphlets of Lewis Carroll ; v. 2)
    Includes bibliographical references and index.
    ISBN 0-930326-09-1
    1. Mathematics.   I. Abeles, Francine F.   II. Title.
III. Series: Carroll, Lewis, 1832–1898. Pamphlets of Lewis Carroll ; v2.
QA3.C295   1994
510—dc20   94-25395

The Pamphlets of Lewis Carroll

SERIES EDITORS
STAN MARX & EDWARD GUILIANO

VOLUME 2

The Mathematical Pamphlets of
Charles Lutwidge Dodgson
and Related Pieces

To Ann and Ernest, steadfast friends

Contents

List of Illustrations	ix
Series Preface	xi
Editor's Preface	xiii
Acknowledgments	xv
Introduction to the Mathematical Pamphlets	1

Geometry — 27

1. Notes on the First Two Books of Euclid (1860) — 35
2. The Enunciations of Euclid, Books I and II (1863) — 42
3. The Fifth Book of Euclid (1868) — 52
4. The Enunciations of Euclid I–VI (1873) — 88

Trigonometry — 117

5. The Formulæ of Plane Trigonometry (1861) — 121
6. Formulæ (Group C) (undated) — 140
7. Simple Facts about Circle-Squaring (1882) — 144
8. Proof Sheets: Propositions I, II (undated) — 148
9. Question 11530 (1893) — 150

Algebra — 157

10. Condensation of Determinants (1866) — 170
11. Algebraical Formulæ for the Use of Candidates for Responsions (1868) — 181
12. Formulæ in Algebra (1868?) — 184
13. Algebraical Formulæ and Rules for the Use of Candidates for Responsions (1870) — 190
14. Algebra (1877) — 195
15. Formulæ (1878) — 197
16. Question 9995 (1889) — 198

Probability — 201

17. Note on Question 7695 [including Question 7695 and its solution (1885)] — 209

18. Response to "Infinitesimal or Zero?" (1886)	213
19. Something or Nothing? (1888)	219
20. Question 9588 (1889)	221

Arithmetic Computation and Theory   227

21. Arithmetical Formulæ and Rules for the Use of Candidates for Responsions (1870)	235
22. Examples in Arithmetic (1874)	240
23. Arithmetic I (1870–74?)	266
24. Arithmetic II (1870–74?)	269
25. Arithmetic (undated)	271
26. Practical Hints on Teaching. Long Multiplication Worked with a Single Line of Figures (1879)	273
27. Divisibility by Seven (1884)	277
28. To Find the Day of the Week for Any Given Date (1887)	280
29. Question 9636 (1888)	283
30. Question 12650 (1895)	285
31. Number-Guessing (1896)	288
32. Question 13614 (1897)	291
33. Brief Method of Dividing a Given Number by 9 or 11 (1897)	293
34. Variant of Item 33 (same title, date)	298
35. Rule for Finding Easter-Day for Any Date till A.D. 2499 (1892–97?)	302
36. Abridged Long Division (1898)	312

Cryptology   325

37. The Key-Vowel Cipher (1858)	336
38. The Matrix Cipher (1858)	339
39. The Alphabet Cipher (1868)	341
40. The Telegraph Cipher (1868)	344

Curriculum   349

41. Circular to Mathematical Friends (1862)	350
42. Proof Sheets Accompanying the Circular (undated)	352
43. A Guide to the Mathematical Student (1864)	372
Selected Bibliography	405
Index	413

Illustrations

Rev. Charles Lutwidge Dodgson. *Courtesy of the Houghton Collection, the Pierpont Morgan Library.* frontispiece

Postcard from Henry J. S. Smith to Dodgson. *Courtesy of the Morris L. Parrish Collection, Princeton University Library.* xviii

Bartholomew Price, F.R.S. *Courtesy of the General Research Division, the New York Public Library, Astor, Lenox and Tilden Foundations.* 4

Henry J. S. Smith, F.R.S. *From Alexander Macfarlane,* Lectures on Ten British Mathematicians of the Nineteenth Century *(New York: John Wiley, 1916).* 5

Isaac Todhunter, F.R.S. *From Macfarlane,* Lectures on Ten British Mathematicians. 15

Letter from Archdeacon Dodgson to his son, Charles Lutwidge Dodgson. *Courtesy of the Morris L. Parrish Collection, Princeton University Library.* 20–21

Augustus De Morgan. *From Macfarlane,* Lectures on Ten British Mathematicians. 23

Frontispiece from Dodgson's edition of Euclid, Books I, II. *Courtesy of the Harry Ransom Humanities Research Center, the University of Texas at Austin.* 26

Frontispiece from Curiosa Mathematica, Part I: A New Theory of Parallels *(London: Macmillan, 1888).* 33

Dodgson's application of the Theorem of Tangents. *Courtesy of the Morris L. Parrish Collection, Princeton University Library.* 116

Proof-sheets: Propositions I, II (item 8). *Courtesy of the Harry Ransom Humanities Research Center, the University of Texas at Austin.* 149

William Spottiswoode, F.R.S. *Courtesy of the General Research Division, the New York Public Library, Astor, Lenox and Tilden Foundations.* — 156

Letter from William Spottiswoode to Dodgson. *Courtesy of the Morris L. Parrish Collection, Princeton University Library.* — 160–61

Formulae (item 15). *Courtesy of David and Denise Carlson, D & D Galleries.* — 197

Letter from Dodgson to Mr. Potts. *Courtesy of the Morris L. Parrish Collection, Princeton University Library.* — 200

A test of divisibility by seven by Charles L. Dodgson's father. *Courtesy of the Weaver Collection, Harry Ransom Humanities Research Center, the University of Texas at Austin.* — 226

Archdeacon Charles Dodgson. *Courtesy of the Gernsheim Collection, Harry Ransom Humanities Research Center, the University of Texas at Austin.* — 228

Number-Guessing (item 31). *Courtesy of the Berg Collection, New York Public Library.* — 289

Telegraph Cipher. *Courtesy of the Houghton Collection, the Pierpont Morgan Library.* — 324

Alphabet Cipher (item 39). *Courtesy of the Harry Ransom Humanities Research Center, the University of Texas at Austin.* — 342

Dodgson's worksheet of algebra grades. *Courtesy of the Weaver Collection, Harry Ransom Humanities Research Center, the University of Texas at Austin.* — 348

Series Preface

Between 1860 and 1897, Charles Lutwidge Dodgson wrote and had printed some 185 pamphlets, booklets, leaflets, form letters, and instruction manuals. They varied in format from a single leaf to booklets of more than 40 pages. Some he signed with his given name, others as Lewis Carroll, and others with some other suitable pseudonym. Some were not signed at all. Taken together, they constitute a unique body of work in the history of English literature.

Their uniqueness is manifested in the broad range of Dodgson's interests, his readiness to challenge any authority, and his unlimited creativity, which flourished up to the time of his death. A churchman who loved the theater (frowned on at the time); a mathematician who could make his subject an instrument of fun; a logician; a photographer; a creator of games, puzzles, and brain twisters—Dodgson had a mind that was constantly at work. In the pamphlets he emerges as a Victorian activist, a University man, and a moral crusader. These works additionally reveal layers of character and personality in this eminent Victorian that are not evident in his other publications.

In his pamphlets and related material even more than in his better-known works—the *Alice* books, *The Hunting of the Snark,* volumes of poetry, books on mathematics and logic, and his prolific letters—one most easily comprehends the seemingly unending limits of his mind.

We have only to look at his very first separate publication, *Rules for Court Circular,* issued in 1860, to catch a glimpse of what would follow. Invented in 1858, according to his diary, the game of Court Circular has some resemblance to Rummy. The pamphlet, a forerunner of many others of similar format, was printed on four pages, in size slightly smaller than 5″ by 8″. Also in 1860, he printed up his *Notes on the First Two Books of Euclid,* as well as a list of photographic portraits he had taken. This pattern of varied interests continued throughout his life. Most years would see three or four items on a wide range of subjects, until the last fifteen years of his life, when even more per year would be issued. Beginning in the mid-1880s his interests fastened on logic, and it was during this later period that all 26 of his pamphlets on that subject were written and printed. His growing interest in a field often combining mathematics and philosophy coincided with his revised edition of *Euclid and His Modern Rivals* and his release of *A Tangled Tale, Curiosa Mathematica,* and the *Sylvie and*

Bruno books, the most seriously philosophical of his stories. Months before he died, he notes on November 12, 1897, in his diary, his intention of bringing out "my Games and Puzzles and Part III of *Curiosa Mathematica* . . . in paper cover," and he was indeed working on these projects in his final days.

Over the years insufficient attention has been paid to Carroll's pamphlets and ephemeral pieces, in contrast to the wealth of interest shown in the *Alice* books. A number of the pamphlets were reprinted in various editions of his collected works, but many more were little known or lost. Dodgson's nephew, Stuart Collingwood, even mentions several in *Life and Letters of Lewis Carroll* that have not been seen in this century.

The Pamphlets of Lewis Carroll is an attempt to provide a scholarly annotated collection of the pamphlets, brought together for the first time. Each book in the series—six are projected—will deal with a particular area: Oxford; Mathematics; Games and Puzzles; Parliamentary Procedure and Practices; Logic; *Alice* and Miscellaneous. A full listing of Carroll's pamphlets will appear in the final volume of this series.

For their enthusiastic assistance in the early days of organizing this series, the Lewis Carroll Society of North America would particularly like to thank Dr. Alexander Wainwright of the Princeton University Library for providing copies of the rare pamphlets in the Parrish Collection; Dr. Morton Cohen and Dr. Selwyn Goodacre for their many suggestions and guidance; Dr. Frank Walker of the Bobst Library at New York University; Mary Lou Ashby of the Pierpont Morgan Library; and Dr. Sandor Burstein and Joseph Brabant for their assistance in providing materials and information. The Society appreciates the patience and enterprise of the many people who are contributing to this project.

Finally, the series general editors must be thanked for their prodigious efforts over many years. From the days when Stan Marx conceived of the project and began securing copies of each pamphlet, through the editing and preparation of the manuscripts for each volume, the general editors have brought dedication and cooperation to this monumental task. General editorship for the series through Volume II was a collaboration between Stan Marx and Edward Guiliano; Stan Marx has assumed the responsibility for Volume III and subsequent volumes.

EDWARD GUILIANO
STAN MARX

Editor's Preface

It was Charles Dodgson's habit to provide his readers with a short preface explaining the purpose of each of his books, and so I have written one as well. The reader of this volume will find Dodgson's pamphlets and other relevant items organized by mathematical subject. Since an evaluation of Dodgson as a mathematician is not possible from these pamphlets alone, each subject begins with a general introduction to Dodgson's writings on the topic, followed by bibliographic information about each of the items included in the section. Each headnote gives the date of the piece, followed by one or more bibliographic references citing its item number, and the source of the item in this volume. The contents of each section are organized chronologically. This approach permits a reassessment of Dodgson's contributions to mathematics and provides complete references to all of the material connected with his pamphlets. The abbreviations listed here refer to the collections housing these items and the sources where they have been mentioned.

ANC	*Alice in North Carroll-ina: Schedule of Events and Exhibition Notes for the Spring Meeting of the Lewis Carroll Society of North America* (May 5–7, 1989).
Berg	The Henry W. and Albert A. Berg Collection of English and American Literature, the New York Public Library.
Bodleian	Bodleian Library, Oxford University.
Brabant	Joseph A. Brabant, private collection.
Carlson	David and Denise Carlson of D & D Galleries, Somerville, New Jersey.
Cohen	*The Letters of Lewis Carroll,* edited by Morton N. Cohen, 2 vols., (1979).
Colindale	Newspaper Library of the British Museum, Colindale Avenue.
Collingwood	*The Life and Letters of Lewis Carroll,* Stuart D. Collingwood, (1898)
Diaries	Entries from C. L. Dodgson's diaries appearing in *The Diaries of Lewis Carroll,* edited by Roger L. Green, 2 vols (1954).

Diaries	Unpublished entries from Dodgson's diaries, housed in the British Library.
Goodacre	Selwyn H. Goodacre, private collection.
HRHRC	Harry Ransom Humanities Research Center, University of Texas at Austin.
LCAT	*Lewis Carroll at Texas: The Warren Weaver Collection and Related Dodgson Materials at the Harry Ransom Humanities Research Center,* compiled by Robert N. Taylor (1985).
LCH	*The Lewis Carroll Handbook,* S. H. Williams, F. Madan, and R. L. Green: revised edition by D. Crutch (1979).
Marx	Stan Marx, private collection.
Morgan	The Arthur A. Houghton, Jr., Collection in the Pierpont Morgan Library.
MQS	*Mathematical Questions and Solutions from the "Educational Times."*
NY Public	General Research Division, Astor, Lenox and Tilden Foundations, the New York Public Library.
NYU	The Alfred C. Berol Collection and the Fales Collection, New York University Library.
Princeton	The Morris L. Parrish Collection of Victorian Novelists, Princeton University Library.
Stern	Jeffery Stern, private collection.

FRANCINE F. ABELES
NEW YORK, N.Y.

Acknowledgments

Assembling the unpublished mathematical items for this volume required the assistance of numerous librarians, collectors, and admirers of Charles Dodgson. I owe a considerable debt to Alexander Wainright, Alison McCuaig, and the staff of the Parrish Collection, Princeton University; John Kirkpatrick, Cathy Henderson, and Seyed Moosavi of the Weaver Collection, Harry Ransom Humanities Research Center, University of Texas; and Mary Palmiter and Rena Rogge of the Nancy Thompson Library, Kean College. I also wish to thank Lynne Fortunato of the Research Libraries and Rodney Phillips and Francis O. Mattson of the Berg Collection at the New York Public Library; Frank Walker of the Berol Collection and Fales Collection at New York University Library; Emily Walhout of the Amory Collection, the Houghton Library at Harvard University; Bernard Crystal of the Rare Book and Manuscript Library at Columbia University; Anna Lou Ashby and Inge Dupont of the Houghton Collection of the Pierpont Morgan Library; Janet West of the Port Washington Public Library; John Wing of the Christ Church Library at Oxford; Richard Guly, Deborah O'Donoghue, and the unnamed research librarians in the British Library and Newspaper Library, London, and Bodleian Library, Oxford, who provided assistance and photocopies of rare material.

Stan Marx, Charles C. Lovett, Joseph A. Brabant, Selwyn H. Goodacre, David and Denise Carlson, and Jeffrey Stern, all intrepid collectors, supplied several rare items. I am indebted to Stan Marx for locating the majority of the items included in this volume, obtaining many of them, and providing supporting material whenever I asked. In London, Richard Swift graciously responded to my pleas for items from the *Educational Times*. I am also indebted to the libraries whose policies allowed me to reproduce texts and photographs they own. I wish to thank their curators, librarians, and technical staff for their conscientious assistance in assembling these materials, particularly Anna Lou Ashby of the Pierpont Morgan Library and Cathy Henderson of the Harry Ransom Humanities Research Center. To my son Edward, who spent many hours in libraries searching for photographs, I am grateful.

Dodgson's wide-ranging mathematical interests prompted me to seek the aid of several mathematicians with areas of expertise closely corresponding to Dodgson's contributions. Eugene Seneta edited the probability section and critically read the one on algebra; Stanley Lipson co-edited the cryptology section. Roger Herz-Fischler generously pored over the geometry and trigonometry sections, offering many suggestions for their improvement, as did Thomas Drucker for the arithmetic computation and theory chapter. Additionally, Martin Gardner, Cy Deavours, William Kruskal, Alfred Hales, and David Robbins provided insights and appropriate references at critical points.

Without financial support for travel and time away from teaching duties, I could not have undertaken this project. That support was provided by the state of New Jersey, through its Research Awards for Travel and Released Time, and my colleagues at Kean College, who enabled me to qualify for these grants. I am particularly indebted to Eileen Kennedy, Mark Lender, and T. Felder Dorn for their trust and encouragement in all the time that I worked on this project.

Lastly, I wish to express my gratitude to some extraordinary people. Morton Cohen has contributed enormously to this work at every stage of its development. Allowing me to use items in his own library, supplying needed references, and reading through the introduction to this volume, he has willingly taken time from his own work to assist mine. My daughters Evelyn and Jennifer applied their typing skills cheerfully to the task of ensuring the bibliography's accuracy, and Evelyn also diligently read page proofs. My husband Ernest provided much-needed emotional and material support and graciously endured countless interruptions to normal family living. And Cornelia B. Wright, my editor, has helped beyond all measure to create a book that is pleasurable to read.

F.F.A.

The Mathematical Pamphlets of
Charles L. Dodgson
and Related Pieces

Postcard from Henry J. S. Smith to Charles L. Dodgson about a series summation problem, dated 11 March 1874.

Introduction to the
Mathematical Pamphlets

So long as you do not assert the *existence* of such Lines, you are merely defining an imaginary concept; & your conclusions have no *practical* value, until you do assert their real existence.

"What right, then, have you," your reader may well ask, "to assume (as you evidently do by your use of the phrase 'the amount of twisting') that these two twistings have the *same* amount?"[1]

By writing about ordinary mathematical topics in uncommon ways, Charles L. Dodgson presented unusual methods for recognizing truth in mathematics. He was acutely aware of subtle differences in the meanings of words, developing standards to enable readers to distinguish conventional interpretations where logical fallacies can lurk, from literal interpretations that may convey no intelligible meanings. For Dodgson, mathematical objects had to be precisely defined in a practical sense, possessing qualities whose presence or absence could be proved in particular cases by applying logical rules. He required that these objects exist *in rerum natura*, and also applied this standard to axioms.

The mathematical work he produced exhibits an extraordinary consistency, the mathematical and logical sides of him bound almost as one. Although humorous statements make their appearance in unlikely places and a certain lightness pervades the prose, as if to suggest he is not being completely serious, much of his writing has a pedantic flavor.

Dodgson's overriding concern for precise statements often imparted a labored quality to his writings. This ironically had the unintended result of interfering with their intended meaning, so that many of his proposed methods were considered impractical and those contributions that might have influenced the development of certain topics in mathematics were overlooked by the scientific community of his day.

---

1. From an unpublished Dodgson manuscript, "Pairs of Lines treated on Direction-Theory," in the Parrish Collection, Princeton University Library. The first quote is from a section of the manuscript dated 18 February 1890, p. 4; the second from a section dated 18 March 1891, p. 14.

By the middle of the nineteenth century, science and its applications had begun to redefine the values on which traditional British society was based. Darwin's *On the Origin of Species* (1859) was responsible for moving scientific theories and definitions to the center of aesthetic, ethical, and social discussions. Victorian periodicals provided a forum for debating scientific developments and their consequences, particularly concerning traditions and religion. Periodicals such as the *Athenaeum* and newspapers like the *Pall Mall Gazette* included reports of professional meetings of scientists, popular scientific lectures, debates between men of science on technical topics, reviews of science textbooks, and even scientific gossip. Underlying these activities was the assumption that culture was a single entity—one in which scientific and humanistic values formed an organic whole.

In 1869, James Joseph Sylvester, a distinguished Cambridge mathematician teaching at the Royal Military Academy in Woolwich, addressed the British Association for the Advancement of Science at Exeter with a paper discussing the background and methods of modern mathematics in the context of the role of faith and authority in science and in religion. In a long footnote to his essay, Sylvester wrote,

> If an Aristotle, or Descartes, or Kant assures me that he recognizes God in the conscience, I accuse my own blindness if I fail to see with him. If Gauss, Cayley, Riemann, Schläfli, Salmon, Clifford, Kronecker, have an inner assurance of the reality of transcendental space [four-dimensional geometry], I strive to bring my faculties of mental vision into accordance with theirs.... I acknowledge two separate sources of authority—the collective sense of mankind, and the illumination of privileged intellects.[2]

In his essay, Sylvester made it abundantly clear that the publication of Darwin's *On the Origin of Species* did not end the need to articulate a reasoned and consistent understanding of the relationship between man and God. To the Oxford mathematician Charles Dodgson, recognizing the presence of God in the collective sense of mankind was a simple matter; believing in the reality of four-dimensional space, even when espoused by the most eminent mathematicians of his time, quite another.

Dodgson was not a member of any of the newly founded scientific societies of his time like the London Mathematical Society or the British Association. He did correspond with other mathematicians, seeking their comments about his work or opinions on mathematical topics in which he was interested. He was not a member of the Royal Society, a distinction given to just fifteen men annually, although his paper on determinants was

---

2. J. J. Sylvester, *Nature*, vol. 1 (1869), 238.

presented to that group in 1866. Aside from Henry J. S. Smith, Savilian Professor of Geometry at Oxford, whose work as a mathematician was matched by his reputation as an excellent speaker and popular political and social figure, his mathematical colleagues at Oxford were Edward Frank Sampson, a close friend and fellow–mathematical lecturer at Christ Church; Bartholomew Price, Sedleian Professor of Natural Philosophy, a specialist in the calculus and his coach when Dodgson was an undergraduate, whose friendship he valued; John Cook Wilson, Professor of Logic, with whom Dodgson carried on lively disputations over points in geometry and logic; Thomas Fowler, Wykeham Professor of Logic, a classmate and friend; and Robert Edward Baynes, student and tutor of Christ Church and Lee's Reader in Physics. In the Parrish collection are six letters to Dodgson from William Spottiswoode, an Oxford mathematician. In one of these, Spottiswoode remarked that he had written to Sylvester to enlist his aid in answering one of Dodgson's questions about tetrahedra. Sylvester responded to the letter of 6 November 1867 three days later. Dodgson got to meet one of the most brilliant English mathematicians, Arthur Cayley, on the occasion of Cayley's receiving an honorary degree (Doctor of Civil Law) from Oxford. In his diary entry for 7 June 1864 Dodgson does not mention whether they discussed matrices, a topic of common interest.

Dodgson pursued his mathematical interests in a solitary fashion, preferring to work things out by himself rather than using the material of others. He rarely acknowledged specific sources in his written work, probably because he treated them in such new and different ways that they were no longer recognizable. But he was always careful never to claim the work of another as his own and was particularly critical of anyone else who did. Dodgson's mathematical publications fall into two broad categories: those that concern the teaching of students and those that reflect his personal interests. Both exhibit a practical quality, an emphasis on calculations and on methods by which problems can be solved. His intentions are more algorithmic than theoretical, and ideas having once appeared have the habit of returning in different guise.

Whether his particular mathematical style was primarily idiosyncratic or perhaps was seriously influenced by the buffeting that clerical scientists experienced in late Victorian England is a matter of conjecture. Certainly attitudes toward clergymen occupying scientific posts began changing at mid-century, becoming increasingly more hostile as the British scientific establishment sought greater professionalization. By the third quarter of the century, the scientific and clerical vocations were considered to be quite separate. The number of Anglican clergymen who presided over mathematical sections of the British Association for the Advancement of Science

Bartholomew Price, F.R.S. (1818–1898), Sedleian Professor of Natural Philosophy and Master of Pembroke College, Oxford.

was fifteen for the years 1831 to 1865. This number dropped to just two for the remaining thirty-five years of the century. A similar pattern prevailed in the Royal Society, where the average percentage of clerical members declined from 8.9 percent in the period 1849 to 1869, to 4.1 percent in the last twenty years of the nineteenth century.[3] For someone as shy and inward looking as Dodgson, such an atmosphere would have been suf-

---

3. Frank M. Turner, "The Victorian Conflict between Science and Religion: A Professional Dimension," *Isis* 69, no. 248 (1978), 367.

Henry J. S. Smith, F.R.S. (1826–1883), Fellow of Balliol and Corpus Christi Colleges and Savilian Professor of Geometry, Oxford.

ficient to prevent him from seeking out colleagues in the professional associations.

Moreover, for Dodgson the scientific and clerical were not at all in opposition to each other. They were very much integrated in his thoughts and beliefs. In an extraordinary set of six letters written in 1897, all addressed to an unidentified agnostic, Dodgson laid out this connection. In the first of the three of these letters that have been published,

> In the Agnostic view of Christianity, it seems to be expected, sometimes, that Christians should be able to *prove* what they believe, by

arguments which a reasonable man *must* accept as valid, whatever his *wishes* may be.

This is the case with (say) mathematics. If a proposition of Euclid were put before a man, able to understand it, but very anxious *not* to believe it, he would not be able to help himself: he *must* believe it.

This is *not* the case with Christianity. Some of its beliefs are what would be called in Science "Axioms," and are quite incapable of being *proved,* simply because *proof* must rest on something already granted: but this does not exist in the case of Axioms: if there were anything already granted, which could be used in proving them, they would not be "Axioms," but "Theorems." The existence of Free Will is an Axiom of this kind. Consequently, if, in any discussion between two persons, one accepts some Axiom needed in the discussion, and the other does not, there is no more to be said: further discussion is useless.

The other beliefs of Christianity are mostly, if not wholly, believed as a *balance of probabilities:* one who is resolved *not* to believe never finds himself *compelled* to believe: there is always room for *moral* causes to come in, such as humility, truthfulness, and, above all, the resolution to *do what is right*.[4]

In this first letter, Dodgson contrasts the structure of a mathematical subject, Euclidean geometry, with that of Christianity. As structures they both rest on axioms, but their respective developments, after the initial set of axioms is assumed, are quite different. The geometry unfolds in the form of propositions and theorems, which forces the reader to believe because the subject is derived logically from the axiom base and previously proved theorems. Christianity, on the other hand, consists mostly of probable truths that can never be established as true. Dodgson attaches great importance to accepting the complete set of axioms of a structure: without that agreement there can be no meaningful exchange of ideas. This exactly parallels his own experience with the non-Euclidean geometries: he does not accept a non-Euclidean parallel postulate, and there the matter ends.

In the second letter that follows, Dodgson reveals his method of establishing for himself whether something is true: you must work it out on your own, using the available facts. Imagine, for a moment, that Euclidean geometry is the "theory" referred to in the second paragraph of the letter, rather than Christianity. Remarkably, the description fits it perfectly! Even

---

4. *The Letters of Lewis Carroll,* edited by Morton N. Cohen (New York: Oxford University Press, 1979), vol. II, 1122.

more remarkable is the ease with which one can interchange Euclidean geometry and Christianity throughout the letter.

> All this shows what God expects of you, namely that you should consider the evidence, accessible to you, as to its truth or falsehood, and thus reach a state of definite belief or disbelief....
>
> Now here is a *theory*, which is before your eyes, believed in by many around you, whom you feel to be thoughtful, intelligent and anxious to know the truth—a theory which is consistent with itself, *and which accounts for the known facts of life*. An Agnostic, who declares "this theory is *not*, the most probable of all," may reasonably be asked "what theory is *more* probable?"
>
> ... Now there are certain undeniable *facts*, which have to be accounted for on *some* theory, and which *this* theory *does* account for. Some of these are as follows:
>
> ... (3) The Christian church is now one of the greatest facts in the world, and an incalculable influence for good: and its doctrines are found, by experience, to be a perfect remedy for sin, a perfect means of reaching the sense that the load of guilt is taken away, and a perfect support and comfort in suffering.
>
> So far, then, the probabilities are enormous that this theory is true: and enormous probabilities are what we live by.[5]

In the third letter below, Dodgson establishes the grounds on which a theory that in the past has accounted for all the known facts can be abandoned. In doing so he reaffirms the critical role of self-verification in accepting or rejecting a theoretical structure. Dodgson maintains that without going through this process, whether by choice or by not being in a position to carry it through, it is really not possible to deny the previously held position.

> But *your* position is this. You *had* reasons, very good reasons, for *affirming* the truth of Christianity. You now *deny* it. This is not a reasonable thing to do, unless you have *stronger* reasons for *denying* it than you had for *affirming* it. What we have now to do is, I believe, to look into these reasons.
>
> ... I think I may fairly conclude that these reasons were *not* worked out by *you*.... You have, I imagine, simply accepted these reasons, second-hand, from some infidel writer, without being able to verify his statements for yourself....
>
> Now I want to put clearly before you what it is that the infidel

---

5. *Ibid.*, 1142–44.

writer . . . is bound to prove to you, before you can be reasonably expected to listen to him.

The facts of history have to be accounted for *somehow*. Christianity offers a theory which fully accounts for them. Before putting aside this theory, you should find a *better*.[6]

Early in 1885, Dodgson became acquainted with Edith Rix, a nineteen-year-old mathematically gifted young woman. They struck up a friendship that included him advising her on which college she should attend. Responding to a letter written by her mother, Dodgson suggests that rather than going to Cambridge, Edith should try Oxford.

But why not Oxford? There is I believe *no* ground for thinking that, because her turn is for Mathematics, she ought to go to Cambridge. Oxford teaching is, I am sure, all that she can *possibly* need, for a great many years to come.

. . . Lady Margaret Hall is the one conducted on Church principles: Somerville Hall professes no particular form of religious persuasion. Do not however suppose that, because I am a clergyman, and recommend the former of the two, that it would make any abatement in the friendship. . . .

The "Honour Mathematics" at Lady Margaret Hall are taught by a Christ Church man.[7]

In this letter Dodgson clearly states his own belief that the "mix" of religion and science is not only desirable, but that the academic subject, here mathematics, does not suffer from the association. As it turned out, Edith Rix did not choose to attend any college at Oxford, but enrolled instead in Newnham College, Cambridge.

Dodgson's own journey in the cause of his church is marked with conflict as well as commitment. Much of the story is well known. He did not choose to be ordained a priest but took deacon's orders instead. Some of his concerns were that his slight stammer would be a cause of embarrassment in delivering sermons. The general expectations placed on a parish priest were uncompromisingly rigid; besides, the Bishop of Oxford, Samuel Wilberforce, who ordained him, had decided that habitual attendance at the theater was inappropriate for a priest. Dodgson had a passion for the theater as well as a deep feeling for his church and its principles. In a letter written in 1885, long after he had resigned his lectureship (but not his studentship), he explained the actions he had taken to his cousin, W. M. Wilcox.

6. *Ibid.*, 1151.   7. *Ibid.*, vol. I, 565.

INTRODUCTION 9

Meanwhile, I will tell you a few facts about myself. . . . When I was about 19, the Studentships at Christ Church were in the gift of the Dean and Chapter. . . . Dr. Pusey [Canon at Christ Church; he and Dodgson's father had been students together at Christ Church] . . . told me he would like to nominate me, but had made a rule to nominate *only* those who were going to take Holy Orders. I told him that was my intention, and he nominated me. That was a sort of "condition," no doubt: but I am sure, if I had told him, when the time came to be ordained, that I had changed my mind, he would not have considered it as in any way a breach of contract.

When I reached the age for taking Deacon's Orders, I found myself established as the Mathematical Lecturer, and with no sort of inclination to give it up and take parochial work: and I had grave doubts whether it would not be my duty *not* to take orders. I took advice on this point . . . and came to the conclusion that, so far from educational work (even Mathematics) being unfit occupation for a clergyman, it was distinctly a *good* thing that many of our educators should be men in Holy Orders.

And a further doubt occurred. I could not feel sure that I should ever wish to take *Priest's* Orders. And I asked Dr. Liddon [Student of Christ Church and a close friend] whether he thought I should be justified in taking Deacon's Orders as a sort of experiment, which would enable me to try how the occupations of clergyman suited me, and *then* decide whether I would take full Orders. He said "most certainly"—and that a Deacon is in a totally different position from a Priest: and much more free to regard himself as *practically* a layman. So I took Deacon's Orders in that spirit. . . .[8]

This is an extraordinary letter. We can sense the layers of conflict he experienced as he arrived at the decision to become a deacon rather than a priest. He seemed to be of two minds in relation to what his church expected of him. Acknowledging the parochial nature of the Anglican church, he concludes that educational work, although inappropriate for a clergyman, is not inappropriate for a deacon. But he pushes this point even further, deciding that it would be desirable for a member of the church to take up this important work. In a stroke he has healed the rift within himself between the church and professional mathematics by considering his educational duties as his clerical obligation. He also allows us to glimpse the doubts he harbors about whether his personal life and the life of a cleric, at any level, are compatible. He does not mention the theater

8. *Ibid.,* 602–3.

here, but surely it was an important consideration. So was his career as a mathematician at Christ Church. Life as a don suited him, and he had no intention of relinquishing it.

Dodgson's work in mathematics fits the "blueprint" implied in this letter. He was unflagging in the care and energy he devoted to what he considered to be his duty: the preparation of Christ Church men to pass examinations in mathematics. The quality of the work in mathematics that he produced in carrying out this endeavour reflects the obligation of his commitment. It does not, however, capture the spirit of his creativity. That was relegated to situations that stimulated his mathematical thinking outside the domain of the mathematical lectureship.

Charles Dodgson produced his first mathematical pamphlet, *Notes on the First Two Books of Euclid*, in 1860, the same year he wrote *Syllabus of Plane Algebraical Geometry*. He continued to write about geometry until 1895 when he published the fourth edition of *Curiosa Mathematica, Part I: A New Theory of Parallels*. Dodgson's writings on geometrical topics dominate his mathematical output. As mathematical lecturer at Christ Church, Oxford University, his primary task was to prepare students for the examinations that determined pass or honours degrees. Examinations controlled every aspect of a student's career; Euclid's geometry was the centerpiece because it was thought to be the deductive study of spatial truth, with Euclid's axioms and theorems describing physical space exactly. By extension, Euclidean geometry stood as the model for the attainment of absolute truths by the power of human reasoning generally. At Oxford and other British universities Euclid was read as an exercise to strengthen intellectual power, not as an instrument to solve practical problems. Euclidean geometry was the primary element in British mathematical education and Euclid's ancient text, *The Elements*, was its vehicle. But there were difficulties, as the huge number of editions of Euclid's text (more than 200 appeared in the nineteenth century) suggests. Many of these editions were unadorned Euclid: no exercises or hints for proving theorems and certainly no problems for solution were provided for students. It was gradually becoming apparent that Euclid's text was not appropriate for school use, but the universities refused to give it up, fearing chaos in the examination system.

At Oxford, the first of three examinations for the B.A. degree was called Responsions (also known as "Little-go"), and the second, Moderations. The last step was a final examination in the appropriate school, such as the School of Mathematics and Physics. According to the examination statute of 1850, a pass in two schools was required for the B.A. degree. For an honours degree, one of the final two schools had to be the Classical one, otherwise known as a Greats school, hence the name "Great-go" for that examination. Beginning with Moderations, a student elected either the pass or honours course in one or both schools for the degree, which determined the difficulty as well as the scope of the examinations. Students were referred to as either passmen or classmen, obtaining their degrees in three or four years, respectively.

Students obtained instruction in subject matter in professorial lectures and in the lectures given by tutors. During Dodgson's tenure at Christ Church, one of the most important educational issues was whether the professorial lectures, where the student was a passive recipient of information, provided the mental discipline of the tutorial system, in which the textbook was central to the development of critical thinking ability.

As the Mathematical Lecturer at Christ Church, Dodgson was responsible for enabling his students to pass the examinations that he also helped to construct. In his diary entry of 11 November 1863, he wrote, ". . . we [Bartholomew Price] settled the days of examination & also that I am to set the 2 Pass papers in Algebra, & the Honour papers in Geometry and Algebra, Algebraical Geometry, & Differential Calculus."[9]

The mathematics required of Oxford students differed widely as a function of their chosen course of study. The number of passmen who selected the Mathematical School as their second school was very small. The majority of passmen took their degrees with only a rudimentary knowledge of arithmetic and algebra and no formal study of geometry beyond the first two books of Euclid, which were tested in Responsions. Additional mathematics was not considered to be part of the general education of the average Oxford student, but rather a subject for specialized study. The course of reading required for mathematical honours at Moderations was the only one that included differential and integral calculus and the calculus of finite differences.[10]

Mathematics had been added to the requirements for a public school gentleman's education only in the first half of the nineteenth century. As reported in the *Public Schools Commission* [Clarendon] *Report* in 1864, just three hours of weekly mathematical instruction was provided at some of the best of the public schools like Eton, Harrow, and Rugby. Mathematics was considered distinctly inferior to the classics. Now, having gained a foothold, mathematics had to conform to classical ideals. What better choice for stimulating continuity of attention, coherence of thought, and

---

9. Diaries, 11 November 1863.

10. See T. D. Acland, *The Discouragement of Elementary Mathematics in General Education at Oxford* (Oxford and London: James Parker and Co., 1867). The arguments set forth in this pamphlet are the contents of a letter sent by its author to the Vice-Chancellor of the University in a letter dated 25 February 1867.

confidence in reasoning than Euclid's *Elements!* This classical text represented the essence of the world of Plato and Aristotle and combined both humanistic and formal traditions of the most admired of cultures, that of the ancient Greeks.[11]

When the Association for the Improvement of Geometrical Teaching, founded in 1871, advocated the displacement of Euclid, Dodgson ridiculed the idea in his book, *Euclid and his Modern Rivals,* first published in 1879. Written in the form of a drama in four acts, the dialogue is witty and satirical. In a short review of the book appearing in the *Scotsman* on 15 May 1879, the writer captured Dodgson's reasons for favoring Euclid's text.

> It [*Euclid and his Modern Rivals*] is an attempt to demonstrate the inutility and positive mischief of having a large number of text-books for the teaching of elementary geometry. Mr. Dodgson does not take his stand on any narrow principle, or oppose modern attempts to supersede Euclid merely from a dislike to innovation. What he tries . . . to show is, that none of the recent text-books contain anything of importance that is not in Euclid; that the axioms and definitions which have been proposed in addition to, or in substitution for, those laid down by him will not, for the most part, stand the test of close examination; and that in simplicity and logical sequence of method he has the advantage of all his rivals.[12]

Dodgson's great concern was that the order and numbering of Euclid's axioms and theorems, as well as his method of treating lines, angles and parallels, remain unchanged. He had no objection to improvements in proofs or to the addition of new problems and theorems within the existing structure. He explained that his reasons were not wholly motivated by the needs of examiners.

> If examinations were done away with altogether, there would still be abundant reason for adopting the same logical sequence in all sciences where some logical sequence is essential: otherwise any attempt at communication between one mathematical student and another is merely a revival of the scenes enacted around the Tower of Babel.[13]

---

11. See W. H. Brock, "Geometry and the Universities: Euclid and His Modern Rivals, 1860–1900," *History of Education* 4 (1975), 21–35.

12. C. L. Dodgson, *Supplement to Euclid and his Modern Rivals* (London: Macmillan and Co., 1885), 340–41.

13. *Ibid.,* 356. Dodgson appended this comment to the extract from a paper that appeared in the *Educational Times* on 1 March 1880.

Dodgson linked his religious beliefs and his work in geometry through the way he perceived natural theology and the nature of mathematical truth. In his time, mathematics was considered uniquely capable of generating truths from axioms that captured the nature of reality; Dodgson's fundamental concern with rigor, i.e., precise definitions, the proper order of reasoning from one theorem to another, etc., is understandable. His need to study and develop logical rules for reasoning correctly and to expose fallacious arguments further reflects this conviction. The introduction of the non-Euclidean geometries in mid-century threatened to upset the legitimacy of mathematics, geometry particularly, in establishing universal truths. Dodgson unequivocally rejected these unwanted intruders, preferring instead to shore up Euclid as a text in *Euclid and his Modern Rivals* and as a doctrine in *Curiosa Mathematica, Part I: A New Theory of Parallels* (1888).

H. J. S. Smith, one of the most brilliant, creative mathematicians of his era, was a staunch admirer of Euclid in the educational setting. As president of the Mathematical and Physical Section of the British Association he had this to say in an address to the section on 18 September 1873 about the attempts to improve the teaching of geometry:

> For some years past this Section has appointed a committee to aid in the improvement of geometric teaching in this country. . . . [The committee has held] a middle course between the views of the innovators who would uphold the absolute monarchy of Euclid . . . and the radicals who would dethrone him altogether. One thing at least they have not forgotten, that geometry is nothing if it be not rigorous and that the whole educational value of the study is lost if strictness of demonstration be trifled with. . . .
>
> Geometry is hard, just as Greek is hard, and one reason why Geometry and Greek are such excellent educational subjects is precisely that they are hard. . . .
>
> The work undertaken by the Association for the improvement of geometrical teaching is still far from complete; and even when it is complete it must be left to hold its own against the criticism of all comers before it can acquire such an amount of public confidence as would justify us in recommending its adoption by the great teaching and examining bodies of the country.[14]

---

14. *The Collected Mathematical Papers of Henry John Stephen Smith,* ed. J. W. L. Glaisher (London: Oxford University Press, 1894; reprint, New York: Chelsea Publishing Co., 1965), Appendix I, 686–87.

Smith had a great deal to say about the substance of Euclid as well. In this same address he spoke about two controversial aspects of Euclid's work: the parallel axiom and Book V of *The Elements,* on proportion.

> That Euclid's treatment of the doctrine of parallels is an example of perfect rigorousness, is an assertion which sounds almost paradoxical, but which I, nevertheless, believe to be true. Euclid has based his theory on an axiom . . . which, it may be safely said, no unprejudiced mind has ever accepted as self-evident. And this unaxiomatic axiom Euclid has chosen to state, without wrapping it up or disguising it, . . . but in its crudest shape, as if to warn his reader that a great assumption was being made. . . . for it is one of the triumphs of modern geometry to have shown that the . . . [parallel axiom] is so far from being an axiom . . . that we cannot at this moment be sure whether it is absolutely and rigorously true, or whether it is a very close approximation to the truth.
>
> Again, the doctrine of proportion as laid down in the fifth book of Euclid, is, probably, still unsurpassed as a masterpiece of exact reasoning; although the cumbrousness of the forms of expression . . . has led to the total exclusion of this part of the elements from the ordinary course of geometrical education. A zealous defender of Euclid might add with truth that the gap thus created in the elementary teaching of mathematics has never been adequately supplied.[15]

Listening to this speech were two leading mathematicians of the second half of the nineteenth century, a young faculty member at the University of Göttingen, Felix Klein, and Arthur Cayley, Sadlerian Professor of Mathematics at Cambridge. Their work was central to the development of both projective and non-Euclidean geometries. In 1859 Cayley had published his "Sixth Memoir upon Quantics" in the *Philosophical Transactions of the Royal Society,* in which he drew together projective geometry and the algebra of invariants. The latter discipline was begun in 1841 by George Boole and extended by J. J. Sylvester, who succeeded Smith as Savilian Professor of Geometry in 1883, and George Salmon, Professor of Mathematics at Trinity College in Dublin. In 1873, Klein was able to build on Cayley's work to place the non-Euclidean geometries within the projective framework. In addition to these men, William K. Clifford, Thomas Archer Hirst, and Olaus Henrici (the last of the "rivals" criticized by Dodgson in *Euclid and his Modern Rivals*), all professors of mathematics at University College London, and the Oxford mathematician, William Spottiswoode, took up the cause of projective geometry. They were

---

15. *Ibid.,* 686–87.

Isaac Todhunter, F.R.S. (1820–1884), Fellow of St. John's College, Cambridge.

also members of various scientific groups like the Royal Society—Spottiswoode was its president from 1878 to 1883—the British Association for the Advancement of Science, and the London Mathematical Society.[16]

Dodgson's focus on Euclidean geometry in his own writings has led several scholars to believe that he did not know of the existence of non-Euclidean geometry, but there is ample evidence that he did. Between 1860 and 1895 many publications dealing with the subject appeared in English, French, and German by mathematicians such as H. Richard Baltzer, Hermann von Helmholtz, Arthur Cayley, W. K. Clifford, R. S. Ball, George Bruce Halsted, Robert Tucker, and O. Henrici. Dodgson was familiar

16. For a discussion of the impact of projective geometry, the reader is referred to Joan L. Richards, "Projective Geometry and Mathematical Progress in Mid-Victorian Britain," *Studies in the History and Philosophy of Science* 17 (1968), 310.

with Baltzer's *Theorie und Anwendung der Determinanten*. He specifically mentioned Baltzer in the introduction to his own book on determinants. Von Helmholtz's work was discussed in the popular and influential scientific journals *Nature* and *Mind*. From 1887 Dodgson himself published articles on various topics in these two journals. Cayley, Henrici, and Ball were active in the British Association, and Henrici also published in *Nature*. Clifford's translation of Riemann's now-famous "On the Hypotheses Which Lie at the Bases of Geometry" appeared in *Nature* in 1873. Robert Tucker, secretary of the London Mathematical Society, who had corresponded with Dodgson in connection with his new theory of parallels, was the editor of Clifford's mathematical papers (1882).[17]

Dodgson also owned Halsted's *Elements of Geometry* (1885).[18] In the second edition of *Euclid and his Modern Rivals*, Dodgson's best known geometrical work, he included as the first appendix an extract of an essay by Isaac Todhunter on elementary geometry published in 1873. Todhunter was a founding member of the London Mathematical Society, served on the council of the Royal Society, and was an influential mathematical educator. Todhunter wrote in his essay,

> I may say that I have taught geometry both Euclidean and non-Euclidean, that my own early studies and prepossessions were towards the latter, but that my testimony would now be entirely in favour of the former.
>
> ... The object of some of our reformers is not to teach the same subject with the aid of a different text-book, but to teach something very different from what is found in Euclid, under the common name of geometry.[19]

The lack of actual evidence whether Dodgson did know of the existence of the non-Euclidean geometries has led mathematical commentators like Eric Temple Bell and Warren Weaver to conclude that he did not. Presuming that Dodgson was unacquainted with his well-regarded Oxford mathematical colleague, H. J. S. Smith, Bell wrote, "Carroll could have got straightened out on geometry in one hour, if he had consulted Smith."[20] But Dodgson *did* know Smith! The two were frequent fellow

---

17. Robert Tucker, "Mr. Dodgson on Parallels," *Nature* 39 (1889), 175.

18. What is known about the contents of Dodgson's personal library can be found in *Lewis Carroll's Library,* edited by Jeffrey Stern, Carroll Studies No. 5, LCSNA (Silver Spring, Md., 1981), especially pp. 29, 47.

19. Lewis Carroll, *Euclid and his Modern Rivals,* 2d ed. (London: Macmillan, 1885; reprint, New York: Dover, 1973), 229, 234.

20. Florence Becker Lennon, *Lewis Carroll* (London: Cassell, 1947), 266.

train travelers. The Parrish Collection has correspondence between them from 1863 and 1875 in which Dodgson queries Smith on several mathematical topics, but not on geometry. Even more bluntly, Weaver decided that "non-Euclidean geometry, with its revolutionary consequences for mathematics and science, was not dreamt of in Dodgson's philosophy."[21]

Curiously, the evidence refuting Weaver's statements on this matter lay buried in a seldom-referred-to volume, the *Supplement to Euclid and his Modern Rivals,* issued in April 1885. The *Supplement,* paged continuously with the first edition, contains eight reviews of the original publication as well as additional substantive material. When the second edition was published later in the same year, the substantive material from the *Supplement* was incorporated; the eight reviews were not. In one of these, from the *Saturday Review* of 10 May 1879, Dodgson clearly stated his views about non-Euclidean geometry. Discussing an equivalent of Playfair's axiom, the reviewer wrote,

> Through a given point outside a straight line only one parallel can be drawn to it. This at once raises the question, what business have you to assume that *any* parallel can be drawn? . . . The assumption . . . is not made by Euclid. And we may further observe that it is not such a small one as it looks, especially in the light of modern geometrical speculations.

At this point in the review Dodgson commented, "But who in the world wants to make this assumption?" The reviewer continues,

> For it results from the work of Lobatschewsky and others that our actual geometry is not an elucidation of eternal and immutable and unique relations, but is rather in the nature of a purely physical science. . . . A consistent geometry (though of course inapplicable to our real experience) can be, and has been, founded on the categorical denial of Playfair's axiom. Euclid's geometry is the science, not of space absolutely, but of a particular kind of space; and in this view the doctrine of parallels lays down very characteristic and important properties of that kind of space.[22]

It is very clear that Dodgson knew of but did not accept non-Euclidean geometry because he could not reject the idea that geometry was not an absolute science. Dodgson fervently believed that mathematics produced results of absolute certainty, that axioms had to refer to meaningful ideas,

---

21. Warren Weaver, "Lewis Carroll: Mathematician," *Scientific American* 194 (1956), 118.
22. *Supplement to Euclid and his Modern Rivals,* 338.

that the rules of acquiring new results must have logical certainty, and that those results defined reality. These attitudes, far from being unusual, were held by the majority of the British mathematical community. Even mathematicians who accepted the non-Euclidean geometries, like Arthur Cayley and Henry J. S. Smith, did so without challenging Euclid's definition of reality. Non-Euclidean geometries were accepted as consistent mathematical theories, but not as depictions of real space.

Three aspects of Dodgson's personality provide additional insight into his staunch defense of Euclid. Dodgson was an artist as well as a mathematician. To him, mathematics was the art of building logical structures. His artist's eye enjoyed using visual forms—but figures have to *look* right. Intuitively, we know what a straight line looks like; the non-Euclidean geometries distort that view. Euclid's geometry was preferable because Euclid used logic on figures based on ordinary geometric intuition. It was not faulty mathematical ability that led him to ignore one of the great discoveries of his time, but a restricted philosophical outlook, one that bound together his aesthetic values with demonstrable truths.

In *Curiosa Mathematica, Part I*, Dodgson takes on Euclidean foundational issues involving the parallel postulate.[23] In the preface, he states that his purpose in writing the book is to present a more "axiomatic" parallel postulate, one that will enable him to separate the Euclidean theorems into two classes: those that are absolutely true and those that are only approximately true. By "axiomatic" Dodgson really means intuitively reasonable. To him Euclid's 12th (parallel) axiom [two lines unequally inclined to a transversal will meet] is unacceptable because it involves the notions of infinities and infinitesimals, notions Dodgson wants to expunge. Theorems only approximately true involve infinitesimal errors. His substitute parallel axiom is an unusual, eye-appealing one: in every circle, the inscribed equilateral hexagon is greater than any of the segments lying outside it. In fact, any equilateral polygon can take the place of the hexagon.

In Appendix II Dodgson explains the basis for wanting to exclude infin-

---

23. Dodgson published two of the three parts of a series of books he called *Curiosa Mathematica*. Part I deals primarily with Euclid's parallel postulate and topics related to it. Much better known is Part II, "Pillow Problems Thought Out During Sleepless Nights," first published in 1893. The fourth edition was reprinted by Dover Publications in 1958 as one of two books bound as one under the title *Pillow Problems and a Tangled Tale*. In 1897 Dodgson was working on Part III, devoted to numbers, but never completed it. Fragments appear in *The Lewis Carroll Picture Book*, edited by Stuart Dodgson Collingwood, published in 1899. It was reprinted in 1961 by Dover under two titles, *Diversions and Digressions of Lewis Carroll* and *The Unknown Lewis Carroll*. Aspects of Parts II and III of *Curiosa Mathematica* are discussed throughout this volume.

itesimals. He points out that in Euclid Book X Proposition I, a property called the Archimedean axiom is assumed which is equivalent to the denial of the existence of infinitesimals. This axiom guarantees that if n is a positive integer and PQ, RS are any two line segments, n times PQ will exceed RS, a property not true for infinitesimals. The Archimedean axiom is essential for defining ratios of segments, and the Euclidean theory of proportionality developed in Book V requires that proportional segments have this property.

Dodgson acknowledged that any attempt to remedy the defect in Euclid's axiom, or any substitute, leads to the "bewildering region of Infinities and Infinitesimals."[24] He also provided additional evidence that he was well aware of non-Euclidean geometry when he discussed objections to his own substitute axiom, on the supposition of the denial of the Euclidean parallel postulate. Referring to the failure of his axiom under this supposition he wrote,

> This phenomenon, however, does not appal me so much as might be expected: for I have often observed it to occur that, when Theorem α logically leads to Theorem β, then, on the supposition of Theorem β *not* being true, it may be proved that Theorem α also is not true. Hence this objection, if worth anything, *proves too much:* since it upsets the whole edifice of Logic itself: and, if you tell me, on such grounds as these, that *I* cannot prove what I assert, I may fairly retort upon you, that *you* cannot prove anything *at all!*[25]

Again, Dodgson is stressing the logical integrity of Euclid's geometry which for him, when coupled with its intuitive appeal about the nature of space, made the Euclidean system unassailable.

We are still left with Dodgson's refusal to accept infinities and infinites-

---

24. C. L. Dodgson, *Curiosa Mathematica, Part I: A New Theory of Parallels* (London: Macmillan, 1888), x:v. Dodgson was never enthusiastic about infinitesimals and infinities and preferred to avoid them whenever possible. His (erroneous) idea of an infinitesimal seemed to be that of a positive quantity smaller than any other positive quantity but not really zero. As Dodgson's papers in Packet 47 of the Parrish Collection show, his confusion about infinitesimals in no way interfered with his ability to calculate derivatives. In fairness to Dodgson, problems concerning orders of infinity and infinitesimals were not settled until R. Dedekind and G. Cantor completed their work on the theory of infinite aggregates in the second half of the nineteenth century.

25. *Ibid.*, xvii. Here Dodgson is supposing that a theorem, I.32 (The sum of the three angles of a triangle equals two right angles), provable from the Euclidean parallel postulate, is not true, and that it can then be proved that the relationship of magnitude between the hexagon and the segment in his axiom is not true, i.e., in an infinitely large circle, the hexagon is less than the segment in area.

Letter from Archdeacon Charles Dodgson to his son, Charles L. Dodgson, concerning a geometrical series summation problem, dated 6 January 1864. Father and son corresponded frequently on topics of a mathematical nature.

imals. Florian Cajori observed that infinitesimals had largely been eliminated in Great Britain in the eighteenth century as "undesirable, unreal, and mischief-making." But they returned in the nineteenth century "to flourish for a time as never before" through the work of D'Alembert, Clairaut, Lagrange, Laplace, Legendre, and others who used Leibniz's notation in calculus, which involved infinitesimals: "How thoroughly the infinitesimal invaded certain parts of British territory is seen in Price's

[Dodgson's mentor and friend] large work on the Infinitesimal Calculus. . . . From the standpoint of rigor, the British treatment of the calculus [which utilized Newton's more Euclidean geometric method of fluxions] was in advance of the Continental."[26]

26. Florian Cajori, *A History of the Conceptions of Limits and Fluxions in Great Britain from Newton to Woodhouse* (Chicago: Open Court, 1919), 278.

In a book in manuscript form, which he nearly completed, Dodgson presents a theory of parallels in which he distinguishes parallel lines that are equidistant from parallel lines that are asymptotic. He began this work, "The Direction-Theory as Applied to Pairs of Lines," early in 1890 and continued to work on it through the spring of 1891. He meant this book to be the third in the series that began with *Euclid and His Modern Rivals* and continued with *Curiosa Mathematica, Part I*.[27] In Appendix III of *Curiosa Mathematica, Part I*, Dodgson points out that two equidistant lines have the property that they will not meet no matter how far they are produced, but that using this property to define parallel lines implies a different meaning of parallelism: "It may be proved [with this property], that, given a Line and a Point not on it, a whole 'pencil' of Lines may be drawn, through the Point, and not meeting the given Line."[28]

However, Dodgson maintained his rejection of this notion of parallelism because such lines cannot be recognized by the practical geometric test that they make equal angles with all transversals.[29]

Dodgson wrote on many topics besides geometry: probability, algebra, arithmetic computation and theory, trigonometry, and cryptology. His work, particularly on nongeometric topics, displays a high level of creativity. Underlying his choice of subjects is the need to get beneath the usual assumptions, to question their validity, and if necessary to turn them upside down. He was a powerful manipulator of probabilities in finite sample spaces, but when he tried problems that touched on deep issues in the foundations of probability, issues that were in the process of being resolved in the mathematical community, he had difficulty. He wrote on linear algebra and on a wide set of topics—in what we now recognize as the mathematics of politics—where he was able to apply linear algebraic techniques to real situations. His best work, particularly the first proof of an important theorem in the development of linear algebra and a method for computing a determinant that has led to surprisingly important consequences in enumerative combinatorics, was done in these areas. Motivated by a sense of fairness, he attempted to solve problems that are more

---

27. This manuscript is in the Parrish Collection in the Princeton University Library.

28. *Curiosa Mathematica, Part I*, 51.

29. Contrary to Warren Weaver's statement on page 16-2 of his typescript introduction to Dodgson's mathematical pamphlets in the Parrish collection in the Princeton University Library, Dodgson avoided a frontal attack on geometrical infinitesimals. The fact is that his geometric point of view obscured a clear understanding of infinitesimals, which has to be based on functions. Augustin Cauchy provided this understanding and clarified the relationship between an infinitesimal and a limit. The 25 volumes of his complete works were published in Paris between 1882 and 1932.

Augustus De Morgan (1806–1871), Professor of Mathematics, University College, London (now the University of London). Founder and first president of the London Mathematical Society.

of our time than his, such as knotty problems in the design of tournaments and schemes for proportional representation—problems that were not resolved until well into this century.

On the other hand, publications on symbolical algebra are conspicuously absent. Introduced by the British mathematician George Peacock in the 1830s, this new algebra was viewed as a branch of mathematics dealing with undefined symbols governed by abstract operations. Augustus De Morgan, with whom Dodgson shared many similar mathematical views, was able to support both symbolical algebra as well as Euclid's geometry. Dodgson could not.

Around mid-century, algebraic methods also suggested that spaces of dimension greater than three could be utilized to solve problems. The issue of the conceivability of higher dimensions opened another crack in the mathematical armor. It was an issue Dodgson avoided. As Helena Pycior notes, "Dodgson ignored symbolical algebra because it was inimical to the traditional view of mathematics. . . . Algebraic meaninglessness and arbitrariness went directly counter to his deep-seated belief in the absolute certainty or truth of mathematics."[30]

Concerned about practicality and the elimination of error in computations, his work on number-theoretic topics is both useful and clever. One of the most novel items appears in a publication on trigonometry in which he gives a series for approximating π that in his lifetime was exceeded in computational accuracy only by the work of the great mathematician, C. F. Gauss. Finally, as a dilettante cryptologist, he originated some ingenious ciphers that were virtually unbreakable in his time, and also formed the first mathematically constructed cipher. What is abundantly apparent in all of his work is his considerable talent as a calculator and popularizer of useful mathematics.

The mind of Charles Dodgson, the mathematician, contained elements dynamically in opposition to each other. He was bold and skeptical, questioning what appeared to be paradoxical or fallacious reasoning, yet he could not escape the bounds of strictly logical thinking. He experimented with unique notation and with a prose style that included humor and dramatic elements, quite unlike that of any other mathematician of his time. He wrote on topics within traditional mathematics while struggling with aspects outside it, a notion that suggests he was hampered by convention of a nonmathematical nature. He could not appreciate the more experimental parts of mathematics, those that relied more on inductive rather than deductive methods, involving levels of abstraction that broke with common sense and with tradition. But he questioned, extended, and generalized conventional mathematics in surprising and unexpected ways.

30. Helena Pycior, "At the Intersection of Mathematics and Humor: Lewis Carroll's *Alices* and Symbolical Algebra," *Victorian Studies* 28, no. 1 (1984), 162.

Geometry

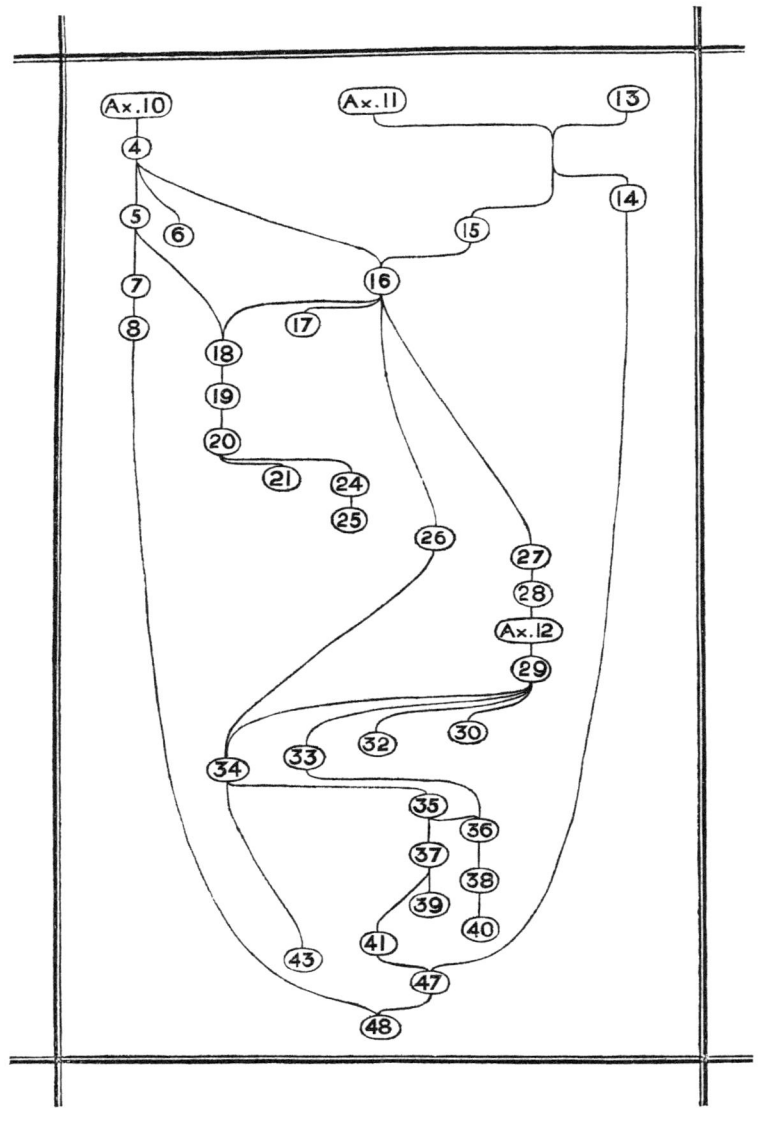

Frontispiece from Euclid, Books I, II, *edited by C. L. Dodgson (Oxford, 1875), illustrating the relationships among Euclid's axioms and theorems.*

Introduction

Dodgson published five books, many of which appeared in several editions, and five extant pamphlets on geometrical topics. The books are *Syllabus of Plane Algebraical Geometry* (1860); *Euclid, Book V* (1874); *Euclid, Books I, II* (unpublished private edition, 1875; published 1882); *Euclid and His Modern Rivals* (1879); and *Curiosa Mathematica, Part I* (1888). The pamphlets, which are reproduced here, with the exception of *Supplement to "Euclid and His Modern Rivals"* (1885), include *Notes on the First Two Books of Euclid* from 1860 (item 1), *The Enunciations of Euclid I, II* from 1863 (item 2), *The Fifth Book of Euclid* from 1868 (item 3), and *The Enunciations of Euclid I–VI* from 1873 (item 4).

The themes of the books and the pamphlets are similar; Dodgson was trying to correct what he saw as gaps, inconsistencies, and inaccuracies in existing texts on Euclidean geometry. He also wanted to help students by providing clarifications, additional definitions, and approaches to proving theorems. He preferred an algebraical approach for the treatment of ratio and proportion in Book V. In this he was completely in agreement with Henry J. S. Smith, who in his 1873 address to the British Association had complained that the contents of Book V were not being taught because of the way in which Euclid had expressed his ideas. Dodgson was aware of this problem early, 1855 by his own account.[1]

The most popular version of Euclid's *Elements* in England in Dodgson's time was Simson's *Euclid*. Although other editions of Euclid's *Elements* in English existed, none were as widely accepted; Simson captured the spirit of Euclid, and his book was the standard geometry textbook of England. His version of Euclid was excellent because it was based on an authoritative Latin translation, not because Simson himself was a suberb textual critic of Euclid. Robert Simson had been professor of mathematics in the University of Glasgow in the eighteenth century. He was considered an excellent interpreter as well as critic of Euclid. His first edition of Euclid's *Elements* appeared in 1756; the twenty-sixth appeared in 1844. Simson's arrangement of Euclid's axioms and definitions was different from Euclid's text:[2] the famed Euclidean parallel postulate, number five in Euclid,

---

1. See the diary entry for 31 May 1855 on p. 29.
2. The comparison made here is between Simson's edition as revised by Isaac Todhunter in 1862 and the authoritative text of Euclid's *Elements* as restored by J. L. Heiberg between 1883 and 1888.

was axiom twelve in Simson; Euclid's fourth postulate that all right angles are equal to each other was Simson's eleventh axiom; and the definition of parallel lines, number twenty-three in Euclid, was number thirty-five in Simson.

Algebraic symbolism, introduced in England in the seventeenth century, did not appear in geometry texts for many years because of the concern that students would confuse Euclid's geometrical concepts with arithmetical ideas. For example, geometrical equality is not the same as numerical equality; when the symbol = is used in geometrical reasoning it must be understood as referring to geometrical equality.

The Greek geometers had no algebraical notation; instead they used a geometrical algebra of remarkable ingenuity. For example, since the geometrical ratio of two magnitudes (continuous quantities) of the same kind (lines, angles, planes, etc.) can be represented by the magnitudes themselves, there is no need to refer to a third magnitude of the same kind against which to compare the original two. The ratio of two magnitudes is their relation to each other, and no standard unit is necessary. This is demonstrated in Euclid's Book V on the theory of proportion; all its propositions dealing with the equality or inequality of ratios are demonstrated without any reference to the numerical measures of those ratios. The advantages of this geometrical method are considerable. First, it is independent of the commensurability or incommensurability of the magnitudes themselves; second, the constructions can be done in terms of straight lines although they apply to magnitude generally. However, the method is not a simple one. It is simpler to express the ratio of two magnitudes of the same kind by two numbers, but this is only possible if the two magnitudes are commensurable.

In Book VII, Euclid dealt with the numerical theory of proportions. There he gave a numerical definition of proportion, number 20: "Numbers are *proportional* when the first is the same multiple, or the same part, or the same parts, of the second that the third is of the fourth."[3]

In Book V, the general definition of proportion is provided in definitions 5 and 6:

> Magnitudes are said to *be in the same ratio,* the first to the second and the third to the fourth, when, if any equimultiples whatever be taken of the first and third, and any equimultiples whatever be taken of the second and fourth, the former equimultiples alike exceed, are alike

---

3. Thomas L. Heath, *The Thirteen Books of Euclid's Elements,* 2d ed. (New York: Dover Publications, 1956), vol. 2, 278.

equal to, or alike fall short of, the latter equimultiples respectively taken in corresponding order. Let magnitudes which have the same ratio be called *proportional*.[4]

This definition can be applied to all geometrical magnitudes because it only involves multiplying magnitudes by numbers. Euclid's deductions in Book V and beyond are heavily dependent on this definition of proportionals.

In 1868, Dodgson published *The Fifth Book of Euclid Treated Algebraically* (item 3). His aim was to put the two treatments of proportion together, an approach that can only be applied to measurable quantities. He used Simson's organization of Euclid as the basis for his pamphlet.

Presumably there were other pamphlets and treatises on the algebraic treatment of ratio and proportion circulating at the time. Their common source was probably Augustus De Morgan's 1841 articles in *Penny Cyclopaedia*, vol. XIX.[5] Item 3 bears a resemblance to an 1858 pamphlet with almost the same title that is sometimes ascribed to Dodgson. However, Thomas Fowler, tutor of Lincoln College and a classmate of Dodgson, is the probable author of the 1858 pamphlet.[6] The quality of the writing is much more fluid and the terminology much less precise than in Dodgson's 1868 pamphlet.

Dodgson's authorship of the 1858 pamphlet has been inferred from a diary entry of 31 May 1855 and from the fact that he on occasion referred to himself as a tutor, which he was not (a tutor taught several different subjects, a lecturer only one). In the diary entry he wrote, "During the last month I have again written out, in an improved form, *The Fifth Book of Euclid proved Algebraically,* and have made considerable progress in my treatise on Algebraic Geometry."[7] Why ten years elapsed before he actually published his pamphlet remains unexplained.

Dodgson begins the preface to his pamphlet with an apology for not having included incommensurable magnitudes, which he—and anyone else who was knowledgeable—understood to be at the heart of Book V. He states his reasons for the omission to be that his publication is intended only for students interested in a pass degree and that the usual way that the required material on measurable quantities is presented is too lengthy.

4. *Ibid.*, 114.
5. *Ibid.*, 116.
6. S. H. Williams and F. Madan, *The Lewis Carroll Handbook*, revised by R. L. Green and D. Crutch (Folkestone: Wm. Dawson & Sons, 1979), 12–13.
7. *The Diaries of Lewis Carroll,* edited by R. L. Green (New York: Oxford University Press, 1954), vol. I, 51.

He also thought that nonmeasurable quantities would be too difficult for passmen.

The organization of the pamphlet is unfortunate. In attempting to sharply distinguish between Euclid's approach and an algebraic one Dodgson uses a two-column presentation, placing the relevant geometric and algebraic material side by side. He uses copious footnotes to compare, clarify, and illustrate various aspects of that material, mixing Euclid's text with Simson's and adding some of his own modifications. The result through the first thirteen pages is precise but confusing; however, the hints for proving theorems and the proofs themselves that follow are straightforward. Dodgson's contribution is his methodical approach to laying out the development of the numerical part of ratio and proportion. He thereby made accessible to ordinary students some of the concepts in Euclid's Books V and VI, free of the original geometrical method used to present them.

In his diary entry of 1 November 1882 Dodgson wrote, "Began today preparing the second edition . . . of *Euclid, I, II* by erasing '+', '=', '>' and '<' signs, which seem likely to make the book less useful, as all algebraical signs are supposed to be forbidden in Cambridge examinations."[8] He was referring to the book he first produced in 1875, which he obviously hoped would enjoy popularity beyond Oxford. This was preceded by two pamphlets, *Notes on the First Two Books of Euclid* from 1860 (item 1) and *The Enunciations of Euclid, Books I and II* from 1863 (item 2). The substance of Euclid's first two books is principally the properties of triangles and parallelograms in Book I and the geometry of areas in Book II. In the 1860 pamphlet, Dodgson exhibits a preference for algebraic notation in proofs of propositions. He provides a proof of the second part of I,16 and of II,13. Why does he choose these two? In Todhunter's edition of Simson's *Euclid,* the proof of the second part of proposition I,16 is omitted. (The enunciation of I,16 is: In any triangle, if one side of a triangle is produced, the exterior angle is greater than either of the interior opposite angles.) Most likely, this is the reason Dodgson includes it. Curiously, this proposition is not universally true.[9] Dodgson's construction is, of course, in Euclidean space and thus meets the criterion for the validity of the proof.

Dodgson provides a simpler proof of proposition II,13 (In every triangle, the square on the side subtending an acute angle is less than the

---

8. *Ibid.,* vol. II, 410.

9. Thomas Heath remarks that in a Riemannian space, which is infinite in extent but finite in size, for the proof to be valid ck must always be inside angle bcg so that angle bck will be less than angle bcg. This will not always be the case in a Riemannian space. *The Thirteen Books of Euclid's Elements,* vol. 1, 280.

squares on the sides containing that angle, by twice the rectangle contained by either of these two sides, and the straight line intercepted between the perpendicular let fall on it from the opposite angle, and the acute angle), in which he relies heavily on algebraic notation while retaining the geometric flavor of the proof. It was through Simson's editions of Euclid that algebraical symbols became accepted.

The other two notes on propositions appear to be clarifications of Simson's alterations of Euclid's proofs. In I,35 (Parallelograms on the same base, and between the same parallels, are equal to one another), Dodgson tries to make clear to the student that a triangle is not being taken twice from the same trapezium, but rather that there are really two congruent trapezia from which two congruent triangles are being taken. In II,7 (If a straight line is divided into any two parts, the squares on the whole line, and on one of the parts, are equal to twice the rectangle contained by the whole and that part, together with the square on the other part), he provides clarification of the diagram designed to appeal to the underlying algebraic meaning of the proposition that the square on the difference of two straight lines equals the sum of the squares on the straight lines minus twice the rectangle contained by them.

Commenting on the *Notes,* Roger Herz-Fischler remarks that they are not self-contained but were meant to be read with one of Simson's editions of Euclid. Herz-Fischler points out that the definitions of point, angle, and triangle are neither those of Euclid nor of Simson, but are rather Dodgson's explanations to students about the meaning of these terms.[10]

The 1863 pamphlet, *The Enunciations of Euclid, Books I and II* (item 2) is an outline of what a student would be expected to know for examination purposes. Dodgson's presentation shows his concern with precision. Moreover, he wanted students to recognize fine distinctions of meaning, differentiating between *define* and *explain, term* and *phrase,* and the like.

In his most ambitious presentation of Euclid's geometry for Oxford students, *The Enunciations of Euclid I–VI* (item 4), Dodgson takes issue with two prevailing customs strongly enough to comment on them. In Proposition 11 of Book I, he omits the Corollary (Two straight lines cannot have a common segment) for the reasons that it is not Euclid's; its proof is illogical and Euclid has actually assumed the Corollary in Proposition 4. This Corollary was added by Simson and was in dispute. Todhunter, for example, thought it should follow Proposition 13, but he added that it is also possible to place it after Proposition 5 or even after Proposition 1 because Euclid *tacitly* assumed there that two straight lines cannot have a

---

10. Roger Herz-Fischler, personal communication.

segment in common where they both meet.[11] Dodgson thinks that the assumption was made no earlier than Proposition 4.

Dodgson's second important change was to move the Corollaries following Proposition 15 in Book I to Proposition 13 because he considered Proposition 13 sufficient for their proof. There is some question whether the two Corollaries were given originally by Euclid, but they are included in Simson's textbook. Heath comments that the balance of manuscript authority seems to be against the genuineness of the first Corollary, the only one of the two that Euclid included.[12] However Proclus, whose commentary on Book I of Euclid is one of the two chief sources of information on the history of Greek geometry (the other being the *Collection* of Pappus), does include it. Proclus adds that the second Corollary follows from the first. There does not appear to be a precedent for switching the Corollaries to Proposition 13.

Dodgson also constructed two geometrical fallacies in the form of theorems and a geometrical puzzle that appear in *The Lewis Carroll Picture Book*, as well as twenty-two geometrical Pillow Problems that were published in his *Curiosa Mathematica, Part II*. Twenty of these dealt with plane geometry and two with algebraical geometry—in modern terms, analytic geometry. As one might expect from the author of the *Alice* books, Dodgson's approach in constructing these problems was backward.

> I generally wrote down the *answer*, first of all: and *afterwards* the question and its solution. For example, in No. 70, the very first words I wrote down were as follows:—
>
> "(1) down back-edge; up again; down again; and so on; (2) about ·7 of the way down the back-edge; (3) about 18° 18′; (4) about 14°."
>
> These answers are not quite correct; but at least they are *genuine*, as the results of *mental* work *only*. "A poor thing, Sir, but mine own!"[13]

Dodgson wrote more books and pamphlets on geometry than on any other topic in mathematics.[14] Only his two books, *Euclid and His Modern*

---

11. *The Elements of Euclid*, edited by Isaac Todhunter (London: J. M. Dent & Sons, 1933), 257.

12. Heath, vol. 1, 278–79.

13. C. L. Dodgson, *Curiosa Mathematica, Part II: Pillow Problems*, 4th ed. (1895; reprint, New York: Dover Publications, 1958), xiii. Dodgson generalized the puzzle sometime between 1890 and 1893. For a complete discussion the reader should consult Warren Weaver's "Lewis Carroll and a Geometrical Paradox," *American Mathematical Monthly* 45 (1938), 234–36.

14. A recent discovery by David Singmaster is an exchange of correspondence between Dodgson and Richard A. Proctor of New York on Euclid's theory of parallels

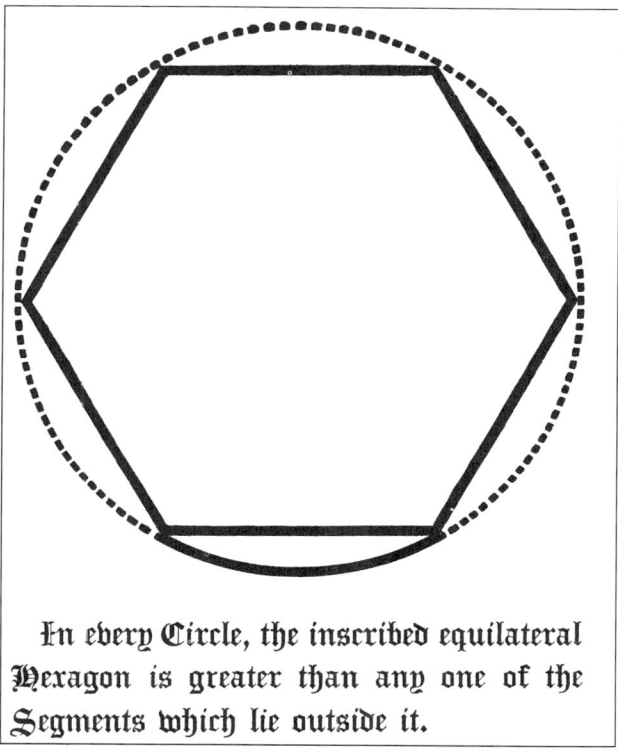

Frontispiece, with the statement of Charles L. Dodgson's alternative Euclidian parallel axiom, from Curiosa Mathematica, Part I: A New Theory of Parallels *(1888).*

Rivals and *Curiosa Mathematica, Part I: A New Theory of Parallels* merit serious consideration. In *Euclid and His Modern Rivals* he questioned assumptions that were not in accordance with Euclid's principles, drew important fine distinctions missed by other experts, and exposed the fallacy of "direction" theory, a method used in textbooks where "direction" is considered to be an undefined notion, and showed it to be an idea derived

---

which appeared in *Knowledge* 6 (1884). It includes Proctor's "Chats about Geometrical Measurement" (337–39), Dodgson's criticisms of Proctor's arguments in Letter 1495 (390–91), and Proctor's response in Letter 1542 (531). I am grateful to Martin Gardner for this information.

from properties of parallel lines.[15] In *Curiosa Mathematica, Part I* Dodgson introduced a novel equivalent of Euclid's parallel postulate: in any circle, the inscribed hexagon is greater than any one of the segments that lie outside it.[16] Although he explored non-Euclidean thoughts concerning infinities and infinitesimals, for example, that space has a limit and that it has none, he found them to be beyond reason. For Dodgson the fundamental assumptions about the universe were Euclidean. How could he accept the non-Euclidean theorem that the area of a triangle is finite when all its sides are infinite? It defies "reason" and must, therefore, be "nonsense."

15. Heath describes several of these instances: the mistake made by some editors in regarding I,7 as having no other purpose than to prove I,8, whereas Dodgson shows it supplements I,1 and I,22 and is therefore an equivalent of III,10 (v. 1, p. 261); the failure to see that Playfair's axiom not only involves Euclid's parallel postulate, but also involves more than that which is unnecessary (v. 1, p. 313); the lack of evidence in VI,33 that Euclid recognized angles greater than two right angles (v. 2, pp. 275–76). In *Euclid and His Modern Rivals,* the main protagonist of "direction" theory is J. M. Wilson. Dodgson devotes forty-seven pages to this "rival's" book, *Elementary Geometry* (2d ed., 1869).

16. In the third and fourth editions of *Curiosa Mathematica, Part I,* the hexagon is replaced by a tetragon (square).

1. *Notes on the First Two Books of Euclid*
DESIGNED FOR CANDIDATES FOR RESPONSIONS

[1860: LCH 25, LCAT 332: Princeton]

This pamphlet was published anonymously, but there is no doubt that Dodgson is the author. It is divided into four sections: "Notes on the Definitions," "Additional Definitions," "Notes on the Propositions," and "List of Abbreviations." Clearly meant to help students pass the first of three examinations for the B.A. degree, the pamphlet contains explanations, modifications, and clarifications of important concepts, terms, and methodology that a candidate would need to know to answer examination questions based on Euclid, Books I and II.

NOTES ON THE DEFINITIONS

A POINT. It must not be supposed that a Point has *only* negative qualities, in which case it would be identical with absolute *nothing*. It has the positive quality of *position*, and it is this which enables us to distinguish one Point from another.

A PLANE ANGLE. It should be observed that an angle is *not* the point where two lines intersect; it is *not* the lines themselves; it is *not* the space between them; but it *is* the "inclination," (or "bending,") of the two lines to each other.

A TRIANGLE. A Triangle may be said to consist of 6 parts, viz., 3 sides and 3 angles. In order to prove one Triangle equal in all respects to another, it must be first given, (or proved) that 3 of these 6 parts in the one are equal to the corresponding 3 in the other. The various cases are:—

Prop. IV. Two sides and the included angle of the one equal to two sides and the included angle of the other.

Prop. VII. The three sides of the one equal to the three sides of the other.

Prop. XXVI. Two angles and one side of the one equal to two angles and one side of the other.

ADDITIONAL DEFINITIONS
I.

A Postulate is something to be done, for which no proof is given. (From *postulatum,* because it is *demanded* that the reader should grant it to be possible.)

II.

An Axiom is something to be believed, for which no proof is given. (From ἀξίωμα, because it is *thought worthy* of belief.)

III.

A Proposition is something either to be done, or to be believed, for which a proof is given. (From *propositum,* because it is *put before* the reader.) Propositions therefore are of 2 kinds, Problems, and Theorems.

IV.

A Problem is something to be done, for which a proof is given. (From πρόβλημα, for the same reason as the last.)

V.

A Theorem is something to be believed, for which a proof is given. (From θεώρημα, because it is to be *considered,* as to its being true or false.)

VI.

The Enunciation of a Proposition states (1) what is given, (2) what is required to be done or proved true: these two parts of the Enunciation are called THE Data, and THE Quæsita.

VII.

The Hypothesis of a Proposition is the same as its Data: it is a word used only with reference to Theorems. (From ὑπόθεσις, because it is *supposed* to be true.)

VIII.

A Corollary is something proved in the course of Proposition, which it was not the object of the Proposition to prove. (From *corolla,* because it is a sort of ornament, or *garland,* of the Proposition.)

IX.

Two Propositions are said to be converse to each other when that which is given in the first is to be done or believed in the second, and that which is to be done or believed in the first is given in the second. For example, the Propositions, *"if 2 sides of a triangle be equal, the opposite angles are equal,"* and *"if 2 angles of a triangle be equal, the opposite sides are equal,"* are converse to each other.

X.

"A fortiori," (or *"much more then,"*) is a phrase used when the conclusion has a *stronger* claim to be believed than any of the facts from which it is proved. For example, "3 is greater than 2, and 4 is greater than 3; *much more then* is 4 greater than 2."

XI.

REDUCTIO AD ABSURDUM is the name of a form of argument, in which something is proved true, by showing that, if it were *not* true, an absurdity, (or impossibility,) would follow. For example, Prop. XXVII. *"If a straight line, falling on 2 other straight lines, make the alternate angles equal to each other; these 2 straight lines shall be parallel,"* is proved true, by showing that, if they were *not* parallel, an absurdity would follow.

The following scheme may assist the reader in understanding the first five Definitions given above.

Things set before us in Geometry may be of 2 kinds, viz.:

(without proofs)	(with proofs)
	i.e. PROPOSITIONS.

(to be done)	(to be believed)	(to be done)	(to be believed)
i.e. POSTULATES.	i.e. AXIOMS.	i.e. PROBLEMS.	i.e. THEOREMS.

A Proposition may be divided into the following parts;—

I. General Enunciation.
    (1) Data.
    (2) Quæsita.
II. Particular Enunciation.
    (1) Data.
    (2) Quæsita.
III. Construction.
IV. Proof.
V. Particular Conclusion.
VI. General Conclusion. (Only found in Theorems.)

Take, for example, Prop. VI of Book I.

I. General Enunciation.
    (1) Data.
If two angles of a triangle be equal to each other,
    (2) Quæsita.
the sides also which subtend the equal angles, shall be equal to each other.

II. Particular Enunciation.
    (1) Data.
Let ABC be a △ having ∠ ABC = ∠ ACB,
    (2) Quæsita.
then side AB shall = side AC.

III. Construction.
For, if AB ≠ AC, one of them is > the other;
let AB be > AC;
from BA cut off BD = AC, and join DC.

IV. Proof.
Then, in △s DBC, ABC,
∵ DB = AC, and BC is common, and ∠ DBC = ∠ ABC;
∴ base DC = base AB,
and △ DBC = △ ABC,
the < = the >, which is absurd;
∴ AB is not ≠ AC;

V. Particular Conclusion.
that is, AB = AC.

VI. General Conclusion.
Therefore, if two triangles, &c.   Q.E.D.

NOTES ON THE PROPOSITIONS

When two Propositions are converse to each other, the *second* has generally no "construction," and is proved by a "reductio ad absurdum."

Book I. Prop. XVI. The second part may be proved thus:—

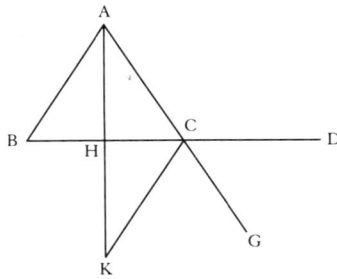

Bisect BC in H, join AH, and produce it to K, making HK = AH, and join CK;

∵   AH = HK, and BH = HC, and ∠ AHB = ∠ KHC, (being vertical angles);

∴   CK = AB, and ∠ KCH = ∠ ABH, i.e. ∠ ABC;

but ∠ GCB > ∠ KCB;

∴   ∠ GCB > ∠ ABC;

and ∠ ACD = ∠ GCB, (being vertical angles);

∴   ∠ ACD > ∠ ABC.   Q.E.D.

Book I. Prop. XXXV. Here the taking away of the triangles from the trapezium is meant to be done thus:—

From the trapezium ABCF take the triangle FDC, and observe that there remains the parallelogram ABCD; then replace this triangle, so as to make the trapezium complete again; then take away from it the triangle EAB, and observe that there remains the parallelogram EBCF. Since, in the two subtractions, equal triangles are taken away, the remainders are equal.

Book II. Prop. VII. It should be observed, that on the space CGKB there are supposed to be *two* squares, one lying over the other. One of these is supposed to be added to the complement AG, the other to the complement GE.

NOTES ON EUCLID, I AND II

Book II. Prop. XIII. This may be proved more simply thus:—

        ∵ (in fig. 1.) BC is divided into 2 parts in D,
and ∵ (in fig. 2.) BD is divided into 2 parts in C,
∴ (in both) $BC^2 + BD^2 = 2BC.BD + CD^2$;
add to each AD^2;
∴ $BC^2 + BD^2 + AD^2 = 2BC.BD + CD^2 + AD^2$;
but $BD^2 + AD^2 = AB^2$,
and $CD^2 + AD^2 = AC^2$;
∴ $BC^2 + AB^2 = 2BC.BD + AC^2$;
∴ $AC^2 < BC^2 + AB^2$ by $2BC.BD$.
Also ∵ (in fig. 3.) $AB^2 = BC^2 + AC^2$,
add to each BC^2;
∴ $BC^2 + AB^2 = 2BC^2 + AC^2$;
∴ $AC^2 < BC^2 + AB^2$ by $2BC^2$, i.e. by $2BC.BC$.
Therefore in any triangle, &c.   Q.E.D.

LIST OF ABBREVIATIONS

/AB	*means*	the line AB
∠ BAC	"	the angle BAC
L	"	right angle
⊥	"	at right angles to
⊙	"	circle
△	"	triangle
∥	"	parallel to
∦	"	not parallel to
=	"	equal to
≠	"	not equal to
>	"	greater than
<	"	less than
≯	"	not greater than
≮	"	not less than
□	"	parallelogram
AB^2	"	the square described on AB
AB.AC	"	the rectangle contained by AB, AC
∵	"	because
∴	"	therefore
Q.E.F.	"	quod erat faciendum
Q.E.D.	"	quod erat demonstrandum

N.B. The symbol for "less than" must be carefully distinguished from that for "angle."

The symbols for "greater than" and "less than" may be distinguished from each other by remembering that the *greater* end of the symbol is placed next the *greater* quantity. Thus "A > B" shows that A is the greater; "A < B" that B is the greater.

The symbols for "because" and "therefore" may be distinguished thus: when "because" is used, the pyramid (∵) is balanced on the point, to show that the argument is still unsettled; but when "therefore" is used, the pyramid (∴) is placed on the base, to show that the argument is settled.

2. The Enunciations of the Propositions and Corollaries, Together with Questions on the Definitions, Postulates, Axioms, &c. in Euclid, Books I and II

[1863: LCH 34: Bodleian]

This pamphlet, published anonymously, contains many corrections in Dodgson's hand; he incorporated them into the much larger pamphlet on Euclid's Books I–VI that he published ten years later (see item 4). The present pamphlet presumably lists all of the material needed to answer the geometry questions on the Responsions examination. An outline for the pamphlet in Dodgson's hand is in the Stephen Rudin Collection. The authors of the *Lewis Carroll Handbook* suggest that Dodgson modified a vulgus of this work that had existed from 1823 and had been published in a common form in 1862. Following Collingwood in his edition of Lewis Carroll's diaries, they also suggest that Dodgson's entry for 17 November 1862, "I think an interesting collection might be made of axioms tacitly assumed by Euclid," and his entry for 19 October 1863, "Took to the University Press the MS. of a small book of enunciations etc. for Euclid I,II," are both referring to this pamphlet.[1] The second certainly does but not the first. Aside from indirect references in *Curiosa Mathematica,* Part I, Dodgson only deals with Euclid's tacitly assumed axioms in three publications: in the preface to *The Enunciations of Euclid I–VI* (item 4), where he mentions the corollary to I, 2 as having been tacitly assumed by Euclid in I, 4; in *Euclid and His Modern Rivals* in his description of one of the rivals, Cuthbertson; and in *The Fifth Book of Euclid* (item 3), where he lists the four axioms tacitly assumed by Euclid that Simson supplied.

Oxford: T. Combe, E. P. Hall and H. Latham, 1863.

1. *The Diaries of Lewis Carroll,* edited by Roger L. Green (New York: Oxford University Press, 1954), vol. I, 189, 206.

EUCLID.
Book I.

Define the following terms:—

1. Point.
2. Line.
3. Right Line.
4. Superficies.
5. Plane Superficies.
6. Plane Angle.
7. Plane Rectilineal Angle.
8. Right Angle.
9. Obtuse Angle.
10. Acute Angle.
11. Term, or Boundary.
12. Figure.
13. Circle.
14. Diameter of a Circle.
15. Semicircle.
16. Rectilineal Figure.
17. Trilateral Figure, or Triangle.
18. Quadrilateral Figure.
19. Multilateral Figure, or Polygon.
20. Equilateral Triangle.
21. Isosceles Triangle.
22. Scalene Triangle.
23. Right-angled Triangle.
24. Obtuse-angled Triangle.
25. Acute-angled Triangle.
26. Square.
27. Oblong.
28. Rhombus.
29. Rhomboid.
30. Trapezium.
31. Parallel Straight Lines.

32. Parallelogram.
33. Diameter, or Diagonal, of a Parallelogram.

Write out the Postulate concerning—

34. The drawing of a straight Line.
35. The producing of a terminated straight Line.
36. The describing of a Circle.

Write out the Axiom concerning—

37. Things equal to the same thing.
38. Equals added to equals.
39. Equals taken from equals.
40. Equals added to unequals.
41. Equals taken from unequals.
42. Things double of the same thing.
43. Things which are halves of the same thing.
44. Magnitudes which coincide with one another.
45. A straight Line meeting two straight Lines, so as to make the two interior angles on the same side of it taken together less than two right angles.

Explain the following terms:—

46. Definition.
47. Postulate.
48. Axiom.
49. Proposition.
50. Problem.
51. Theorem.
52. Corollary.
53. *"A fortiori."* Give an instance.
54. *"Reductio ad absurdum."* Give an instance.

55. When are Propositions said to be converse to each other? Give an instance.
56. Into what parts may a Proposition be divided?

Prop. i. Probl. To describe an equilateral triangle on a given finite straight line.

Prop. ii. Probl. From a given point to draw a straight line equal to a given straight line.

Prop. iii. Probl. From the greater of two given straight lines to cut off a part equal to the less.

Prop. iv. Theor. If two sides of one triangle be equal to two sides of another, each to each, and the angles included by those sides equal; the bases or *third sides* shall be equal; as also the triangles; and their other angles shall be equal; viz. those to which the equal sides are opposite.

Prop. v. Theor. The angles at the base of an isosceles triangle are equal; and if the equal sides be produced, the angles on the other side of the base shall be equal.

Cor. Every equilateral triangle is also equiangular.

Prop. vi. Theor. If two angles of a triangle be equal to each other; the sides also, which subtend, or are opposite to, the equal angles, are equal to each other.

Cor. Every equiangular triangle is also equilateral.

Prop. vii. Theor. On the same base, and on the same side of it, there cannot be two triangles which have their sides terminated in one extremity of the base equal to each other, and at the same time their sides terminated in the other extremity equal.

Prop. viii. Theor. If two triangles have two sides of the one equal to two sides of the other, each to each, and likewise their bases equal; the angle contained by the two sides of the one shall be equal to the angle contained by the two sides equal to them of the other.

Prop. ix. Probl. To bisect a given rectilineal angle; that is, to divide it into two equal angles.

Prop. x. Probl. To bisect a given finite straight line; that is, to divide it into two equal parts.

Prop. xi. Probl. To draw a straight line at right angles to a given straight line from a given point in the same.

Cor. By the help of this Problem it may be proved that two straight lines cannot have a common segment.

PROP. XII. PROBL. To draw a straight line perpendicular to a given straight line of unlimited length, from a given point without it.

PROP. XIII. THEOR. The angles, which one straight line makes with another on one side of it, are either two right angles, or are together equal to two right angles.

PROP. XIV. THEOR. If, at a point in a straight line, two other straight lines on opposite sides of it make the adjacent angles together equal to two right angles; these two straight lines shall be in one and the same straight line.

PROP. XV. THEOR. If two straight lines cut one another; the vertical, or *opposite,* angles shall be equal.

COR. 1. Hence the angles at the point of intersection of two straight lines are together equal to four right angles.

COR. 2. And consequently all the angles made by any number of lines meeting in one point, are together equal to four right angles.

PROP. XVI. THEOR. If one side of a triangle be produced; the exterior angle is greater than either of the interior opposite angles.

PROP. XVII. THEOR. Any two angles of a triangle are together less than two right angles.

PROP. XVIII. THEOR. The greater side of every triangle is opposite to the greater angle.

[i.e. If two sides of a triangle be unequal, the angle opposite to the greater side is greater than the angle opposite to the lesser side.]

PROP. XIX. THEOR. The greater angle of every triangle is subtended by the greater side, or *has the greater side opposite to it.*

[i.e. If two angles of a triangle be unequal, the side opposite to the greater angle is greater than the side opposite to the lesser angle.]

PROP. XX. THEOR. Any two sides of a triangle are together greater than the third.

PROP. XXI. THEOR. If from the ends of the side of a triangle two straight lines are drawn to a point within the triangle; these two straight lines shall be less than the other two sides of the triangle, but shall contain a greater angle.

PROP. XXII. PROBL. To make a triangle, the sides of which shall be equal to three given straight lines, but any two whatever of these must be greater than the third.

PROP. XXIII. PROBL. At a given point in a given straight line to make an angle equal to a given rectilineal angle.

PROP. XXIV. THEOR. If two triangles have two sides of the one equal to two sides of the other, each to each, but the angle contained by two sides of one of them greater than the angle contained by the two sides equal to them of the other; the base of that which has the greater angle shall be greater than the base of the other.

PROP. XXV. THEOR. If two triangles have two sides of the one equal to two sides of the other, each to each, but the base of the one greater than the base of the other; the angle contained by the sides of that which has the greater base shall be greater than the angle contained by the sides equal to them of the other.

PROP. XXVI. THEOR. If two triangles have two angles of the one equal to two angles of the other, each to each, and one side equal to one side; viz. either the sides adjacent to the equal angles, or opposite to the equal angles in each: then shall the other sides be equal, each to each, and also the third angle of the one to the third angle of the other.

PROP. XXVII. THEOR. If a straight line falling upon two other straight lines makes the alternate angles equal to each other; those two straight lines shall be parallel.

PROP. XXVIII. THEOR. If a straight line falling upon two other straight lines, makes the exterior angle equal to the interior and opposite angle on the same side of the line, or makes the interior angles on the same side together equal to two right angles; the two straight lines shall be parallel to each other.

[This contains two Propositions:

(α) If a straight line, falling upon two other straight lines, makes the exterior angle equal to the interior and opposite angle on the same side of the line; the two straight lines shall be parallel to each other.

(β) If a straight line, falling upon two other straight lines, makes the interior angles on the same side together equal to two right angles; the two straight lines shall be parallel to each other.]

PROP. XXIX. THEOR. If a straight line fall upon two parallel straight lines; it makes the alternate angles equal, and the exterior angle equal to the interior and opposite upon the same side, and the two interior angles upon the same side together equal to two right angles.

PROP. XXX. THEOR. Straight lines which are parallel to the same straight lines are parallel to one another.

PROP. XXXI. PROBL. To draw a straight line through a given point parallel to a given straight line.

PROP. XXXII. THEOR. If a side of any triangle be produced, the exterior angle is equal to the two interior and opposite angles; and the three interior angles of every triangle are equal to two right angles.

COR. 1. All the interior angles of any rectilineal figure, together with four right angles, are equal to twice as many right angles as the figure has sides.

COR. 2. All the exterior angles of any rectilineal figure are equal to four right angles.

PROP. XXXIII. THEOR. The straight lines which join the extremities of equal and parallel straight lines towards the same parts are themselves equal and parallel.

PROP. XXXIV. THEOR. The opposite sides and angles of parallelograms are equal to one another; and the diameter bisects them, that is, divides them into two equal parts.

PROP. XXXV. THEOR. Parallelograms on the same base, and between the same parallels, are equal to each other.

PROP. XXXVI. THEOR. Parallelograms upon equal bases, and between the same parallels, are equal to each other.

PROP. XXXVII. THEOR. Triangles on the same base, and between the same parallels, are equal.

PROP. XXXVIII. THEOR. Triangles upon equal bases, and between the same parallels, are equal.

PROP. XXXIX. THEOR. Equal triangles upon the same base, and upon the same side of it, are between the same parallels.

PROP. XL. THEOR. Equal triangles upon equal bases in the same straight line, and towards the same parts, are between the same parallels.

PROP. XLI. THEOR. If a parallelogram and a triangle be upon the same base, and between the same parallels; the parallelogram is double of the triangle.

PROP. XLII. PROBL. To describe a parallelogram, which shall be equal to a given triangle, and have one of its angles equal to a given rectilineal angle.

PROP. XLIII. THEOR. The complements of the parallelograms, which are about the diameter of any parallelogram, are equal to one another.

PROP. XLIV. PROBL. To a given straight line to apply a parallelogram which shall be equal to a given triangle, and have one of its angles equal to a given rectilineal angle.

PROP. XLV. PROBL. To describe a parallelogram equal to a given rectilineal figure, and having an angle equal to a given rectilineal angle.

COR. Hence it is manifest, how to a given straight line to apply a parallelogram, which shall have an angle equal to a given rectilineal angle, and shall be equal to a given rectilineal figure.

PROP. XLVI. PROBL. To describe a square upon a given straight line.

COR. Hence every parallelogram, that has one right angle, has all its angles right angles.

PROP. XLVII. THEOR. In any right-angled triangle, the square which is described upon the side subtending the right angle, is equal to the squares described upon the sides which contain the right angle.

PROP. XLVIII. THEOR. If the square described on one side of a triangle be equal to the squares described on the other two; the angle contained by those sides is a right angle.

Book II.

Define the following terms:—

1. Rectangle.
2. Complements of a Parallelogram.
3. Gnomon.

PROP. I. THEOR. If there be two straight lines, one of which is divided into any number of parts; the rectangle contained by the two straight lines is equal to the rectangles contained by the undivided line, and the several parts of the divided line.

PROP. II. THEOR. If a straight line be divided into any two parts; the rectangles contained by the whole and each of the parts are together equal to the square of the whole line.

PROP. III. THEOR. If a straight line be divided into any two parts; the rectangle contained by the whole and one of the parts is equal to the rectangle contained by the two parts, together with the square of the aforesaid part.

PROP. IV. THEOR. If a straight line be divided into any two parts; the square of the whole line is equal to the squares of the two parts together with twice the rectangle contained by the parts.

COR. Hence the parallelograms about the diameter of a square are likewise squares.

PROP. V. THEOR. If a straight line be divided into two equal parts, and also into two unequal parts; the rectangle contained by the unequal parts, together with the square of the line between the points of section, is equal to the square of half the line.

COR. From this proposition it is manifest that the difference of the squares of two unequal lines is equal to the rectangle contained by their sum and difference.

PROP. VI. THEOR. If a straight line be bisected, and produced to any point; the rectangle contained by the whole line produced, and the part of it produced, together with the square of half the line bisected, is equal to the square of the line made up of the half and the part produced.

PROP. VII. THEOR. If a straight line be divided into any two parts; the squares of the whole line and of one of the parts are equal to twice the rectangle contained by the whole and that part, together with the square of the other part.

PROP. VIII. THEOR. If a straight line be divided into any two parts; four times the rectangle contained by the whole line and one of the parts, together with the square of the other part, is equal to the square of the line made up of the whole and that part.

PROP. IX. THEOR. If a straight line be divided into two equal, and also into two unequal parts; the squares of the two unequal parts are together double of the square of half the line, and of the square of the line between the points of section.

PROP. X. THEOR. If a straight be bisected, and produced to any point; the square of the whole line thus produced, and the square of the part of it produced, are together double of the square of half the line, and of the square of the line made up of the half and the part produced.

PROP. XI. PROBL. To divide a given straight line into two parts, so that the rectangle contained by the whole and one of the parts shall be equal to the square of the other part.

PROP. XII. THEOR. In obtuse-angled triangles, if a perpendicular be drawn from either of the acute angles to the opposite side produced; the square of the side subtending the obtuse angle is greater than the squares of the sides which contain it by twice the rectangle contained by the side upon which, when produced, the perpendicular falls, and the straight line intercepted without the triangle between the perpendicular and the obtuse angle.

PROP. XIII. THEOR. In every triangle, the square of the side subtending any of the acute angles is less than the squares of the sides containing that angle, by twice the rectangle contained by either of these sides, and the straight line intercepted between the perpendicular let fall upon it from the opposite angle and the acute angle.

PROP. XIV. PROBL. To describe a square which shall be equal to a given rectilineal figure.

3. The Fifth Book of Euclid Treated Algebraically, so Far as It Relates to Commensurable Magnitudes, with Notes.

[1868: LCH 64, LCAT 349: HRHRC]

This work is similar to the *Notes* pamphlet, item 1, in that Dodgson includes explanations, modifications, and proofs of what passmen needed to know about Euclid's theory of ratios for Moderations, the second examination required for the baccalaureate degree at Oxford colleges. Despite the title, he also does discuss incommensurables, it being virtually impossible to tease the two completely apart.[1] In his diary entry of 16 January 1868, Dodgson wrote, "During my stay at Ripon [Dodgson's father was Canon of Ripon Cathedral and Dodgson stayed there during the first three months of the year], I have written almost all of the pamphlet on Euclid V by Algebra, with notes, and have gone on with the MS. on Geometric Conic Sections."[2] It is unclear whether the references to Algebra in the text and footnotes refer to a particular book or are meant to be algebraical equivalents of Euclid's definitions.

PREFACE

The theory of Incommensurable Magnitudes, without which the whole subject of Geometrical Proportion is so incomplete as

---

Oxford and London: James Parker and Co., 1868.
1. Dodgson remarks in footnote [11] on page 56 that Euclid's fourth definition excludes the case of one magnitude being infinitely greater than the other. However, it does not exclude finite incommensurables—they have this same property as commensurables.
2. *The Diaries of Lewis Carroll,* edited by Roger L. Green (New York: Oxford University Press, 1954), vol. II, 265.

Dodgson used the symbols *, †, ‡, and § for footnotes, which are replaced here by the bracketed numbers.

to be, from a logical point of view, utterly valueless, is nevertheless omitted, as far as possible, from the following treatise.

My reasons for this omission are two: first, that I believe it to be much too abstruse a subject for the ordinary Pass Examination; secondly, that it is not required in it. The exemption is a most necessary one, though the effect of it is to reduce the Vth and VIth Books, in the form in which they are now learned and accepted in the Schools, to a logical absurdity.

Whether it would not be preferable to substitute for these Books an equivalent quantity of Algebra, perhaps as far as Permutations and Combinations, is a question I do not here enter on. To supply, in its shortest form, that knowledge of the subject which is at present required and accepted in the Schools, is my object in putting forth this treatise. I hope that it may not be long wanted.

Euclid.

Definitions.

Algebra.

Definitions.

I.

To multiply a magnitude by an integer *is to take so many of it as the integer denotes.*

II.

To divide a magnitude by an integer *is to make of it so many equal magnitudes as the integer denotes, and to take one of them.*

Axiom I.

To multiply a magnitude by a fraction *is to make of it so many equal magnitudes as the denominator denotes, and to take so many of them as the numerator denotes.*

II.

A greater magnitude is said to be a **multiple** of a less, when the less measures the greater; that is, "when the less is contained a certain number of times exactly in the greater."[1]

I.

A less magnitude is said to be a part of a greater, when the less measures the greater; that is, "when the less is contained a certain number of times exactly in the greater."[3]

III.

If a magnitude be multiplied by an integer, the magnitude so obtained is called a **multiple** *of the first.* [2]

IV.

If a magnitude be divided by an integer, or (which is the same thing) be multiplied by a fraction whose numerator is unity, the magnitude so obtained is called a **part,** *or* **measure,** *of the first.*[4]

V.

If a magnitude be multiplied by a fraction, the magnitude so obtained is called a **fraction** *of the first.*[5]

VI.

If a magnitude be multiplied by a number, integral or fractional, the number is said to **denote what multiple, part, or fraction, the magnitude so obtained is of the first.**[6]

[1] *Euc. Def. I, II.* These definitions are incomplete, since they do not include the case of *equal* magnitudes. If we multiply, or divide, any magnitude by unity, we obtain the same magnitude again: thus any magnitude is a *multiple* of itself, and also a *part* of itself.

[2] *Alg. Def. III.* Thus, if "six inches" be multiplied by 3, we obtain "eighteen inches," which is a *multiple* of "six inches."

[3] See note [1].

[4] *Alg. Def. IV.* Thus, if "six inches" be divided by 3, or multiplied by $\frac{1}{3}$, we obtain "two inches," which is a *part* of "six inches."

[5] *Alg. Def. V.* Thus, if "six inches" be multiplied by $\frac{2}{3}$, we obtain "four inches," which is a *fraction* of "six inches."

[6] *Alg. Def. VI.* Thus, the numbers 3, $\frac{1}{3}$, $\frac{2}{3}$, are said to denote what multiple, part, or fraction, each of the magnitudes, "eighteen inches," "two inches," "four inches," is of "six inches."

VII.

If 2 or more magnitudes be each multiplied by the same number, the magnitudes so obtained are called **equimultiples** *of the first magnitudes.*[7]

VIII.

If 2 or more magnitudes be such that a certain magnitude is a measure of each of them, it is called a common measure *of them: and they are said to be* **commensurable**.[8]

Proposition I.

If 2 magnitudes be commensurable, each is a multiple, part, or fraction, of the other.

For the common measure may be obtained from the first by dividing by an integer;

and the second may be obtained from that common measure by multiplying by an integer;

hence the second magnitude may be obtained from the first by multiplying by the number, integral or fractional, formed by taking as a denominator the first integer and as a numerator the second;

and, if this fraction, when reduced to lowest terms, have for its denominator unity, the second is a multiple of the first;

---

[7] *Alg. Def. VII.* Thus, if "two shillings," "three miles," "half-an-ounce," be each multiplied by 4, we obtain "eight shillings," "twelve miles," "two ounces"; and these are *equimultiples* of the first magnitudes.

[8] *Alg. Def. VIII.* Thus, "tenpence" is a *common measure* of "half-a-crown" and "four and twopence": and these sums are said to be *commensurable*.

if it have for its numerator unity, the second is a part of the first;

in all other cases, the second is a fraction of the first.

Similarly, it may be proved that the first is a multiple, part, or fraction of the second.

Therefore, if 2 magnitudes, &c.

Q.E.D.[9]

III.

Ratio is a mutual relation of two magnitudes of the same kind to one another, in respect of quantity.[10]

IV.

Magnitudes are said to have a ratio to each other when the less can be multiplied so as to exceed the other.[11]

IX.

Ratio *is a certain relation which a magnitude has to another of the same kind, in respect of quantity.*

X.

A magnitude is said to have a ratio to another of the same kind when either they are equal or the less can be multiplied so as to exceed the other.

XI.

If 2 magnitudes be commensurable, the number, integral or fractional, denoting what multiple, part, or fraction, the one is of the other, is called **the ratio of the one to the**

|9| *Alg. Prop. I.* Thus, each of the two sums, "half-a-crown" and "four and twopence," is a multiple part, or fraction of the other.

For "tenpence," the common measure, may be obtained from "half-a-crown" by dividing by 3;

and "four and twopence" may be obtained from "tenpence," by multiplying by 5;

hence "four and twopence" may be obtained from "half-a-crown," by multiplying by $\frac{5}{3}$;

hence, in this case, "four and twopence" is a *fraction* of "half-a-crown."

|10| *Euc. Def. III.* This is too vague to be of practical use. Euclid gives a more exact definition afterwards (see Def. V.) which is meant to include *incommensurable* magnitudes. As we are going to treat of *commensurable* magnitudes only, we may substitute a simpler definition for that of Euclid. (See Alg. Def. XI.)

|11| *Euc. Def. IV.* This is to exclude the case of one magnitude being *infinitely greater* than the other.

other, *or the ratio which the one has to the other.*[12]

V.

The first of four magnitudes is said to have the same ratio to the second, that the third has to the fourth, when any equimultiples whatever of the first and third being taken, and any equimultiples whatever of the second and fourth, if the multiple of the first be less than that of the second, the multiple of the third is also less than that of the fourth, and if the multiple of the first be equal to that of the second, the multiple of the third is also equal to that of the fourth, and if the multiple of the first be greater than that of the second, the multiple of the third is also greater than that of the fourth.

XII.

If there be 4 magnitudes such that the first and second are incommensurable, as also are the third and fourth, and such as to satisfy Euclid's Vth Definition; then the first is said to **have the same ratio to the second that the third has to the fourth.**

Axiom II.

If 2 magnitudes be incommensurable, that which is called "the ratio of the one to the other" cannot be expressed as a number.

VI.

Magnitudes which have the same ratio are called **proportionals.**

Definition XIII.

If there be 2 or more pairs of magnitudes such that the ratio of the first to the second in each pair is the same, or if there be 3 or more magnitudes such that the ratio of the first to the second, of the second to the third, and so on, is the same: in either case the

---

[12] *Alg. Def. XI.* Thus, $\frac{5}{3}$ is the *ratio* of "four and twopence" to "half-a-crown." Again, $\frac{5}{3}$ is the *ratio* of "fifteen miles" to "nine miles."

[Euclid uses the phrase "continual proportionals," but does not define it.] magnitudes are called **proportionals**, and in the latter case they are called **continual proportionals**.[13]

Axiom III.

If 3 or more magnitudes be proportionals, and if one of the ratios be a number; so also is each of the others. And if one of the ratios be between 2 incommensurable magnitudes (so that it cannot be expressed as a number); so also is each of the others.

Convention I.

If 2 or more magnitudes be commensurable, let it be agreed to select a common measure of them, and let the numbers denoting what multiples they respectively are of that common measure be said **to represent** *them.*[14]

Proposition II.

If 2 commensurable magnitudes be represented by numbers, and if a number, integral or fractional, be formed by taking as a numerator the one of these numbers and as a de-

---

[13] *Alg. Def. XIII.* Thus, "four and twopence," "half-a-crown," "fifteen miles," "nine miles," are *proportionals*. Thus also, "four and twopence," "half-a-crown," "one and sixpence" are *continual proportionals*.

From this and the preceding two Definitions we may gather that there are two cases, and only two, in which a set of four magnitudes are called "proportionals."

One is, where the first and second are commensurable, as also are the third and fourth; and where the *number,* integral or fractional, expressing the ratio of the first to the second, is the same as that expressing the ratio of the third to the fourth.

The other is, where the first and second are *in*commensurable, as also are the third and fourth; and where the four magnitudes fulfil the conditions enumerated in Euclid's Fifth Definition.

[14] *Alg. Conv. I.* Thus, "four and twopence," "half-a-crown," "one and sixpence," have as common measures "twopence," "a penny," &c.: if we select "twopence" these sums are represented by the numbers 25, 15, 9; if "a penny," by 50, 30, 18.

nominator the other: this number is the ratio of the one magnitude to the other.

For the selected common measure may be obtained from that other magnitude by dividing by the number taken as a denominator;

and the one magnitude may be obtained from the common measure by multiplying by the number taken as a numerator;

hence the one magnitude may be obtained from the other by multiplying by this number, integral or fractional;

∴ this number denotes what multiple, part, or fraction, the one is of the other; (Def. VI.

∴ *this number is the ratio of the one magnitude to the other.*

(Def. VIII.

Therefore, *if* 2 *commensurable magnitudes*, &c. Q.E.D.[15]

Conventions (continued).

II.

Let *"the ratio of* a *to* b*"* *be represented by* "a:b."

III.

Let *"the ratio of* a *to* b *is the same as the ratio of* c *to* d," *or* "a, b, c, d *are proportionals" be represented by* "a:b::c:d."[16]

---

[15] *Alg. Prop. II.* Thus, if "four and twopence" and "half-a-crown" be represented by the numbers 25, 15, the ratio of "four and twopence" to "half-a-crown" is $\frac{25}{15}$;

if they be represented by 50, 30, the ratio is $\frac{50}{30}$;

and these results coincide, since each fraction, when reduced to its lowest terms, becomes $\frac{5}{3}$.

[16] *Alg. Conv. III.* This would be read "as *a* is to *b*, so is *c* to *d*."

Axioms (continued).

IV.

If a:b::c:d, *then*
$$\frac{a}{b} = \frac{c}{d}.$$

V.

If $\frac{a}{b} = \frac{c}{d}$, *then*
a:b::c:d.

VII.

When of the equimultiples of four magnitudes, taken as in the fifth definition, the multiple of the first is greater than the multiple of the second, but the multiple of the third is not greater than the multiple of the fourth, then the first is said to have to the second a greater ratio than the third has to the fourth; and the third is said to have to the fourth a less ratio than the first has to the second.

VIII.

Analogy, or proportion, is the similitude of ratios.

IX.

Proportion consists in three terms at least.[17]

X.

When three magnitudes are proportionals, the first is said to have to the third the **duplicate ratio** of that which it has to the second.[18]

---

[17] *Euc. Def. IX.* This is properly an Axiom.
[18] *Euc. Def. X.* Thus, "four and twopence" is said to have to "one and sixpence" the *duplicate ratio* of that which it has to "half-a-crown."

XI.

When four magnitudes are continued proportionals, the first is said to have to the fourth, the **triplicate ratio** of that which it has to the second, and so on, **quadruplicate,** &c. increasing the denomination still by unity, in any number of proportionals.[19]

Definition A.

When there are any number of magnitudes of the same kind, the first is said to have to the last of them the ratio **compounded** of the ratio which the first has to the second, and of the ratio which the second has to the third, and of the ratio which the third has to the fourth, and so on unto the last magnitude.[20]

XII.

In proportionals, the antecedent terms are said to be **homologous** to one another; as also the consequents to one another.[21]

XIII.

Permutando, or *alternando,* by permutation or alternately. This word is used when there are four proportionals, and it is inferred that the first is to the third, as the second is to the fourth.

(Prop. XVI.

---

[19] *Euc. Def. XI.* The 4 magnitudes "eight shillings," "four shillings," "two shillings," "one shilling," are continual proportionals: hence, "eight shillings" is said to have to "one shilling," the *triplicate ratio* of that which it has to "four shillings."

[20] *Euc. Def. A.* Thus, if we take at random the 4 magnitudes "half-a-crown," "three shillings," "three and twopence," "three and fivepence"; then "half-a-crown" is said to have to "three and fivepence" the ratio *compounded* of 3 ratios, namely, the ratio of "half-a-crown" to "three shillings," the ratio of "three shillings" to "three and twopence," and the ratio of "three and twopence" to "three and fivepence."

Again, if we represent 4 magnitudes of the same kind by the symbols "a, b, c, d"; then a is said to have to d the ratio *compounded* of the 3 ratios, $a:b, b:c, c:d$.

That is, $\frac{a}{d}$ is said to be *compounded* of $\frac{a}{b}, \frac{b}{c}, \frac{c}{d}$.

Now $\frac{a}{d}$ is the *product* of $\frac{a}{b} \cdot \frac{b}{c} \cdot \frac{c}{d}$.

Hence we see that, if a set of ratios can be expressed as numbers, the ratio said to be *compounded* of them is the *product* of these numbers.

[21] *Euc. Def. XII.* Thus, in the proportion $a:b::c:d$, a and c are *homologous;* and so are b and d.

XIV.

Invertendo, by inversion; when there are four proportionals, and it is inferred, that the second is to the first as the fourth is to the third. (Prop. *B.*

XV.

Componendo, by composition; when there are four proportionals, and it is inferred that the first together with the second, is to the second, as the third together with the fourth, is to the fourth. (Prop. XVIII.

XVI.

Dividendo, by division; when there are four proportionals, and it is inferred, that the excess of the first above the second, is to the second, as the excess of the third above the fourth, is to the fourth. (Prop. XVII.

XVII.

Convertendo, by conversion; when there are four proportionals, and it is inferred, that the first is to its excess above the second, as the third to its excess above the fourth. (Prop. *E.*

XVIII.

Ex æquali (sc. distantiâ), or *ex æquo,* from equality of distance: when there is any number of magnitudes more than two, and as many others, such that they are proportionals when taken two and two of each rank, and it is inferred, that the first is to the last of the first rank of magnitudes, as the first is to the last of the others: "Of this there are the two following kinds, which arise from the different order in which the magnitudes are taken, two and two."

XIX.

Ex æquali, from equality. This term is used simply by itself, when the first magnitude is to the second of the first rank, as the first to the second of the other rank; and as the second is to the third of the first rank, so is the second to the third of the other;

and so on in order: and the inference is as mentioned in the preceding definition; whence this is called ordinate proportion.
(Prop. XXII.

XX.

Ex æquali in proportione perturbatâ seu inordinatâ, from equality in perturbate or disorderly proportion. This term is used when the first magnitude is to the second of the first rank, as the last but one is to the last of the second rank; and as the second is to the third of the first rank, so is the last but two to the last but one of the second rank; and as the third is to the fourth of the first rank, so is the third from the last to the last but two of the second rank; and so on in a cross order: and the inference is as in the XVIIIth definition. (Prop. XXIII.

Axioms.

I.

Equimultiples of the same, or of equal magnitudes, are equal to one another.

II.

Those magnitudes, of which the same or equal magnitudes are equimultiples, are equal to one another.

III.

A multiple of a greater magnitude is greater than the same multiple of a less.

IV.

That magnitude, of which a multiple is greater than the same multiple of another, is greater than that other magnitude.

HINTS ON PROVING PROPOSITIONS.

Most of the Propositions of the Vth Book may be proved by observing the following three rules:—

1. Express the "data" in algebraical language and reduce it to its simplest form.

2. Do the same with the "conclusion."

3. Seek for a connecting link between these two results. In many cases this is obvious, but, where it is not so, obtain, from the data, values of some of the magnitudes in terms of others, then take the terms of the conclusion one by one, and substitute in them the values obtained from the data.

These rules will be easily understood by applying them to an example. Let us take Prop. II. "If the first magnitude be the same multiple of the second that the third is of the fourth, and the fifth the same multiple of the second that the sixth is of the fourth; then the first and fifth together shall be the same multiple of the second that the third and sixth together are of the fourth."

1. We represent the magnitudes by $a, b, c, d, e, f,$ making $a = mb$, so that $c = md$, and making $e = nb$, so that $f = nd$. This *cannot* be reduced to a simpler form.

2. The "conclusion" is that $(a + e)$ is the same multiple of b as $(c + f)$ is of d. And this also cannot be reduced to a simpler form.

3. No connecting link is obvious: we have therefore to obtain, from the data, values of some of the magnitudes in terms of others; but these we already have, viz. $a = mb, c = md, e = nb, f = nd$. Nothing remains then but to take the terms of the conclusion by one by, and substitute in them these values. Thus $(a + e) = mb + nb) = (m + n) \cdot b$, and $(c + f) = (md + nd) = (m + n) \cdot d$. Hence we see that $(a + e)$ is $(m + n)$ times b, and that $(c + f)$ is $(m + n)$ times d, which proves the "conclusion": we then arrange the proof thus:—

$$\therefore \quad a + e = mb + nb = (m + n) \cdot b;$$
$$\text{and } c + f = md + nd = (m + n) \cdot d. \qquad \text{Q.E.D.}$$

Again, let us take Prop. IV, "If the first of four magnitudes has the same ratio to the second which the third has to the fourth; then any equimultiples whatever of the first and third shall have the same ratio to any equimultiples of the second and fourth, viz. 'the equimultiple of the first shall have the same ratio to that

FIFTH BOOK OF EUCLID 65

of the second, which the equimultiple of the third has to that of the fourth'."

1. We represent the magnitudes by a, b, c, d, so that $a:b::c:d$. We represent the equimultiples of a, c, by A, C, making $A = ma$, so that $C = mc$; and we represent the equimultiples of b, d, by B, D, making $B = nb$, so that $D = nd$. We then reduce the proportion, $a:b::c:d$, to its simplest form, viz. $\dfrac{a}{b} = \dfrac{c}{d}$.

2. The "conclusion" is $A:B::C:D$; and this, reduced to its simplest form, becomes $\dfrac{A}{B} = \dfrac{C}{D}$.

3. We have now to seek for a link to connect the equation $\dfrac{a}{b} = \dfrac{c}{d}$ with the equation $\dfrac{A}{B} = \dfrac{C}{D}$, and it is tolerably obvious that this may be done by multiplying both sides of the first by $\dfrac{m}{n}$; we thus obtain $\dfrac{ma}{nb} = \dfrac{mc}{nd}$, and thence $\dfrac{A}{B} = \dfrac{C}{D}$: we then arrange the proof thus:—

$\because$ $a:b::c:d$;

$\therefore$ $\dfrac{a}{b} = \dfrac{c}{d}$;

$\therefore$ $\left(\times^{\text{ng}} \text{ by } \dfrac{m}{n}\right), \dfrac{ma}{nb} = \dfrac{mc}{nd}$;

$\therefore$ $\dfrac{A}{B} = \dfrac{C}{D}$;

$\therefore$ $A:B::C:D.$ Q.E.D.

But suppose this connecting link were not obvious. We then obtain, from the data, values of some of the magnitudes in terms of others, thus: $\dfrac{a}{b} = \dfrac{c}{d} = k$; $\therefore a = kb, c = kd$. Now the terms of the conclusion are $\dfrac{A}{B}, \dfrac{C}{D}$; taking these one by one, and substituting in them the values obtained from the data, we get $\dfrac{A}{B} = \dfrac{ma}{nb} = \dfrac{mkb}{nb} = \dfrac{mk}{n}$, and $\dfrac{C}{D} = \dfrac{mc}{nd} = \dfrac{mkd}{nd} = \dfrac{mk}{n}$. Having thus proved that $\dfrac{A}{B}$ and $\dfrac{C}{D}$ are each equal to the same thing, viz. $\dfrac{mk}{n}$, we see that they are equal to each other, and the conclusion follows as before: we then arrange the proof thus:—

$$\because \quad a:b::c:d;$$
$$\therefore \quad \frac{a}{b} = \frac{c}{d} = k \text{ (say)};$$
$$\therefore \quad a = kb,\ c = kd;$$
$$\text{now} \quad \frac{A}{B} = \frac{ma}{nb} = \frac{mkb}{nb} = \frac{mk}{n};$$
$$\text{and} \quad \frac{C}{D} = \frac{mc}{nd} = \frac{mkd}{nd} = \frac{mk}{n};$$
$$\therefore \quad \frac{A}{B} = \frac{C}{D};$$
$$\therefore \quad A:B::C:D. \qquad \text{Q.E.D.}$$

The method of proving the proposition by introducing "k," and thus reducing the equation "$\dfrac{a}{b} = \dfrac{c}{d}$" to the form "$a = kb$, $c = kd$," is generally rather longer than the other, and should not be adopted without first seeking for a shorter connecting link.

The student is recommended to attempt to prove the propositions for himself before consulting the solutions here given.

Proposition I.

If any number of magnitudes be equimultiples of as many, each of each; what multiple soever any one of them is of its part, the same multiple shall all the first magnitudes be of all the others.

Let the first set be represented by A, B, C, &c., and the second by a, b, c, &c.; and let $A = ma$, so that $B = mb$, $C = mc$, &c.
Then
$$A + B + C + \&c. = ma + mb + mc + \&c.,$$
$$= m \cdot (a + b + c + \&c.).$$
Q.E.D.

Proposition II.

If the first magnitude be the same multiple of the second that the third is of the fourth, and the fifth the same multiple of the second that the sixth is of the fourth; then the first and fifth together shall be the same multiple of the second that the third and sixth together are of the fourth.

Let them be represented by a, b, c, d, e, f; and let $a = mb$, so that $c = md$; and let $e = nb$, so that $f = nd$.

Then $\quad a + e = mb + nb = (m + n) \cdot b$;
and $\quad c + f = md + nd = (m + n) \cdot d$. Q.E.D.

Corollary to Prop. II.

From this it is plain that, if any number of magnitudes be multiples of another, X; and as many others be the same multiples of x, each of each: then the whole of the first is the same multiple of X, that the whole of the last is of x.

Let the first set be represented by A, B, C, &c., with X; and the other set by a, b, c, &c., with x; and let $A = \alpha X, B = \beta X, C = \gamma X$, &c.; so that $a = \alpha x, b = \beta x, c = \gamma x$, &c.
Then

$$A + B + C + \&c. = \alpha X + \beta X + \gamma X + \&c.$$
$$= (\alpha + \beta + \gamma + \&c.) \cdot X;$$
and
$$a + b + c + \&c. = \alpha x + \beta x + \gamma x + \&c.$$
$$= (\alpha + \beta + \gamma + \&c.) \cdot x.$$
Q.E.D.

Proposition III.

If the first be the same multiple of the second which the third is of the fourth; and if of the first and third there be taken equimultiples, these shall be equimultiples, the one of the second, and the other of the fourth.

Let them be represented by a, b, c, d; and let $a = mb$, so that $c = md$; and let the new equimultiples of a, c, be represented by A, C; and let $A = na$, so that $C = nc$.

Then $A = na = nmb$;
and $C = nc = nmd$. Q.E.D.

Proposition IV.

If the first of four magnitudes has the same ratio to the second which the third has to the fourth; then any equimultiples whatever of the first and third shall have the same ratio to any equimultiples of the second and fourth, viz., "the equimultiple of the first shall have the same ratio to that of the second, which the equimultiple of the third has to that of the fourth."

Let them be represented by a, b, c, d; and let $a:b::c:d$; and let the equimultiples of a, c, be represented by A, C, and those of b, d, by B, D; and let $A = ma$, so that $C = mc$; and let $B = nb$, so that $D = nd$.

Because $a:b::c:d$;

$$\therefore \frac{a}{b} = \frac{c}{d};$$

$\therefore (\times^{\text{ng}}$ by $\frac{m}{n})$, $\frac{ma}{nb} = \frac{mc}{nd}$;

$$\therefore \frac{A}{B} = \frac{C}{D};$$

$\therefore A:B::C:D$. Q.E.D.

Corollaries to Prop. IV.

Likewise, if the first has the same ratio to the second, which the third has to the fourth, then also any equimultiples whatever of the first and third shall have the same ratio to the second and fourth; and in like manner, the first and the third shall have the same ratio to any equimultiples whatever of the second and fourth.

$$\because \frac{a}{b} = \frac{c}{d};$$

$\therefore (\times^{\text{ng}}$ by m), $\dfrac{ma}{b} = \dfrac{mc}{d}$;

$$\therefore \frac{A}{b} = \frac{C}{d};$$

$\therefore A:b::C:d.$

Again,

$$\because \frac{a}{b} = \frac{c}{d};$$

$\therefore (\div^{\text{ng}}$ by n), $\dfrac{a}{nb} = \dfrac{c}{nd}$;

$$\therefore \frac{a}{B} = \frac{c}{D};$$

$\therefore a:B::c:D.$

Proposition V.

If one magnitude be the same multiple of another, which a magnitude taken from the first is of a magnitude taken from the other; the remainder shall be the same multiple of the remainder that the whole is of the whole.

Let them be represented by A, B, and the magnitudes to be taken from them by a, b; and let $A = mB$, so that $a = mb$.

Then $A - a = mB - mb = m \cdot (B - b)$.   Q.E.D.

Proposition VI.

If two magnitudes be equimultiples of two others, and if equimultiples of these be taken from the first two; the remainders are either equal to these others, or equimultiples of these.

Let the first set be represented by A, B, and the second by a, b; and let $A = ma$, so that $B = mb$; and let the new equimultiples of a, b, be represented by α, β; and let $\alpha = na$, so that $\beta = nb$.

Then $A - \alpha = ma - na = (m - n) \cdot a$;
and $B - \beta = mb - nb = (m - n) \cdot b$;
and if $(m - n) = 1$, then $(A - \alpha) = a$,
and $(B - \beta) = b$;
but if $(m - n) \neq 1$, then $(A - \alpha)$, $(B - \beta)$ are equimultiples of a, b. Q.E.D.

Proposition A.

If the first of four magnitudes has to the second the same ratio which the third has to the fourth; then, if the first be greater than the second, the third is also greater than the fourth; if equal, equal; and if less, less.

Let them be represented by a, b, c, d; and let $a:b::c:d$.

Then $\dfrac{a}{b} = \dfrac{c}{d}$;

now let $a > b$;

∴ ($\div$^{ng} by b), $\dfrac{a}{b} > 1$;

∴ $\dfrac{c}{d} > 1$;

∴ ($\times$^{ng} by d), $c > d$.

Similarly if $a = b$, or if $a < b$. Q.E.D.

Proposition B. (*Invertendo.*)

If four magnitudes are proportionals, they are proportionals also when taken inversely.

Let them be represented by a, b, c, d; and let $a:b::c:d$.

Then $\dfrac{a}{b} = \dfrac{c}{d}$;

∴ ($\div^{ng}$ 1 by each), $\dfrac{b}{a} = \dfrac{d}{c}$.

∴ $b:a::d:c.$ Q.E.D.

Proposition C.

If the first be the same multiple or part of the second, that the third is of the fourth; the first is to the second, as the third is to the fourth.

This is included in the Definition of proportionals.

Proposition D.

If the first be to the second, as the third to the fourth; and if the first be a multiple or part of the second; the third is the same multiple or part of the fourth.

It is evident that the ratio of the first to the second is a number;

(Alg. Def. XI.

∴ the ratio of the third to the fourth is a number;

(Alg. Ax. III.

and these numbers are the same;   (Alg. Def. XIII.

∴ if the first be a multiple or part of the second, the third is the same multiple or part of the fourth. Q.E.D.

Proposition VII.

Equal magnitudes have the same ratio to the same magnitude; and the same has the same ratio to equal magnitudes.

Let the equal magnitudes be represented by a, b, and the other by x.

Because $a = b$;

∴ ($\div^{ng}$ by x), $\dfrac{a}{x} = \dfrac{b}{x}$;

∴ $a:x::b:x.$

Again, $\because a = b$;

$\therefore$ ($\div^{\text{ng}} x$ by each), $\dfrac{x}{a} = \dfrac{x}{b}$;

$\therefore x:a::x:b.$ Q.E.D.

Proposition VIII.

Of unequal magnitudes, the greater has a greater ratio to the same than the less has; and the same magnitude has a greater ratio to the less, than it has to the greater.

Let the unequal magnitudes be represented by a, b, and the other by x; and let $a > b$.

Because $a > b$;

$\therefore$ ($\div^{\text{ng}}$ by x), $\dfrac{a}{x} > \dfrac{b}{x}$;

$\therefore a:x$ is a greater ratio than $b:x$.

Again, $\because a > b$;

$\therefore$ ($\div^{\text{ng}}$ by ab), $\dfrac{1}{b} > \dfrac{1}{a}$;

$\therefore$ ($\times^{\text{ng}}$ by x), $\dfrac{x}{b} > \dfrac{x}{a}$;

$\therefore x:b$ is a greater ratio than $x:a$. Q.E.D.

Proposition IX.

Magnitudes which have the same ratio to the same magnitude, are equal to one another; and those to which the same magnitude has the same ratio, are equal to one another.

Let the first magnitudes be represented by a, b, and the other by x.

And first, let $a:x::b:x$;

$\therefore \dfrac{a}{x} = \dfrac{b}{x}$;

$\therefore$ ($\times^{\text{ng}}$ by x), $a = b$.

Secondly, let $x:a::x:b$;

FIFTH BOOK OF EUCLID

$$\therefore \quad \frac{x}{a} = \frac{x}{b};$$

$\therefore$ ($\div^{\text{ng}}$ by x), $\dfrac{1}{a} = \dfrac{1}{b};$

$\therefore$ ($\times^{\text{ng}}$ by ab), $b = a.$ Q.E.D.

Proposition X.

That magnitude which has a greater ratio than another has to the same magnitude, is the greater of the two; and that magnitude to which the same has a greater ratio than it has to another magnitude, is the lesser of the two.

Let them be represented by a, b, with x; and first let $a:x$ be a greater ratio than $b:x$.

Then $\dfrac{a}{x} > \dfrac{b}{x};$

$\therefore$ ($\times^{\text{ng}}$ by x), $a > b.$

Secondly, let $x:a$ be a greater ratio than $x:b$.

Then $\dfrac{x}{a} > \dfrac{x}{b};$

$\therefore$ ($\div^{\text{ng}}$ by x), $\dfrac{1}{a} > \dfrac{1}{b};$

$\therefore$ ($\times^{\text{ng}}$ by ab), $b > a;$

$\therefore \quad a < b.$ Q.E.D.

Proposition XI.

Ratios that are the same to the same ratio, are the same to one another.

Let the ratios be represented by $a:b$, $c:d$, with $x:y$; and let $a:b::x:y$, and $c:d::x:y$.

Then $\dfrac{a}{b} = \dfrac{x}{y},$ and $\dfrac{c}{d} = \dfrac{x}{y};$

$\therefore \quad \dfrac{a}{b} = \dfrac{c}{d};$

$\therefore \quad a:b::c:d.$ Q.E.D.

Proposition XII.

If any number of magnitudes be proportionals, as one of the antecedents is to its consequent, so shall all the antecedents taken together be to all the consequents.

Let them be represented by a, b, c, &c.; and let $a:b::c:d::e:f::$ &c.

Then $\dfrac{a}{b} = \dfrac{c}{d} = \dfrac{e}{f} =$ &c. $= k$ (say);

$\therefore\ a = kb, c = kd, e = kf$, &c.

$\therefore\ \dfrac{a + c + e + \&c.}{b + d + f + \&c.} = \dfrac{kb + kd + kf + \&c.}{b + d + f + \&c.}$

$= \dfrac{k(b + d + f + \&c.)}{b + d + f + \&c.} = k = \dfrac{a}{b};$

$\therefore\ a:b::(a + c + e + \&c.):(b + d + f + \&c.).$ Q.E.D.

Proposition XIII.

If the first has to the second the same ratio which the third has to the fourth, but the third to the fourth a greater ratio than the fifth to the sixth; the first shall also have to the second a greater ratio than the fifth to the sixth.

Let them be represented by a, b, c, d, e, f; and let $a:b::c:d$, but let $c:d$ be a greater ratio than $e:f$.

Then $\dfrac{a}{b} = \dfrac{c}{d}$, and $\dfrac{c}{d} > \dfrac{e}{f};$

$\therefore\ \dfrac{a}{b} > \dfrac{e}{f};$

$\therefore\ a:b$ is a greater ratio than $e:f$. Q.E.D.

Corollary to Prop. XIII.

And if the first have a greater ratio to the second, than the third has to the fourth, but the third the same ratio to the fourth, which the fifth has to the sixth; it may be demonstrated, in like manner, that the first has a greater ratio to the second, than the fifth has to the sixth.

Let $a:b$ be a greater ratio than $c:d$; but let $c:d::e:f$.
Then $\dfrac{a}{b} > \dfrac{c}{d}$, and $\dfrac{c}{d} = \dfrac{e}{f}$;
∴ $\dfrac{a}{b} > \dfrac{e}{f}$;
∴ $a:b$ is a greater ratio than $e:f$.   Q.E.D.

Proposition XIV.

If the first has to the second the same ratio which the third has to the fourth: then if the first be greater than the third, the second shall be greater than the fourth; if equal, equal; and if less, less.

Let them be represented by a, b, c, d; and let $a:b::c:d$.
Then $\dfrac{a}{b} = \dfrac{c}{d} = k$(say);
∴ $a = kb, c = kd$;
now let $a > c$;
∴ $kb > kd$;
∴ ($\div^{\text{ng}}$ each by k), $b > d$.
Similarly if $a = c$, or if $a < c$.   Q.E.D.

Proposition XV.

Magnitudes have the same ratio to each other which their equimultiples have.

Let them be represented by a, b, and their equimultiples by A, B; and let $A = ma$, so that $B = mb$.
Then $\dfrac{A}{B} = \dfrac{ma}{mb} = \dfrac{a}{b}$;
∴ $A:B::a:b$.   Q.E.D.

Proposition XVI.

If four magnitudes of the same kind be proportionals, they shall also be proportionals when taken alternately.

Let them be represented by a, b, c, d; and let $a:b::c:d$.

Then $\dfrac{a}{b} = \dfrac{c}{d} = k$ (say);

∴ $a = kb, c = kd;$

now $\dfrac{a}{c} = \dfrac{kb}{kd} = \dfrac{b}{d}$;

∴ $a:c::b:d.$

Q.E.D.

Proposition XVII. (*Dividendo.*)

If magnitudes, taken jointly, be proportionals, they shall also be proportionals when taken separately; that is, if two magnitudes together have to one of them the same ratio which two others have to one of these, the remaining one of the first two shall have to the other the same ratio which the remaining one of the last two has to the other of these.

let them be represented by *a, b, c, d;* and let $a+b:b::c+d:d.$

Then $\dfrac{a+b}{b} = \dfrac{c+d}{d};$

∴ $\dfrac{a}{b} + 1 = \dfrac{c}{d} + 1;$

∴ $\dfrac{a}{b} = \dfrac{c}{d};$

∴ $a:b::c:d.$

Q.E.D.

Proposition XVIII. (*Componendo.*)

If magnitudes, taken separately, be proportionals, they shall also be proportionals when taken jointly; that is, if the first be to the second, as the third to the fourth, the first and second together shall be to the second as the third and fourth together to the fourth.

Let them be represented by *a, b, c, d;* and let $a:b::c:d.$

Then $\dfrac{a}{b} = \dfrac{c}{d};$

∴ (adding 1 to each), $\dfrac{a}{b} + 1 = \dfrac{c}{d} + 1;$

$$\therefore \frac{a+b}{b} = \frac{c+d}{d};$$

$\therefore\ a+b:b::c+d:d.$ Q.E.D.

Proposition XIX.

If a whole magnitude be to a whole as a magnitude taken from the first is to a magnitude taken from the other, the remainder shall be to the remainder as the whole to the whole.

Let the first magnitudes be represented by A, B; and the magnitudes to be taken from them by a, b; and let $A:B::a:b$.

Then $\dfrac{A}{B} = \dfrac{a}{b} = k$ (say);

$\therefore\ A = kB,\ a = kb;$

$$\therefore \frac{A-a}{B-b} = \frac{kB-kb}{B-b} = \frac{k(B-b)}{B-b} = k = \frac{A}{a};$$

$\therefore\ A-a:B-b::A:a.$ Q.E.D.

Corollary to Prop. XIX.

If the whole be to the whole, as a magnitude taken from the first is to a magnitude taken from the other; the remainder shall likewise be to the remainder, as the magnitude taken from the first to that taken from the other. The demonstration is contained in the preceding.

Proposition E. (*Convertendo*.)

If four magnitudes be proportionals, they are also proportionals by conversion; that is, the first is to its excess above the second, as the third to its excess above the fourth.

Let them be represented by a, b, c, d; and let $a:b::c:d$.

Then $\dfrac{a}{b} = \dfrac{c}{d} = k$ (say);

$\therefore\ a = kb,\ c = kd;$

$$\therefore \frac{a}{a-b} = \frac{kb}{kb-b} = \frac{k}{k-1};$$

and $\dfrac{c}{c-d} = \dfrac{kd}{kd-d} = \dfrac{k}{k-1}$;

∴ $\dfrac{a}{a-b} = \dfrac{c}{c-d}$;

∴ $a:a-b::c:c-d.$ Q.E.D.

Proposition XX.

If there be three magnitudes, and other three, which taken two and two, have the same ratio; if the first be greater than the third, the fourth shall be greater than the sixth; if equal, equal; and, if less, less.

Let the first set be represented by A, B, C, and the other set by a, b, c; and let $A:B::a:b$, and $B:C::b:c$.

Then $\dfrac{A}{B} = \dfrac{a}{b}$, and $\dfrac{B}{C} = \dfrac{b}{c}$;

∴ $\dfrac{A}{B} \cdot \dfrac{B}{C} = \dfrac{a}{b} \cdot \dfrac{b}{c}$;

∴ $\dfrac{A}{C} = \dfrac{a}{c}$.

Now let $A > C$;

∴ ($\div^{\text{ng}}$ each by C), $\dfrac{A}{C} > 1$;

∴ $\dfrac{a}{c} > 1$;

∴ ($\times^{\text{ng}}$ each by c), $a > c$.

Similarly if $A = C$, or if $A < C$. Q.E.D.

Proposition XXI.

If there be three magnitudes, and other three which have the same ratio, taken two and two, but in a cross order; if the first magnitude be greater than the third, the fourth shall be greater than the sixth; if equal, equal; and, if less, less.

Let the first set be represented by A, B, C, and the other set by a, b, c; and let $A:B::b:c$, and $B:C::a:b$.

Then $\dfrac{A}{B} = \dfrac{b}{c}$, and $\dfrac{B}{C} = \dfrac{a}{b}$;

∴ $\dfrac{A}{B} \cdot \dfrac{B}{C} = \dfrac{b}{c} \cdot \dfrac{a}{b}$;

∴ $\dfrac{A}{C} = \dfrac{a}{c}$.

Now let $A > C$;

∴ (÷^{ng} each by C), $\dfrac{A}{C} > 1$;

∴ $\dfrac{a}{c} > 1$;

∴ (×^{ng} each by c), $a > c$.

Similarly if $A = C$, or if $A < C$. Q.E.D.

Proposition XXII. (*Ex æquali.*)

If there be any number of magnitudes, and as many others, which, taken two and two in order, have the same ratio; the first shall have to the last of the first magnitudes the same ratio which the first of the others has to the last.

Let the first set be represented by A, B, C, &c., X, Y, Z, and the second by a, b, c, and x, y, z; and let
$A:B::a:b,\ B:C::b:c,$ &c., $X:Y::x:y,\ Y:Z::y:z.$

Then $\dfrac{A}{B} = \dfrac{a}{b}$;

$\dfrac{B}{C} = \dfrac{b}{c}$,

&c.

$\dfrac{X}{Y} = \dfrac{x}{y}$;

$\dfrac{Y}{Z} = \dfrac{y}{z}$;

∴ $\dfrac{A}{B} \cdot \dfrac{B}{C} \ldots \ldots \dfrac{X}{Y} \cdot \dfrac{Y}{Z} = \dfrac{a}{b} \cdot \dfrac{b}{c} \ldots \ldots \dfrac{x}{y} \cdot \dfrac{y}{z}$;

$$\therefore \frac{A}{Z} = \frac{a}{z};$$
$$\therefore A:Z::a:z.$$
Q.E.D.

Proposition XXIII. (*æquali in proportione perturbatâ.*)

If there be any number of magnitudes, and as many others, which, taken two and two in a cross order, have the same ratio; the first shall have to the last of the first magnitudes the same ratio which the first of the others has to the last.

Let the first set be represented by *A, B, C,* &c., *X, Y, Z,* and the second by *a, b, c,* &c., *x, y, z;* and let

$A:B::y:z, B:C::x:y$, &c., $X:Y::b:c, Y:Z::a:b$.

Then $\dfrac{A}{B} = \dfrac{y}{z};$

$\dfrac{B}{C} = \dfrac{x}{y};$

&c.

$\dfrac{X}{Y} = \dfrac{b}{c};$

$\dfrac{Y}{Z} = \dfrac{a}{b};$

$$\therefore \frac{A}{B} \cdot \frac{B}{C} \cdots \cdots \frac{X}{Y} \cdot \frac{Y}{Z} = \frac{y}{z} \cdot \frac{x}{y} \cdots \cdots \frac{b}{c} \cdot \frac{a}{b};$$

$$\therefore \frac{A}{Z} = \frac{a}{z};$$

$$\therefore A:Z:a:z.$$
Q.E.D.

Proposition XXIV.

If the first has to the second the same ratio which the third has to the fourth, and the fifth to the second the same ratio which the sixth has to the fourth; the first and fifth together shall have the same ratio to the second which the third and sixth together have to the fourth.

Let them be represented by *a, b, c, d, e, f;* and let $a:b::c:d$, and $e:b::f:d$.

Then $\dfrac{a}{b} = \dfrac{c}{d}$, and $\dfrac{e}{b} = \dfrac{f}{d}$;

$\therefore \dfrac{a+e}{b} = \dfrac{c+f}{d}$;

$\therefore a+e:b::c+f:d.$ Q.E.D.

Corollaries to Prop. XXIV.

1.

If the same hypothesis be made as in the proposition, the excess of the first and fifth shall be to the second as the excess of the third and sixth to the fourth. The demonstration of this is the same with that of the proposition, if division be used instead of composition.

2.

The proposition holds true of two ranks of magnitudes, whatever be their number, of which each of the first rank has to the second magnitude the same ratio that the corresponding one of the second rank has to a fourth magnitude; as is manifest.

Proposition XXV.

If four magnitudes of the same kind be proportionals, the greatest and least of them together are greater than the other two together.

Let the pair containing the greatest magnitude be taken first, and the ratios inverted if necessary, so as to make the greatest magnitude first of the four: and let the magnitudes, so arranged, be represented by a, b, c, d, so that $a:b::c:d$.

Then $\dfrac{a}{b} = \dfrac{c}{d} = k$ (say);

$\therefore a = kb$, and $c = kd$.

Now $a > b$; $\therefore kb > b$; $\therefore k > 1$; $\therefore kd > d$; $\therefore c > d$; again, $a > c$; $\therefore kb > kd$; $\therefore b > d$; again, $a > d$;

$\therefore d$ is the least.

Since $k > 1$, let $k = 1+h$ (say);
∴ $a = b + hb, c = d + hd$.
Now $a + d$ will be $> b + c$;
if $b + hb + d > b + d + hd$;
   i.e. if $hb > hd$;
      i.e. if $b > d$, which is true;
∴ $a + d > b + c$.　　　　　　　　　　Q.E.D.

Proposition F.

Ratios which are compounded of the same ratios, are the same to one another.

Let the first set of ratios be represented by $a:b$, $b:c$, &c., $x:y$, $y:z$, so that the ratio compounded of them is $a:z$; and let the second set be represented by $A:B$, $B:C$, &c., $X:Y$, $Y:Z$, so that the ratio compounded of them is $A:Z$; and let
$$a:b::A:B,$$
$$b:c::B:C,$$
$$\&c.,$$
$$x:y::X:Y,$$
$$y:z::Y:Z.$$

Then $\dfrac{a}{b} = \dfrac{A}{B}$;

$\dfrac{b}{c} = \dfrac{B}{C}$,

&c.

$\dfrac{x}{y} = \dfrac{X}{Y}$,

$\dfrac{y}{z} = \dfrac{Y}{Z}$,

∴ $\dfrac{a}{b} \cdot \dfrac{b}{c} \cdots \dfrac{x}{y} \cdot \dfrac{y}{z} = \dfrac{A}{B} \cdot \dfrac{B}{C} \cdots \dfrac{X}{Y} \cdot \dfrac{Y}{Z}$.

∴ $\dfrac{a}{z} = \dfrac{A}{Z}$;

∴ $a:z::A:Z$.　　　　　　　　　　Q.E.D.

Proposition G.

If several ratios be the same to several ratios, each to each; the ratio which is compounded by ratios which are the same to the first ratios, each to each, shall be the same to the ratio compounded of ratios which are the same to the other ratios, each to each.

Let the first set of ratios be represented by $a:b$, $c:d$, &c. $k:l$, $m:n$, and the second set by
$$A:B, C:D, \&c., K:L, M:N;$$
also let the ratios, which are to be compounded, and which are the same as the first set, be presented by $r:s$, $s:t$, &c., $x:y$, $y:z$, so that the ratio compounded of them is $r:z$;

also let the ratios, which are to be compounded, and which are the same as the second set, be represented by
$$R:S, S:T, \&c., X:Y, Y:Z,$$
so that the ratio compounded of them is $R:Z$; and let

$$\left.\begin{array}{l} a:b::A:B, \\ c:d::C:D, \\ \&c. \\ k:l::K:L, \\ m:n::M:N; \end{array}\right\} \quad \left.\begin{array}{l} a:b::r:s, \\ c:d::s:t, \\ \&c. \\ k:l::x:y, \\ m:n::y:z; \end{array}\right\} \quad \left.\begin{array}{l} A:B::R:S, \\ C:D::S:T, \\ \&c. \\ K:L::X:Y, \\ M:N::Y:Z. \end{array}\right\}$$

Then

$$\left.\begin{array}{l} \dfrac{a}{b}=\dfrac{A}{B}, \\ \dfrac{c}{d}=\dfrac{C}{D}, \\ \&c. \\ \dfrac{k}{l}=\dfrac{K}{L}, \\ \dfrac{m}{n}=\dfrac{M}{N}. \end{array}\right\} \quad \left.\begin{array}{l} \dfrac{a}{b}=\dfrac{r}{s}, \\ \dfrac{c}{d}=\dfrac{s}{t}, \\ \&c. \\ \dfrac{k}{l}=\dfrac{x}{y}, \\ \dfrac{m}{n}=\dfrac{y}{z}. \end{array}\right\} \quad \left.\begin{array}{l} \dfrac{A}{B}=\dfrac{R}{S}, \\ \dfrac{C}{D}=\dfrac{S}{T}, \\ \&c. \\ \dfrac{K}{L}=\dfrac{X}{Y}, \\ \dfrac{M}{N}=\dfrac{Y}{Z}. \end{array}\right\}$$

hence $\dfrac{r}{s}\cdot\dfrac{s}{t}\ldots\ldots\dfrac{x}{y}\cdot\dfrac{y}{z} = \dfrac{a}{b}\cdot\dfrac{c}{d}\ldots\ldots\dfrac{k}{l}\cdot\dfrac{m}{n}$,

$$= \dfrac{A}{B}\cdot\dfrac{C}{D}\ldots\ldots\dfrac{K}{L}\cdot\dfrac{M}{N},$$

$$= \dfrac{R}{S}\cdot\dfrac{S}{T}\ldots\ldots\dfrac{X}{Y}\cdot\dfrac{Y}{Z};$$

$\therefore\ \dfrac{r}{z} = \dfrac{R}{Z};$

$\therefore\ r:z::R:Z.$ Q.E.D.

Proposition H.

If a ratio which is compounded of several ratios be the same to a ratio which is compounded of several other ratios; and if one of the first ratios, or the ratio which is compounded of several of them, be the same to one of the last ratios, or to the ratio which is compounded of several of them; then the remaining ratio of the first, or, if there be more than one, the ratio compounded of the remaining ratios, shall be the same to the remaining ratio of the last, or, if there be more than one, to the ratio compounded of these remaining ratios.

Let the first set of ratios be represented by $a:b$, $b:c$, &c., $l:m$, $m:n$, so that the ratio compounded of them is $a:n$; and let the second set be represented by $q:r$, $r:s$, &c., $x:y$, $y:z$, so that the ratio compounded of them is $q:z$; and let $a:n::q:z$. Also let the ratio compounded of several of the first set, viz. $a:b$, $b:c$, be the same as the ratio compounded of several of the second set, viz $q:r$, $r:s$, $s:t$; that is, let $a:c::q:t$. Then shall the ratio compounded of the remaining ratios of the first set, viz. $c:d$, $d:e$, &c., $l:m$, $m:n$, be the same as the ratio compounded of the remaining ratios of the second set, viz. $t:u$, $u:v$, &c., $x:y$, $y:z$; that is, the ratio $c:n$ shall be the same as the ratio $t:z$.

Because $a:n::q:z$;

$\therefore\ \dfrac{a}{n} = \dfrac{q}{z};$

$$\therefore \frac{a}{b} \cdot \frac{b}{c} \cdot \frac{c}{d} \cdot \frac{d}{e} \ldots \ldots \frac{l}{m} \cdot \frac{m}{n} =$$
$$\frac{q}{r} \cdot \frac{r}{s} \cdot \frac{s}{t} \cdot \frac{t}{u} \cdot \frac{u}{v} \ldots \ldots \frac{x}{y} \cdot \frac{y}{z};$$
$$\therefore \frac{a}{c} \cdot \frac{c}{n} = \frac{q}{t} \cdot \frac{t}{z};$$

but $a:c::q:t;$

$$\therefore \frac{a}{c} = \frac{q}{t};$$

$$\therefore \frac{\frac{a}{c} \cdot \frac{c}{n}}{\frac{a}{c}} = \frac{\frac{q}{t} \cdot \frac{t}{z}}{\frac{q}{t}}.$$

$$\therefore \frac{c}{u} = \frac{t}{z};$$

$\therefore\ c:n::t:z.$

And similarly for any other case. Q.E.D.

Proposition K.

If there be any number of ratios, and any number of other ratios such, that the ratio which is compounded of ratios which are the same to the first ratios, each to each, is the same to the ratio which is compounded or ratios which are the same, each to each, to the last ratios; and if one of the first ratios, or the ratio which is compounded of ratios which are the same to several of the first ratios, each to each, be the same to one of the last ratios, or to the ratio which is compounded of ratios which are the same, each to each, to several of the last ratios; then the remaining ratio of the first, or, if there be more than one, the ratio which is compounded of ratios which are the same, each to each, to the remaining ratios of the first shall be the same to the remaining ratio of the last, or, if there be more than one, to the ratio which is compounded of ratios which are the same, each to each, to these remaining ratios.

86   ITEM 3

Let the first set of ratios be represented by $a:b$, $c:d$, &c., $v:x$, $y:z$, and the second set by
$$A:B, C:D, \&c., V:X, Y:Z;$$
also let the ratios, which are to be compounded, and which are the same as the first set, be represented by $\alpha:\beta$, $\beta:\gamma$, &c., $\eta:\theta$, $\theta:\kappa$, so that the ratio compounded of them is $\alpha:\kappa$;

also let the ratios, which are to be compounded, and which are the same as the second set, be represented by $\lambda:\mu$, $\mu:\nu$, &c., $\chi:\psi$, $\psi:\omega$, so that the ratio compounded of them is $\lambda:\omega$;

and let
$$\left.\begin{array}{l} a:b::\alpha:\beta, \\ c:d::\beta:\gamma, \\ e:f::\gamma:\delta, \\ \\ \&c. \\ v:x::\eta:\theta, \\ y:z::\theta:\kappa; \end{array}\right\} \quad \left.\begin{array}{l} A:B::\lambda:\mu, \\ C:D::\mu:\nu, \\ E:F::\pi:\varpi, \\ G:H::\nu:\rho, \\ \&c. \\ V:X::\chi:\psi, \\ Y:Z::\psi:\omega; \end{array}\right\} ;$$

also let $\alpha:\kappa::\lambda:\omega$.

Also let the ratio, which is compounded of ratios which are the same to several of the first ratios, each to each, viz. the ratio compounded of $\alpha:\beta$, $\beta:\gamma$, be the same as the ratio, which is compounded of ratios which are the same, each to each, to several of the last ratios, viz. the ratio compounded of $\lambda:\mu$, $\mu:\nu$, $\nu:\pi$; that is, let $\alpha:\gamma::\lambda:\pi$. Then shall the ratio, which is compounded of ratios which are the same, each to each, to the remaining ratios of the first set, viz. the ratio compounded of $\gamma:\delta$, &c., $\eta:\theta$, $\theta:\kappa$, be the same as the ratio, which is compounded of ratios which are the same, each to each, to the remaining ratios of the last set, viz. the ratio compounded of $\pi:\rho$, &c., $\chi:\psi$, $\psi:\omega$, that is, it shall be proved that $\gamma:\kappa::\pi:\omega$.

Because $\alpha:\kappa::\gamma:\omega$;

$\therefore \dfrac{\alpha}{\kappa} = \dfrac{\lambda}{\omega}$;

$\therefore \dfrac{\alpha}{\beta} \cdot \dfrac{\beta}{\gamma} \cdot \dfrac{\gamma}{\delta} \ldots \ldots \dfrac{\eta}{\theta} \cdot \dfrac{\theta}{\kappa}$

$$= \frac{\lambda}{\mu} \cdot \frac{\mu}{\nu} \cdot \frac{\nu}{\pi} \cdot \frac{\pi}{\rho} \ldots \ldots \frac{\chi}{\psi} \cdot \frac{\psi}{\omega};$$

$$\therefore \quad \frac{\alpha}{\gamma} \cdot \frac{\gamma}{\kappa} = \frac{\lambda}{\pi} \cdot \frac{\pi}{\omega};$$

but $\alpha : \gamma :: \lambda : \pi$;

$$\therefore \quad \frac{\alpha}{\gamma} = \frac{\lambda}{\pi};$$

$$\therefore \quad \frac{\dfrac{\alpha}{\gamma} \cdot \dfrac{\gamma}{\kappa}}{\dfrac{\alpha}{\gamma}} = \frac{\dfrac{\lambda}{\pi} \cdot \dfrac{\pi}{\omega}}{\dfrac{\lambda}{\pi}};$$

$$\therefore \quad \frac{\gamma}{\kappa} = \frac{\pi}{\omega};$$

$\therefore \quad \gamma : \kappa :: \pi : \omega$.

And similarly for any other case. Q.E.D.

4. The Enunciations of Euclid I–VI. Together with Questions on the Definitions, Postulates, Axioms, &c.

[1873: LCH 92: Bodleian]

This pamphlet does not bear Dodgson's name, but he certainly is its author. On 26 December 1872 he wrote in his diary, "I have printed (for publication next term) *The Enunciations of Euclid I–VI,* with questions on the Definitions, etc. and have now in the Press the Fifth Book, done by Algebra, in double columns, with hints on the left and the solution on the right."[1] Passmen were required to known the contents of Euclid I–VI for the Moderations examination. The first nineteen pages form an amended and somewhat enlarged version of *The Enunciations, Books I and II* (item 2). The material of Book V includes only quantities that have a common measure as in *The Fifth Book of Euclid Treated Algebraically, so Far as It Relates to Commensurable Magnitudes, with Notes* (item 3). Dodgson omitted Euclid's treatment of incommensurables because he felt it would be too confusing for students.

※

In this edition of the Enunciations, some emendations have been ventured on, two of which require special mention.

(1) The Corollary, usually appended to Proposition 11 of Book I, is omitted, for the following reasons: that it is not Euclid's—that the proof, usually given with it, is illogical—and that Euclid has tacitly assumed it in Proposition 4.

(2) The Corollaries, usually appended to Proposition 15, are here transferred to Proposition 13, which is quite sufficient for their proof.

Oxford: E. B. Gardner, E. Pickard Hall, and J. H. Stacy, 1873.
1. The book Dodgson is referring to is his 1874 publication, *Euclid, Book V,* where again he covered only commensurable magnitudes. *The Diaries of Lewis Carroll,* edited by Roger L. Green (New York: Oxford University Press, 1954), vol. II, 316.

Besides these definite innovations, many of the Enunciations have been abridged, by the omission of superfluous words, and their language made more uniform in some instances where different words had been used to express the same thing. Capital initials have been used for the *chief* subjects having position, such as "Line," "Triangle," in order to distinguish them from the *subordinate* subjects having position, such as "centre of Circle," "diagonal of Parallelogram," and also from subjects having magnitude only, without definite position, such as the squares and rectangles treated of in Book II.

BOOK I.

Define the following terms and phrases:—

1. Point.
2. Line.
3. Right Line.
4. Superficies.
5. Plane Superficies.
6. Plane Angle.
7. Plane Rectilineal Angle.
8. Right Angle.
9. Obtuse Angle.
10. Acute Angle.
11. Term, or Boundary.
12. Figure.
13. Circle.
14. "Diameter of a Circle."
15. Semicircle.
16. Rectilineal Figure.
17. Trilateral Figure, or Triangle.
18. Quadrilateral Figure.
19. Multilateral Figure, or Polygon.
20. Equilateral Triangle.
21. Isosceles Triangle.

22. Scalene Triangle.
23. Right-angled Triangle.
24. Obtuse-angled Triangle.
25. Acute-angled Triangle.
26. Square.
27. Oblong.
28. Rhombus.
29. Rhomboid.
30. Trapezium.
31. Parallel Straight Lines.
32. Parallelogram.
33. "Diagonal of a Parallelogram."

Write out the Postulate concerning—

34. The drawing of a straight Line.
35. The producing of a terminated straight Line.
36. The describing of a Circle.

Write out, and illustrate by numbers, the Axiom concerning—

37. Equals added to equals.
38. Equals taken from equals.
39. Equals added to unequals.
40. Equals taken from unequals.

Write out the Axiom concerning—

41. Things equal to the same thing.
42. Things double of the same thing.
43. Things which are halves of the same thing.
44. Magnitudes which coincide with one another.

45. Under what circumstances does Euclid assert, as an Axiom, that two Lines will meet?

Define the following terms and phrases:—

46. Perpendicular.
47. Vertex.

48. "Circumference of a Circle."
49. "Centre of a Circle."
50. "Centre of a Semicircle."
51. "Radius of a Circle."
52. Definition.
53. Postulate.
54. Axiom.
55. Proposition.
56. Problem.
57. Theorem.
58. Enunciation.
59. *Data* and *quæsita*.
60. Hypothesis.
61. Corollary.

62. Define *"A fortiori."* Give an instance.
63. Define *"Reductio ad absurdum."* Give an instance.
64. When is a Line said to "subtend" an angle?

Draw a Line intersecting two others; and point out pairs of angles which are—

65. —adjacent.
66. —vertical.
67. —alternate.
68. —exterior and "interior and opposite on the same side of the Line."
69. —"two interior angles on the same side of the Line."
70. When is a Parallelogram said to be "applied" to a Line?
71. When is a Parallelogram said to be "about the diagonal" of another?
72. What are "the complements of the Parallelograms about the diagonal of a Parallelogram"?
73. When are Propositions said to be "converse" to each other? Give an instance.

1. PROBL. To describe an equilateral Triangle on a given finite straight Line.

2. PROBL. From a given Point to draw a straight Line equal to a given straight Line.

3. PROBL. From the greater of two given straight Lines to cut off a part equal to the less.

4. THEOR. If two Triangles have two sides of the one equal to two sides of the other, each to each, and the angles included by those sides equal; the bases or *third sides* are equal; as also are the Triangles; and their other angles are equal, viz. those to which the equal sides are opposite.

5. THEOR. The angles at the base of an isosceles Triangle are equal; and, if the equal sides be produced, the angles on the other side of the base are equal.

COR. Every equilateral Triangle is also equiangular.

6. THEOR. If two angles of a Triangle be equal; the sides also, which subtend, or are opposite to, the equal angles, are equal.

COR. Every equiangular Triangle is also equilateral.

7. THEOR. On the same base, and on the same side of it, there cannot be two Triangles which have their sides terminated in one extremity of the base equal, and at the same time their sides terminated in the other extremity equal.

8. THEOR. If two Triangles have two sides of the one equal to two sides of the other, each to each, and likewise their bases equal; the angle contained by the two sides of the one is equal to the angle contained by the two sides, equal to them, of the other.

9. PROBL. To bisect a given rectilineal angle.

10. PROBL. To bisect a given finite straight Line.

11. PROBL. To draw a straight Line at right angles to a given straight Line from a given Point in the same.

12. PROBL. To draw a straight Line perpendicular to a given straight Line of unlimited length, from a given Point without it.

13. THEOR. The angles, which one straight Line makes with another on one side of it, are either two right angles, or are together equal to two right angles.

COR. 1. The angles at the point of intersection of two straight Lines are together equal to four right angles.

COR. 2. All the angles, made by any number of Lines meeting in one Point, are together equal to four right angles.

14. Theor. If, at a Point in a straight Line, two other straight Lines on opposite sides of it make the adjacent angles together equal to two right angles; these two straight Lines are in one and the same straight Line.

15. Theor. If two straight Lines cut one another; the vertical, or *opposite,* angles are equal.

16. Theor. If one side of a Triangle be produced; the exterior angle is greater than either of the interior opposite angles.

17. Theor. Any two angles of a Triangle are together less than two right angles.

18. Theor. The greater side of every Triangle is opposite to the greater angle.

19. Theor. The greater angle of every Triangle is subtended by the greater side, or *has the greater side opposite to it.*

20. Theor. Any two sides of a Triangle are together greater than the third.

21. Theor. If from the ends of one side of a Triangle two straight Lines be drawn to a Point within the Triangle; these two straight Lines are less than the other two sides of the Triangle, but contain a greater angle.

22. Probl. To make a Triangle, the sides of which shall be equal to three given straight Lines, but any two whatever of these must be greater than the third.

23. Probl. At a given Point in a given straight Line to make an angle equal to a given rectilineal angle.

24. Theor. If two Triangles have two sides of the one equal to two sides of the other, each to each, but the angle contained by two sides of one of them greater than the angle contained by the two sides, equal to them, of the other; the base of that which has the greater angle is greater than the base of the other.

25. Theor. If two Triangles have two sides of the one equal to two sides of the other, each to each, but the base of the one greater than the base of the other; the angle, contained by the sides of that which has the greater base, is greater than the angle contained by the sides, equal to them, of the other.

26. Theor. If two Triangles have two angles of the one equal to two angles of the other, each to each, and one side equal to one

side; viz. either the sides adjacent to the equal angles, or opposite to the equal angles in each: the other sides are equal, each to each, and also the third angle of the one to the third angle of the other.

27. THEOR. If a straight Line falling upon two other straight Lines make the alternate angles equal; the two straight Lines are parallel.

28. THEOR. If a straight Line falling upon two other straight Lines make the exterior angle equal to the interior and opposite angle on the same side of the Line, or make the two interior angles on the same side together equal to two right angles; the two straight Lines are parallel.

29. THEOR. If a straight Line fall upon two parallel straight Lines; it makes the alternate angles equal, and the exterior angle equal to the interior and opposite upon the same side; and the two interior angles upon the same side together equal to two right angles.

30. THEOR. Straight Lines which are parallel to the same straight Line are parallel to one another.

31. PROBL. To draw a straight Line through a given Point parallel to a given straight Line.

32. THEOR. If a side of any Triangle be produced, the exterior angle is equal to the two interior and opposite angles; and the three interior angles of every Triangle are equal to two right angles.

COR. 1. All the interior angles of any rectilineal Figure, together with four right angles, are equal to twice as many right angles as the Figure has sides.

COR. 2. All the exterior angles of any rectilineal Figure are equal to four right angles.

33. THEOR. The straight Lines which join the extremities of equal and parallel straight Lines towards the same parts are themselves equal and parallel.

34. THEOR. The opposite sides and angles of a Parallelogram are equal; and the diagonal bisects it.

35. THEOR. Parallelograms on the same base, and between the same parallels, are equal.

36. THEOR. Parallelograms upon equal bases, and between the same parallels, are equal.

37. THEOR. Triangles on the same base, and between the same parallels, are equal.

38. THEOR. Triangles upon equal bases, and between the same parallels, are equal.

39. THEOR. Equal Triangles upon the same base, and upon the same side of it, are between the same parallels.

40. THEOR. Equal Triangles upon equal bases in the same straight Line, and towards the same parts, are between the same parallels.

41. THEOR. If a Parallelogram and a Triangle be upon the same base, and between the same parallels; the Parallelogram is double of the Triangle.

42. PROBL. To describe a Parallelogram equal to a given Triangle, and having an angle equal to a given rectilineal angle.

43. THEOR. The complements of the Parallelograms, which are about the diagonal of any Parallelogram, are equal.

44. PROBL. To a given straight Line to apply a Parallelogram equal to a given Triangle, and having an angle equal to a given rectilineal angle.

45. PROBL. To describe a Parallelogram equal to a given rectilineal Figure, and having an angle equal to a given rectilineal angle.

COR. To a given straight Line to apply a Parallelogram equal to a given rectilineal Figure, and having an angle equal to a given rectilineal angle.

46. PROBL. To describe a Square upon a given straight Line.

COR. A Parallelogram, that has one right angle, is rectangular.

47. THEOR. In a right-angled Triangle, the Square, described upon the side subtending the right angle, is equal to the Squares described upon the sides containing it.

48. THEOR. If the Square described upon one side of a Triangle be equal to the Squares described upon the other two; the angle contained by those sides is a right angle.

BOOK II.

Define the following terms and phrases:—

1. Rectangle.
2. Gnomon.
3. "The square of AB."
4. "The rectangle contained by AB, CD," or "the rectangle AB, CD."

1. THEOR. If there be two straight Lines, one of which is divided into any number of parts; the rectangle contained by the two straight Lines is equal to the rectangles contained by the undivided Line, and the several parts of the divided Line.

2. THEOR. If a straight Line be divided into any two parts; the rectangles contained by the whole and each of the parts are together equal to the square of the whole Line.

3. THEOR. If a straight Line be divided into any two parts; the rectangle contained by the whole and one of the parts is equal to the rectangle contained by the two parts, together with the square of the aforesaid part.

4. THEOR. If a straight Line be divided into any two parts; the square of the whole Line is equal to the squares of the two parts together with twice the rectangle contained by the parts.

COR. The Parallelograms about the diagonal of a Square are likewise Squares.

5. THEOR. If a straight Line be divided into two equal, and also into two unequal parts; the rectangle contained by the unequal parts, together with the square of the Line between the points of section, is equal to the square of half the Line.

COR. The difference of the squares of two unequal Lines is equal to the rectangle contained by their sum and difference.

6. THEOR. If a straight Line be bisected, and produced to any Point; the rectangle contained by the whole Line produced, and the part of it produced, together with the square of half the Line

bisected, is equal to the square of the Line made up of the half and the part produced.

7. THEOR. If a straight Line be divided into any two parts; the squares of the whole Line and of one of the parts are equal to twice the rectangle contained by the whole and that part, together with the square of the other part.

8. THEOR. If a straight Line be divided into any two parts; four times the rectangle contained by the whole Line and one of the parts, together with the square of the other part, is equal to the square of the Line made up of the whole and that part.

9. THEOR. If a straight Line be divided into two equal, and also into two unequal parts; the squares of the unequal parts are together double of the square of half the Line, and of the square of the Line between the points of section.

10. THEOR. If a straight Line be bisected, and produced to any Point; the square of the whole Line thus produced, and the square of the part of it produced, are together double of the square of half the Line, and of the square of the Line made up of the half and the part produced.

11. PROBL. To divide a given straight Line into two parts, so that the rectangle contained by the whole and one of the parts shall be equal to the square of the other part.

12. THEOR. In an obtuse-angled Triangle, if a perpendicular be drawn from either of the acute angles to the opposite side produced; the square of the side subtending the obtuse angle is greater than the squares of the sides containing it by twice the rectangle contained by the side upon which, when produced, the perpendicular falls, and the straight Line intercepted without the Triangle between the perpendicular and the obtuse angle.

13. THEOR. In every Triangle, the square of the side subtending any of the acute angles is less than the squares of the sides containing that angle, by twice the rectangle contained by either of these sides, and the straight Line intercepted between the perpendicular let fall upon it from the opposite angle and the acute angle.

14. PROBL. To describe a Square equal to a given rectilineal Figure.

BOOK III.

1. When are Circles said to be equal?
2. When is a straight Line said to touch a Circle?
3. When are Circles said to touch each other?
4. When are straight Lines said to be equally distant from the centre of a Circle?
5. When is one of two straight Lines said to be further than the other from the centre of a Circle?
6. Define an arc of a Circle.
7. Define "a segment of a Circle."
8. Define "the angle of a segment."
9. Define "an angle in a segment."
10. When is an angle said to "stand upon" part of the circumference of a Circle (i.e. upon an arc)?
11. When is part of the circumference of a Circle (i.e. an arc) said to be "subtended" by an angle?
12. Define "a sector of a Circle."
13. Define "similar segments of Circles."
14. When is a segment of a Circle said to be "alternate" with an angle? Illustrate your answer by a figure.

When is a Line said to fall on—

15. —a convex circumference?
16. —a concave circumference?

Give the name of a rectilineal Figure having—

17. —five sides.
18. —six do [ditto].
19. —seven do.
20. —eight do.
21. —fifteen do.

1. PROBL. To find the centre of a given Circle.
COR. If in a Circle a straight Line bisect another at right

angles, the centre of the Circle is on the Line which bisects the other.

2. THEOR. If any two Points be taken in the circumference of a Circle, the straight Line which joins them falls within the Circle.

3. THEOR. If a straight Line drawn through the centre of a Circle bisect a straight Line in it which does not pass through the centre, it cuts it at right angles; and if it cut it at right angles, it bisects it.

4. THEOR. If in a Circle two straight Lines cut each other, which do not both pass through the centre, they do not bisect each other.

5. THEOR. If two Circles cut each other, they have not the same centre.

6. THEOR. If one Circle touch another internally, they have not the same centre.

7. THEOR. Of all the straight Lines drawn from a Point in a Circle, not the centre, to the circumference, the greatest is that which passes through the centre; and the other part of that diameter is the least; and of the others, that which is the nearer to the one through the centre is greater than the more remote; and only two equal straight Lines can be drawn from the Point to the circumference, one on each side of the least.

8. THEOR. If from a Point without a Circle, straight Lines be drawn to the circumference; of those which fall on the *concave* circumference, the greatest is that which passes through the centre; and of the rest, that which is nearer to the greatest is greater than the more remote; but of those which fall on the *convex* circumference the least is that between the Point without the Circle and the diameter; and of the rest, that which is nearer to the least is less than the more remote; and only two equal straight Lines can be drawn from the Point to the circumference, one on each side of the least.

9. THEOR. If, from a Point within a Circle, more than two equal straight Lines be drawn to the circumference, that Point is the centre of the Circle.

10. THEOR. One Circle cannot cut another in more than two Points.

11. THEOR. If one Circle touch another internally, the straight Line which joins their centres, being produced, passes through the Point of contact.

12. THEOR. If two Circles touch each other externally, the straight Line which joins their centres passes through the Point of contact.

13. THEOR. One Circle cannot touch another in more Points than one, whether it touch internally or externally.

14. THEOR. Equal straight Lines in a Circle are equally distant from the centre; and those which are equally distant from the centre are equal.

15. THEOR. The diameter is the greatest straight Line in a Circle; and of all others, that which is nearer to the centre is greater than the more remote; and the greater is nearer to the centre than the less.

16. THEOR. The straight Line drawn at right angles to the diameter of a Circle from the extremity of it, falls without the Circle; and no straight Line can be drawn between that straight Line and the circumference from the extremity, so as not to cut the Circle; or, which is the same thing, no straight Line can make so great an acute angle with the diameter at its extremity, or so small an angle with the straight Line which is at right angles to it, as not to cut the circle.

COR. The straight Line which is drawn at right angles to the diameter of a Circle from the extremity of it, touches the Circle; and it touches it only in one Point. Also, there cannot be more than one straight Line touching the Circle in the same Point.

17. PROBL. To draw a straight Line from a given Point, either without or in the circumference, which shall touch a given Circle.

18. THEOR. If a straight Line touch a Circle, the straight Line drawn from the centre to the Point of contact is at right angles to the Line touching the Circle.

19. THEOR. If a straight Line touch a Circle, and from the Point of contact a straight Line be drawn at right angles to the touching Line, the centre of the Circle is in that Line.

20. THEOR. The angle at the centre of a Circle is double of the

angle at the circumference, upon the same base, that is, upon the same part of the circumference.

21. THEOR. The angles in the same segment of a Circle are equal.

22. THEOR. The opposite angles of any quadrilateral Figure inscribed in a Circle are together equal to two right angles.

23. THEOR. On the same straight Line, and on the same side of it, there cannot be two similar segments of Circles not coinciding with each other.

24. THEOR. Similar segments of Circles upon equal straight Lines are equal.

25. PROBL. A segment of a Circle being given, to describe the Circle, of which it is the segment.

26. THEOR. In equal Circles, equal angles stand upon equal circumferences, whether they be at the centres or circumferences.

27. THEOR. In equal Circles, the angles which stand upon equal circumferences are equal, whether they be at the centres or circumferences.

28. THEOR. In equal Circles, equal straight Lines cut off equal circumferences, the greater equal to the greater, and the less to the less.

29. THEOR. In equal Circles, equal circumferences are subtended by equal straight Lines.

30. PROBL. To bisect a given circumference, or arc.

31. THEOR. In a Circle, the angle in a semicircle is a right angle; but the angle in a segment greater than a semicircle is less than a right angle; and the angle in a segment less than a semicircle is greater than a right angle.

COR. If one angle of a Triangle be equal to the other two, it is a right angle.

32. THEOR. If a straight Line touch a Circle, and from the Point of contact a straight Line be drawn cutting the Circle, the angles made by this Line with the Line touching the Circle are equal to the angles which are in the alternate segments of the Circle.

33. PROBL. Upon a given straight Line to describe a segment of a Circle containing an angle equal to a given rectilineal angle.

34. PROBL. To cut off a segment from a given Circle, which shall contain an angle equal to a given rectilineal angle.

35. THEOR. If two straight Lines in a Circle cut each other, the rectangle contained by the segments of one of them is equal to the rectangle contained by the segments of the other.

36. THEOR. If from any Point without a Circle two straight Lines be drawn, one of which cuts the Circle, and the other touches it; the rectangle contained by the whole Line which cuts the Circle, and the part of it without the Circle, is equal to the square of the Line which touches it.

COR. If from a Point without a Circle there be drawn two straight Lines cutting it, the rectangles contained by the whole Lines and the parts of them without the Circle, are equal.

37. THEOR. If from a Point without a Circle there be drawn two straight Lines, one of which cuts the Circle, and the other meets it; if the rectangle contained by the whole Line, which cuts the Circle, and the part of it without the Circle, be equal to the square of the Line which meets it, the Line which meets the Circle touches it.

BOOK IV.

When is a rectilineal Figure said to be—

1. —inscribed in another?
2. —described about another?
3. —inscribed in a Circle?
4. —described about a Circle?

When is a Circle said to be—

5. —inscribed in a rectilineal Figure?
6. —described about a rectilineal Figure?
7. When is a straight Line said to be placed in a Circle?

1. PROBL. In a given Circle, to place a straight Line equal to a given straight Line, which is not greater than the diameter of the Circle.

2. Probl. In a given Circle to inscribe a Triangle equiangular to a given Triangle.

3. Probl. About a given Circle to describe a Triangle equiangular to a given Triangle.

4. Probl. To inscribe a Circle in a given Triangle.

5. Probl. To describe a Circle about a given Triangle.

Cor. If the centre of the Circle be within the Triangle, each of its angles is less than a right angle; but, if the centre be in one of the sides of the Triangle, the angle opposite to this side is a right angle; and, if the centre be without the Triangle, the angle, opposite to the side beyond which it is, is greater than a right angle. Also, if the given Triangle be acute-angled, the centre of the Circle is within it; if it be right-angled, the centre is in the side opposite to the right angle; and if it be obtuse-angled, the centre is without the Triangle, beyond the side opposite to the obtuse angle.

6. Probl. To inscribe a Square in a given Circle.

7. Probl. To describe a Square about a given Circle.

8. Probl. To inscribe a Circle in a given Square.

9. Probl. To describe a Circle about a given Square.

10. Probl. To describe an isosceles Triangle, having each of the angles at the base double of the third angle.

11. Probl. To describe an equilateral and equiangular Pentagon in a given Circle.

12. Probl. To describe an equilateral and equiangular Pentagon about a given Circle.

13. Probl. To inscribe a Circle in a given equilateral and equiangular Pentagon.

14. Probl. To describe a Circle about a given equilateral and equiangular Pentagon.

15. Probl. To inscribe an equilateral and equiangular Hexagon in a given Circle.

Cor. The side of the Hexagon is equal to the straight Line from the centre, that is, to the semi-diameter of the Circle.

16. Probl. To inscribe an equilateral and equiangular Quindecagon in a given Circle.

BOOK V.

Define the following terms and phrases:—

1. Multiple.
2. Part, or Measure.
3. Equimultiples.
4. Commensurable magnitudes.
5. Ratio. (Give also the algebraical definition.)
6. *"The first of four magnitudes has the same ratio to the second as the third has to the fourth."* (Give also the algebraical definition.)
7. Proportionals.
8. Antecedent and consequent terms.
9. Homologous terms.
10. Continual proportionals.
11. Proportion.
12. Disproportion.
13. Duplicate ratio.
14. Triplicate ratio, &c.
15. A mean proportional.
16. Compound ratio.
17. *"Alternando"* or *"permutando."*
18. *"Invertendo."*
19. *"Componendo."*
20. *"Dividendo."* (Give also the form used in Prop. 17.)
21. *"Convertendo."*
22. *"Ex æquali."*
23. *"In proportione ordinatâ."*
24. *"In proportione perturbatâ."*
25. Equation.
26. Inequality.

Write out, and illustrate by numbers, the Axiom concerning—

27. —equimultiples of the same, or of equal magnitudes.

28. —magnitudes of which equal magnitudes are equimultiples.
29. —equimultiples of unequal magnitudes.
30. —magnitudes of which unequal magnitudes are equimultiples.

31. What condition must be fulfilled by two magnitudes, in order that each of them may be said to have a ratio to the other?
32. Why has Euclid laid down this condition?
33. What is the least number of terms which can be proportionals?
34. Show the identity of the two forms of *"dividendo."*

1. THEOR. If any number of magnitudes be equimultiples of as many, each of each; what multiple soever any one of them is of its part, the same multiple are all the first magnitudes of all the others.

2. THEOR. If the first magnitude be the same multiple of the second that the third is of the fourth, and the fifth the same multiple of the second that the sixth is of the fourth; then the first and fifth together are the same multiple of the second that the third and sixth together are of the fourth.

COR. If any number of magnitudes A, B, C, &c., be multiples of another X; and as many a, b, c, &c., be the same multiples of x, each of each; the whole of the first, viz. $A + B + C +$ &c., is the same multiple of X, that the whole of the last, viz. $a + b + c +$ &c., is of x.

3. THEOR. If the first be the same multiple of the second which the third is of the fourth; and if of the first and third there be taken equimultiples, these are equimultiples, the one of the second, and the other of the fourth.

4. THEOR. If the first of four magnitudes have the same ratio to the second which the third has to the fourth; then any equimultiples whatever of the first and third have the same ratio to any equimultiples of the second and fourth, viz.: "the equimultiple of

the first has the same ratio to that of the second, which the equimultiple of the third has to that of the fourth."

COR. Likewise, if the first have the same ratio to the second, which the third has to the fourth, then also, any equimultiples whatever of the first and third have the same ratio to the second and fourth: And in like manner, the first and the third have the same ratio to any equimultiples whatever of the second and fourth.

5. THEOR. If one magnitude be the same multiple of another, which a magnitude taken from the first is of a magnitude taken from the other; the remainder is the same multiple of the remainder that the whole is of the whole.

6. THEOR. If two magnitudes be equimultiples of two others, and if equimultiples of these be taken from the first two; the remainders are either equal to these others, or equimultiples of them.

A. THEOR. If the first of four magnitudes have to the second the same ratio which the third has to the fourth; then, if the first be greater than the second, the third is also greater than the fourth; if equal, equal; and if less, less.

B. THEOR. If four magnitudes be proportionals, they are proportionals also when taken inversely.

C. THEOR. If the first be the same multiple or part of the second, that the third is of the fourth; the first is to the second, as the third to the fourth.

D. THEOR. If the first be to the second, as the third to the fourth; and if the first be a multiple or part of the second; the third is the same multiple or part of the fourth.

7. THEOR. Equal magnitudes have the same ratio to the same magnitude; and the same has the same ratio to equal magnitudes.

8. THEOR. Of unequal magnitudes, the greater has a greater ratio to the same than the less has; and the same magnitude has a greater ratio to the less, than it has to the greater.

9. THEOR. Magnitudes which have the same ratio to the same magnitude, are equal; and those to which the same magnitude has the same ratio, are equal.

10. Theor. That magnitude which has a greater ratio than another has to the same magnitude, is the greater; and that magnitude to which the same has a greater ratio than it has to another magnitude, is the lesser.

11. Theor. Ratios that are the same to the same ratio, are the same to one another.

12. Theor. If any number of magnitudes be proportionals, as one of the antecedents is to its consequent, so are all the antecedents taken together to all the consequents.

13. Theor. If the first have to the second the same ratio which the third has to the fourth, but the third to the fourth a greater ratio than the fifth to the sixth; the first has to the second a greater ratio than the fifth to the sixth.

Cor. And if the first have to the second a greater ratio than the third has to the fourth, but the third the same ratio to the fourth which the fifth has to the sixth; the first has to the second a greater ratio than the fifth has to the sixth.

14. Theor. If the first have to the second the same ratio which the third has to the fourth: then if the first be greater than the third, the second is greater than the fourth; if equal, equal; and if less, less.

15. Theor. Magnitudes have the same ratio to each other which their equimultiples have.

16. Theor. If four magnitudes of the same kind be proportionals, they are also proportionals when taken alternately.

17. Theor. If magnitudes, taken jointly, be proportionals, they are also proportionals when taken separately; that is, if two magnitudes together have to one of them the same ratio which two others have to one of these, the remaining one of the first two has to the other the same ratio which the remaining one of the last two has to the other of these.

*17. Theor. (*Stated according to the Definition of "Dividendo."*) If the first of four magnitudes have to the second the same ratio which the third has to the fourth; the excess of the first above the second has to the second the same ratio which the excess of the third above the fourth has to the fourth.

18. THEOR. If magnitudes, taken separately, be proportionals, they are also proportionals when taken jointly; that is, if the first be to the second, as the third to the fourth, the first and second together are to the second as the third and fourth together to the fourth.

19. THEOR. If a whole magnitude be to a whole as a magnitude taken from the first is to a magnitude taken from the other, the remainder is to the remainder as the whole to the whole.

COR. If the whole be to the whole, as a magnitude taken from the first is to a magnitude taken from the other; the remainder likewise is to the remainder as the magnitude taken from the first to that taken from the other.

E. THEOR. If four magnitudes be proportionals, they are also proportionals by conversion; that is, the first is to its excess above the second, as the third to its excess above the fourth.

20. THEOR. If there be three magnitudes, and other three, which taken two and two, have the same ratio; if the first be greater than the third, the fourth is greater than the sixth; if equal, equal; and, if less, less.

21. THEOR. If there be three magnitudes, and other three, which have the same ratio, taken two and two, but in a cross order; if the first magnitude be greater than the third, the fourth is greater than the sixth; if equal, equal; and, if less, less.

22. THEOR. If there be any number of magnitudes, and as many others, which, taken two and two in order, have the same ratio; the first has to the last of the first magnitudes the same ratio which the first of the others has to the last.

23. THEOR. If there be any number of magnitudes, and as many others, which, taken two and two in a cross order, have the same ratio; the first has to the last of the first magnitudes the same ratio which the first of the others has to the last.

24. THEOR. If the first have to the second the same ratio which the third has to the fourth, and the fifth to the second the same ratio which the sixth has to the fourth; the first and fifth together have to the second the same ratio which the third and sixth together have to the fourth.

COR. 1. If the same hypothesis be made, the excess of the first and fifth is to the second as the excess of the third and sixth to the fourth.

Cor. 2. The Proposition holds true of two ranks of magnitudes, whatever be their number, of which each of the first rank has to the second magnitude the same ratio that the corresponding one of the second rank has to a fourth magnitude.

25. Theor. If four magnitudes of the same kind be proportionals, the greatest and least of them together are greater than the other two together.

F. Theor. Ratios which are compounded of the same ratios, are the same to one another.

G. Theor. If several ratios be the same to several ratios, each to each; the ratio which is compounded of ratios which are the same to the first ratios, each to each, shall be the same to the ratio compounded of ratios which are the same to the other ratios, each to each.

H. Theor. If a ratio which is compounded of several ratios be the same to a ratio which is compounded of several other ratios; and if one of the first ratios, or the ratio which is compounded of several of them, be the same to one of the last ratios, or to the ratio which is compounded of several of them; then the remaining ratio of the first, or, if there be more than one, the ratio compounded of the remaining ratios, shall be the same to the remaining ratio of the last, or, if there be more than one, to the ratio compounded of these remaining ratios.

K. Theor. If there be any number of ratios, and any number of other ratios such, that the ratio which is compounded of ratios which are the same to the first ratios, each to each, is the same to the ratio which is compounded of ratios which are the same, each to each, to the last ratios; and if one of the first ratios, or the ratio which is compounded of ratios which are the same to several of the first ratios, each to each, be the same to one of the last ratios, or to the ratio which is compounded of ratios which are the same, each to each, to several of the last ratios; then the remaining ratio of the first, or, if there be more than one, the ratio which is compounded of ratios which are the same, each to each, to the remaining ratios of the first shall be the same to the remaining ratio of the last, or, if there be more than one, to the ratio which is compounded of ratios which are the same, each to each, to these remaining ratios.

BOOK VI.

1. Define the "altitude" of a Triangle or Parallelogram.
2. When are rectilineal Figures said to be similar?
3. When are two Triangles said to have two sides of the one "reciprocally proportional" to two sides of the other?
4. When is a straight Line said to be cut in extreme and mean ratio?
5. When are two Lines said to be similarly divided?

1. THEOR. Triangles and Parallelograms of the same altitude are one to another as their bases.

COR. Triangles and Parallelograms of equal altitudes are one to another as their bases.

2. THEOR. If a straight Line be drawn parallel to one of the sides of a Triangle, it cuts the other sides, or those produced, proportionally; and if two sides, or two sides produced, proportionally, the straight Line which joins the Points of section is parallel to the remaining side.

3. THEOR. If the vertical angle of a Triangle be bisected by a straight Line cutting the base, the segments of the base have the same ratio as the other sides of the Triangle; and if the segments of the base have the same ratio as the other sides of the Triangle, the straight Line, drawn from the vertex to the Point of section, bisects the vertical angle.

A. THEOR. If the exterior angle of a Triangle, made by producing one of its sides, be bisected by a straight Line, which also cuts the base produced; the segments between the dividing Line and the extremities of the base have the same ratio as the other sides of the Triangle; and if the segments of the base produced have the same ratio as the other sides of the Triangle, the straight Line, drawn from the vertex to the Point of section, bisects the exterior angle of the Triangle.

4. THEOR. The sides about the equal angles of equiangular Triangles are proportionals; and those which are opposite to the equal angles are homologous.

5. THEOR. If the sides of two Triangles about each of their angles be proportionals, the Triangles are equiangular, and have those angles equal which are opposite to the homologous sides.

6. THEOR. If two Triangles have one angle of the one equal to one angle of the other, and the sides about the equal angles proportionals; the Triangles are equiangular, and have those angles equal which are opposite to the homologous sides.

7. THEOR. If two Triangles have one angle of the one equal to one angle of the other, and the sides about two other angles proportionals; then, if each of the remaining angles be either less, or not less, than a right angle, or if one of them be a right angle, the Triangles are equiangular, and have those angles equal about which the sides are proportionals.

8. THEOR. In a right-angled Triangle, if a perpendicular be drawn from the right angle to the base, the Triangles on each side of it are similar to the whole Triangle, and to one another.

COR. The perpendicular, drawn from the right angle of a right-angled Triangle to the base, is a mean proportional between the segments of the base. Also, each of the sides is a mean proportional between the base and its segment adjacent to that side.

9. PROBL. From a given straight Line to cut off any part required.

10. PROBL. To divide a given straight Line similarly to a given divided straight Line.

11. PROBL. To find a third proportional to two given straight Lines.

12. PROBL. To find a fourth proportional to three given straight Lines.

13. PROBL. To find a mean proportional between two given straight Lines.

14. THEOR. Equal Parallelograms which have one angle of the one equal to one angle of the other, have their sides about the equal angles reciprocally proportional; and Parallelograms, which have an angle of the one equal to an angle of the other, and their

sides about the equal angles reciprocally proportional, are equal.

15. THEOR. Equal Triangles, which have one angle of the one equal to one angle of the other, have their sides about the equal angles reciprocally proportional; and Triangles, which have an angle of the one equal to an angle of the other, and their sides about the equal angles reciprocally proportional, are equal.

16. THEOR. If four straight Lines be proportionals, the rectangle contained by the extremes is equal to the rectangle contained by the means; and if the rectangle contained by the extremes be equal to the rectangle contained by the means, the four straight Lines are proportionals.

17. THEOR. If three straight Lines be proportionals, the rectangle contained by the extremes is equal to the square of the mean; and if the rectangle contained by the extremes be equal to the square of the mean, the three straight Lines are proportionals.

18. PROBL. Upon a given finite straight Line to describe a Rectilineal Figure similar and similarly situated to a given rectilineal Figure.

19. THEOR. Similar Triangles have to one another the duplicate ratio of their homologous sides.

COR. If three straight Lines be proportionals, as the first is to the third, so is any Triangle upon the first to a similar and similarly described Triangle upon the second.

20. THEOR. Similar Polygons may be divided into the same number of similar Triangles, having to one another the same ratio as the Polygons; and the Polygons have to one another the duplicate ratio of their homologous sides.

COR. 1. Similar rectilineal Figures have to one another the duplicate ratio of their homologous sides.

COR. 2. If three straight Lines be proportionals, as the first is to the third, so is any rectilineal Figure upon the first, to a similar and similarly described rectilineal Figure upon the second.

21. THEOR. Rectilineal Figures, which are similar to the same rectilineal Figures, are also similar to one another.

22. THEOR. If four straight Lines be proportionals, the similar rectilineal Figures similarly described upon them are also proportionals; and if the similar rectilineal Figures similarly described

upon four straight Lines be proportionals, those straight Lines are proportionals.

23. THEOR. Equiangular Parallelograms have to one another the ratio which is compounded of the ratios of their sides.

24. THEOR. The Parallelograms about the diameter of any Parallelogram are similar to the whole and to one another.

25. PROBL. To describe a rectilineal Figure similar to one, and equal to another, given rectilineal Figure.

26. THEOR. If two similar Parallelograms have a common angle and be similarly situated, they are about the same diameter.

30. PROBL. To divide a given straight Line in extreme and mean ratio.

31. THEOR. In right-angled Triangles, the rectilineal Figure described on the side opposite to the right angle, is equal to the similar and similarly described Figures on the sides containing it.

32. THEOR. If two Triangles, which have two sides of the one proportional to two sides of the other, be joined at one angle, so as to have their homologous sides parallel; the remaining sides are in a straight Line.

33. THEOR. In equal Circles, angles, whether at the centres or circumferences, have the same ratio which the circumferences have to one another; so also have the sectors.

B. THEOR. If the vertical angle of a Triangle be bisected by a straight Line, which likewise cuts the base; the rectangle contained by the sides of the Triangle is equal to the rectangle contained by the segments of the base, together with the square of the straight Line bisecting the angle.

C. THEOR. If from the vertical angle of a Triangle a straight Line be drawn perpendicular to the base; the rectangle contained by the sides of the Triangle is equal to the rectangle contained by the perpendicular and the diameter of the Circle described about the Triangle.

D. THEOR. The rectangle, contained by the diagonals of a quadrilateral Figure inscribed in a Circle, is equal to the rectangles contained by the opposite sides.

THE END.

Trigonometry

Theorem of Tangents

Given a, b, C

$$\frac{a-b}{a+b} = \tan\frac{A-B}{2} \cdot \tan\frac{C}{2}$$

Let $\frac{a-b}{a+b} = m$, and $\tan\frac{C}{2} = n$

$$\therefore \tan\frac{A-B}{2} = \frac{m}{n}$$

$$\therefore \frac{A-B}{2} = \tan^{-1}\frac{m}{n}$$

If this be a known tangent, $\overline{A-B}$ may be found: also $\overline{A+B}$, whence A, & B —

If not, proceed thus:

$$n = \tan\frac{C}{2} = \tan\frac{180 - \overline{A+B}}{2}$$

$$= \tan\left(90 - \frac{A+B}{2}\right) = \cot\frac{A+B}{2}$$

$$\therefore \tan\frac{A+B}{2} = \frac{1}{n}$$

$$\therefore \frac{A+B}{2} = \tan^{-1}\frac{1}{n}$$

also $\frac{A-B}{2} = \tan^{-1}\frac{m}{n}$

$\therefore$ by addition $A = \tan^{-1}\frac{n(m+1)}{n^2 - m}$

i.e. $\tan A = \frac{mn+n}{n^2-m} = \frac{\sin A}{\cos A}$

$$\therefore \frac{\sin A}{mn+n} = \frac{\cos A}{n^2 - m} = \frac{1}{\sqrt{m^2+n^2} \cdot \sqrt{n^2+1}}$$

$$\therefore \sin A = \frac{mn+n}{\sqrt{m^2+n^2} \cdot \sqrt{n^2+1}}$$

and as a, $\sin A$, & $\sin C$ are now known, the triangle may be solved by the theorem of the Sines.

Q.E.F.

Feb: 1. 1860.

First page, recto, of Charles. L. Dodgson's application of the Theorem of Tangents to solving triangles, dated 1 February 1860.

Introduction

Charles Dodgson published only one pamphlet on trigonometry, *The Formulæ of Plane Trigonometry, Printed with Symbols (Instead of Words) to Express the "Goniometrical Ratios,"* in 1861 (item 5). Of his unpublished writings, a rare, undated item, three pages of trigonometric formulas called "Formulæ (Group C)" (item 6), and fragments of a manuscript on circle-squaring (item 7) deal with trigonometry. His other published works on trigonometry consist of twenty-four Pillow Problems (1872–1891) and Question 11530 (item 9) from the *Educational Times* (1892).[1] Both the 1861 pamphlet and the material on circle-squaring are important, but in entirely different ways. The former represents an (unsuccessful) attempt to influence the development of mathematical symbols, while the latter is a notable contribution to the history of approximation methods for π.

In the preface to the 1861 pamphlet Dodgson expressed two reasons for taking up the task. First, he hoped to elicit comments and suggestions from other mathematicians about the formulae themselves—omissions, better examples, or their mode of presentation—because he was planning a large task, the publication of a collection of "Formulae of Pure Mathematics." Second, and clearly the more important reason, he wished to ascertain whether other mathematicians would adopt the symbols that he proposed to take the place of "sin," "cos," "tan," "cosec," etc. If the new symbols found acceptance, he wanted to use them in his planned "Formulæ" and in another publication he was preparing on analytical geometry that was based on his book, *Syllabus of Plane Algebraical Geometry* (1860). The preface is three and one-half pages long, three of which argue the merits of his symbolic notation. Dodgson was chiefly concerned with the saving of time, space, and labor that the symbols would offer; he argued that they are easily written, suggestive of their meaning, connected with each other, and distinct from existing symbols. But Dodgson's scheme, like that of Jean Bernoulli a century earlier, was never adopted.

In his definitive two-volume work on the history of mathematical no-

---

1. One of the three topics in *Three Problems by Oedipus* (1895), a pamphlet listed in the *Lewis Carroll Handbook* (p. 191) as possibly, but unlikely, by Dodgson can be linked to the circle-squaring dilemma: on the circle and inscribed polygons. However, it is quite clear from the notation and terminology used that Dodgson is not its author.

tations, Florian Cajori gives 1895, the publication date of the fourth edition of *Curiosa Mathematica, Part II: Pillow Problems,* as the entry for Dodgson's contribution to the trigonometric symbols of the nineteenth century.[2] However 1861, the publication date of *The Formulae of Plane Trigonometry,* is the more obvious one for Dodgson's attempt to produce a symbolical trigonometry for writing purposes.

The formulas are divided into three parts: "Goniometry Proper, i.e. the measurement of angles by *angular units,*" "Goniometry by Ratios, i.e. the *indication,* (not *measurement,*) of angles by what are called 'goniometrical ratios,'" and "Trigonometry, i.e. the properties of rectilinear figures." One formula is given in Part I, connecting degree and grad measure and connecting degree and radian measure. Twenty-four formulas are provided in Part II, including number 21, Gregorie's (Gregory's) series. The appearance of the arctangent series in this 1861 publication seems to contradict a statement by Morris Kline that the work of James Gregory (1638–1675) was not known generally until 1939.[3] The nineteen formulas in Part III are divided into ten dealing with triangles, four with quadrilaterals, and five with regular polygons.

In addition to Gregory's series, Dodgson included Euler's and Machin's series to find π. By the middle of the nineteenth century, the value of π, first proved irrational by Johann Lambert in 1761, was accurately known to 500 places. However, it was not until 1882 that Ferdinand Lindemann proved that π was transcendental, thereby settling the ancient Greek problem of squaring the circle. Since π is not algebraic, it is impossible to construct a line segment of length π; therefore one cannot construct a square having area equal to that of a given circle (or a circle with area equal to that of a given square.) Perhaps the intuitive appeal of the task is what made circle-squaring a popular activity among amateur mathematicians in the nineteenth century.

When Augustus De Morgan died in 1871, Dodgson took over the unenviable position of referee for the amateur mathematicians who thought they had indeed squared the circle. Ten years and many, many letters later, Dodgson decided to write a simple treatise presenting the main lines of argument to convince would-be circle-squarers of the impossibility of the task. In July 1881 he began a correspondence with one would-be circle-squarer that lasted several months, over the course of which Dodgson was

---

2. Florian Cajori, *A History of Mathematical Notations* (La Salle, Ill.: Open Court, 1929), vol. II, 170–1. Dodgson used the new symbols in his Pillow Problems.

3. See Morris Kline, *Mathematical Thought from Ancient to Modern Times* (New York: Oxford University Press, 1972), 355. I am grateful to Professor Victor Katz for this observation.

unable to convince him of the folly of his endeavours. In this as well as in other correspondence, the circle-squarers were not using good approximations for π. Dodgson wanted to set the matter straight once and for all by writing a pamphlet on the subject. Never finished, it probably would have been called "Simple Facts For Circle-Squarers." In the introductory chapter (item 7), dated 20 April 1882, he defined the scope of the problem:

> "Circle-Squarers", under which term I include all who have attempted to give an *exact* value to the area of a circle, expressed in terms of the square on its radius.
>
> To measure the area of the circle itself is a complicated matter, and the processes, by which the value 3.14159 has been calculated, are long and abstruse....[4]

The radius is taken as the unit of length so that the area of the circle is π^2 units. A square of equal size would have a side of length $\sqrt{\pi}$. What Dodgson set out to do was

> to prove, by very simple methods (in which I shall make no attempt at measuring the *circle* at all, but shall merely measure certain *rectilinear figures* drawn within it & outside it), that, whatever be the *exact* value of the area, it is at any rate less than 3.1417 times, and greater than 3.1413 times, the square on its radius."[5]

Why did he choose these values for π when he knew sharper ones? In manuscript sheets dated 2 and 13 August 1881 and in undated ones from the Morris L. Parrish Collection of Victorian Novelists in the Princeton University Library, Dodgson tried to calculate π using methods of his own, most likely with the intent of demonstrating them in his pamphlet. One approach that he used involved breaking up a 45° angle ($\pi/4$) into angles with rational tangents. Using different series, he tried to find the number of these angles that would produce the best limit for π. By August 1881, his best approximation, $\pi < 3.1450$, used eight angles. So, between August 1881 and April 1882 he had succeeded in improving this value.

Dodgson was not yet satisfied with his approximation for π. On 1 May 1892, his problem, Question 11530, appeared on page 234 of the *Educational Times*. Here he presented a formula to break up $\tan^{-1} 1/a$ (his notation for the arctangent function) into two angles of the same form. He wanted reactions from other mathematicians to this formula because it enabled him to obtain much better limits for π: 3.141597 and 3.141583.

---

4. See item 7, pp. 145, 146.
5. See item 7, p. 146.

The response by H. J. Woodall, a solution presuming that the angles are the smallest possible, links Dodgson's series with one given by William Wallace, professor of mathematics at the University of Edinburgh, and ultimately with a generalization that had first been given by Carl Friedrich Gauss in the second volume of his *Werke* (1832). Discussing Gauss's work, Woodall stated that the formulas that Gauss found using the theory of numbers were remarkable because π could be accurately and quickly calculated beyond the 707 places that W. Shanks achieved in 1873 (Shanks had made an error in the 527th place which was not discovered until 1945.). Although Dodgson's formula was not quite up to the level of Gauss's, it too broke with the prevailing methodology used by Shanks and others and provided a simpler and more rapidly convergent series approximation for π.[6]

6. Woodall uses the notation A cot x for the function arccotangent x. The substitution of the arccotangent for the arctangent eliminates the fractions in the equations. Although the first appearance of the formula that Dodgson used is credited to Wallace, it is unlikely that Dodgson was aware of it. For a more complete development of approximations of π from arctangent series, see F. Abeles, "Charles L. Dodgson's Geometric Approach to Arctangent Relations for π," *Historia Mathematica* 20 (May 1993): 151–59.

5. *The Formulæ of Plane Trigonometry,*
PRINTED WITH SYMBOLS (INSTEAD OF WORDS) TO EXPRESS THE "GONIOMETRICAL RATIOS"

[1861: LCH 27, LCAT 333: British]

The preface to the pamphlet is dated 11 June 1861. The only reference to this work appears in a letter addressed to Dodgson's sister, Mary Collingwood, on 20 February 1861.

As you ask about my mathematical books I will give you a list of my "Works."

1. Syllabus, etc., etc. (done)
2. Notes on Euclid (done)
3. Ditto on Algebra (done—will be out this week, I hope)
4. Cycle of Examples, Pure Mathematics (about 1/3 done)
5. Collection of formulæ (1/2 done)
6. Collection of Symbols (begun)
7. Algebraical Geometry in 4 vols. (about 1/4 of Vol. I done).

Doesn't it look grand?[1]

The three diary volumes from 18 April 1858 through 8 May 1862 are missing, but Cohen identifies "syllabus" as referring to *A Syllabus of Plane Algebraical Geometry,* published in 1860. In modern terminology, algebraical geometry is more properly called analytic geometry. "Notes" refers to the pamphlet, *Notes on the First Two Books of Euclid* (item 1), from 1860. The third item in Dodgson's list refers to a missing pamphlet from the Hartley collection, *Notes on the First Part of Algebra* (1861); "Cycle" refers to another unavailable pamphlet from the Hartley collection, *General List of Subjects, and Cycle for Working Examples,* published in 1863. Cohen conjectures that "Collection of formulæ" is a reference to *The Formulæ of Plane Trigonometry,* but "Collection of Symbols" probably refers to this pamphlet as well, because the pamphlet's title is somewhat misleading; both symbols and formulas are its main subjects. Dodgson may have developed the two parts separately, combining them later when he saw that it made sense to do so.

Oxford: James Wright, 1861.

1. *The Letters of Lewis Carroll,* edited by Morton N. Cohen (New York: Oxford University Press, 1979), vol. I, 48.

The final item is the most intriguing. In the introduction to *A Syllabus of Plane Algebraical Geometry,* Dodgson says his interest is in "reducing the whole subject to a complete and uniform system, which shall occupy, with regard to Algebraical Geometry, the same position which is occupied by that of Euclid with regard to Pure Geometry."[2]

It seems that Dodgson meant *A Syllabus* to act as a map for the projected four-volume work on analytic geometry. Curiously, he never produced any part of this work, nor did he mention it again.

PREFACE

This Pamphlet is published with two objects: first, to exhibit a specimen of a collection of "Formulæ of Pure Mathematics," which I am preparing for publication; secondly, to suggest the substitution of symbols for the cumbrous expressions "sin," "tan," "cosec," &c., at present employed in Trigonometry.

As to the first of these objects, I am in hopes that, by publishing this specimen by itself, I may receive, from other mathematicians, suggestions both as to the form and the matter of the larger work; and any simple and compendious formulæ, not given in the published treatises, or superseding those which are there given, will be most acceptable for this object.

As to the second, I am anxious to ascertain what probability there is of others' consenting to adopt these, or any similar symbols, before employing them either in the above-mentioned "Formulæ of Pure Mathematics," or in another work, (also in preparation,) on Algebraical Geometry, (on the plan indicated in a Syllabus published in 1860,) into both of which works these symbols would of course enter largely.

Objections may be raised, first, against the introduction of any symbols at all into this subject, and, secondly, against the particular symbols here suggested: these objections I proceed to examine.

---

2. Charles L. Dodgson, *A Syllabus of Plane Algebraical Geometry* (Oxford: James Wright, 1860), x.

First, then, as to the introduction of any symbols at all: if it could be shown that such symbols were likely to confuse those who are used to the existing notation, or to make the books at present in use unintelligible to those who learn only the new notation, I grant that it would be much too late in the history of Trigonometry to attempt any such radical change. But I am sure that no such effects would be produced; the new symbols would still be *called* "sin," "cos," &c., so that the old names would not die out, and those who use the new symbols would find no more difficulty in reading books where the names are given in full, than we do in reading old mathematical books where the words "plus," and "minus" occur. And there can be no question that the use of symbols would save much time, space, and labour; all of which are points of great importance to those who are much engaged in writing or printing Mathematics.

Secondly, as to the particular symbols here suggested: on this point I speak much more doubtfully; and I shall be very willing, if simpler and more appropriate symbols be proposed, to adopt them instead of these; still, as these have not been chosen without some thought, and the trial and rejection of many others, I have great hopes that they may be found sufficiently simple and expressive for the purpose: and perhaps a simple account of the way in which they originated will be the best explanation and defence of them which I can offer.

It seemed necessary, then, that symbols for such a purpose should be—easily written—suggestive (as far as possible) of their meaning—connected with each other—and, above all, distinct from all symbols at present in use. Letters of the alphabet of course suggested themselves first: but besides that all the alphabets, capital and small, Roman, Greek, even Old English and Hebrew, are already largely appropriated in Mathematics, it did not seem that any but initials were sufficiently connected with the words to stand much chance of being remembered. Now the words "sine, secant, cosine, cosecant, cotangent, tangent," have unfortunately only three initials among them; and of these "S" and "C" (whether we take them as capital or small letters,) are already distinctly appropriated in this very subject of Algebraical Geometry—

"S" and "C" signifying, one, the semi-perimeter of a triangle; the other, one of its angles—and "s" and "c" signifying, one, an arc of a curve; the other, one of the sides of a triangle: as to the latter indeed it is sufficient to point out that the phrase "a.cosA+b.cosB+c.cosC" would, under this notation, be written "a.cA+c.cB+c.cC"!

Setting aside, then, all alphabets, and (of course) all purely *mathematical* symbols, such as numerals, &c., no course remained but to *invent* new symbols for this purpose. In setting about this I took as principles—to secure their being easily written, that each should consist of *two* strokes of the pen only—to connect them with each other, that *one* of these two strokes should be the same in all—to make them suggestive of their meaning, that they should represent (as nearly as possible) the *geometrical* lines to which these ratios belong. And as, in modern Trigonometry, each of these ratios involves *two* lines, I thought it better to fall back on the older theory, where the sine, cosine, &c., are all *single* lines.

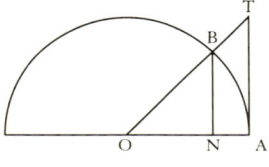

Figure 1.

In the annexed diagram [Fig. 1], *BN* is the *sine* of the angle *AOB*, *ON* its *cosine*, *TO* its *secant*, and *AT* its *tangent*. It occurred to me, then, to construct a set of symbols for these by taking the semicircle itself as the part *common* to all the symbols, and these lines as their distinctive features, altering them only so far as to make each symbol symmetrical. I shall now consider each "goniometrical ratio" by itself.

I. "Sine," ⌒. In this it will be seen that the line *BN* has merely been shifted to the middle of the figure.

II. "Cosine," ⌂. In this the line *ON* has been produced, (which in hasty writing one could hardly avoid doing,) and taken a little *beyond* the curve, to avoid confusion with the existing symbol for "semicircle." Observe here also that the line in this symbol is *at right angles* to that in the last; which may be connected with the fact that these are corresponding ratios of *complemental* angles (i.e. of angles, whose sum is *a right angle*).

III. "Secant," ⌒. In this the line *TO* has been produced downwards, to avoid all chance of confusion with the symbol ⌒.

IV. "Cosecant," ⌒. This I derived from the former, by turning the line round through *a right angle,* on the principle which I have explained in II.

V. "Tangent," ⌒. If the symbol for this were taken from the diagram, it would have a one-sided effect, and it would be difficult to find an analogous symbol for "cotangent," since the three other positions in which it might be placed are pre-occupied by the letters *h,* μ, and *y.* I therefore preferred placing it horizontally on the top of the curve, leaving a little interval between the two, to avoid confusion with the letter π.

VI. "Cotangent," ⌣. I could not derive this from the last, on the principle of II, without destroying its character as *a tangent.* I therefore simply *inverted* the former symbol, which may be taken to indicate the fact that each is the *reciprocal* of the other: so that if $⌒ = \frac{a}{b}, ⌣ = \frac{b}{a}$, which seems to be a very consistent and self-interpreting notation.

VII. "Versed-sine," ⌒. This is of course a combination of the semicircle with the letter *"V"*; it contains one stroke more than the other symbols, but as it is very seldom used, this is of little importance.

I have thus endeavoured to show that the proposed symbols satisfy three of the four requisitions, viz.: that they should be—easily written—suggestive of their meaning—and connected with each other. As to whether they are distinct from existing symbols, it is for the objectors to point out any with which they are liable to be confused; none such have been suggested to me, except the letter Ω, which is something like the new symbol ⌒; but, as the latter is closed, instead of open, below, and much wider in proportion to its height than the former, I do not think there is much danger of either's being mistaken for the other.

I will conclude by putting into the form of definite questions the points to which I wish to draw the attention of those mathematicians into whose hands this may come:

(1) Do you object *in limine* to the introduction of *any symbols whatever* as substitutes for the words sin, cos, &c.? If not,

(2) Can you suggest others, better adapted than these for such a purpose? If not,

(3) Do you so far approve of the symbols here suggested, that, if they were employed in a published work, you would not, *on that account alone,* object to use or to recommend such a work?

Ch. Ch. June 11, 1861.

PRELIMINARY REMARKS.

The subject-matter usually assigned to Plane Trigonometry may be more properly arranged thus:

Part I.

Goniometry Proper.
i.e. the measurement of angles by *angular units*.

Part II.

Goniometry by Ratios.
i.e. the *indication,* (not *measurement,*) of angles by what are called "goniometrical ratios."

Part III.

Trigonometry.
i.e. the properties of rectilinear figures.

Before proceeding to enumerate the Formulæ of the subject, I shall make a few remarks on each of these three Parts.

Part I.

Goniometry Proper.

Two different units are in common use:

(1). a right angle.

(2). the angle which is subtended, (in any circle), by an arc equal to the radius; this can easily be proved to be the same, whatever be the size of the circle, and to be about $\frac{2}{3}$ of a right angle. This angle I propose to call "a radial angle."

When (1) is employed, it is subdivided, either into 90 degress (English measure), or 100 grades (French measure).

When (2) is employed, the following proportion holds good;

No. of radial angles contained : 1 : : arc : radius,

which may be briefly expressed thus; angle $= \dfrac{\text{arc}}{\text{radius}}$.

The quantity "2 right angles" would be algebraically represented—in English measure by "180°"—in French measure by 200^g—and in "radial" measure by "3.14159 . . . . . . radial angles."

Part II.

Goniometry by Ratios.

The student will do well to observe that the statement, to be found in most treatises on Trigonometry, that "if any of the quantities, sin *A,* cos *A,* &c., be given, the angle *A* may be determined," is untrue. The word "angle" is used in various senses; in one of these senses, it could not be determined, even if *all* the "goniometrical ratios" were given, and even in its simplest sense "sin *A*" is *never* sufficient to determine it. I believe that the chief difficulty of Trigonometry is surmounted, when we have once arrived at a clear notion of the different meanings of this word "angle." It may be thought that the following explanation is unnecessarily complicated, but if such complication really exists in the science itself, it surely deserves investigation; we can never get rid of the difficulties of a subject by ignoring them.

First, then, there is the "geometrical angle," or the angle as treated of by Euclid: of this it may be remarked, that it is considered as an *absolute magnitude*, without regard to *direction*, and that it is always measured the *shortest* way round, so that it can never exceed two right angles. Thus, in the annexed figure [Fig. 2], the "geometrical angle" contained between OX and OA is measured (whether from X to A, or from A to X, does not matter) along the arrow marked S, and *not* along that marked L. If OX and OA were in one straight line, the angle might still be called geometrical, though not treated of in Euclid; it would then be *equal* to two right angles, and might of course be measured either way round at pleasure.

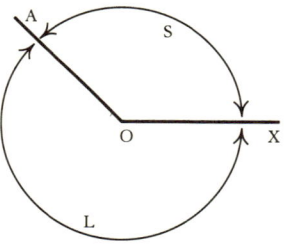

Figure 2.

Secondly, there is the "angle of position": in measuring this, *one* of the lines is supposed to be fixed in position, and the angle to be measured, *from* it, *to* the other, the shortest way round: and as this angle may lie in either of two opposite directions, these are distinguished as "positive" and "negative." Thus, in the annexed figure [Fig. 3], the angles XOA and XOA' are identical when viewed as "geometrical angles" only, since they have the same magnitude: nevertheless OA and OA' have different positions with regard to OX, and these may be distinguished by calling the angle XOA "positive," and XOA' "negative." Observe also, that every possible position of OA, if *above* the line X'OX, may be represented by a positive "geometrical angle," if *below* it, by a negative one, and if coinciding with *OX'*, by a positive or negative one, equal to two right angles; but that in no case need the angle *exceed* two right angles.

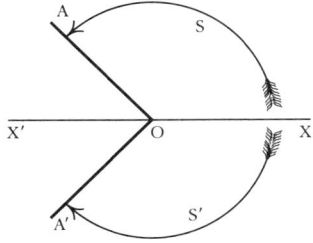

Figure 3.

Thirdly, there is the "angle of revolution": in measuring this, *one* of the lines is (as before) supposed to be fixed, and the angle to be measured *from* it *to* the other; but no longer are we obliged to do this the *shortest* way round, nor even to stop measuring it the first time we come upon the other line: we may go round and round the circle any number of times, the only rule being that we must at last stop on the other line. Thus, in the annexed figure [Fig. 4], the arrows, S_1, S_2, and S_3, are specimens of the various ways in which the "angle of revolution" may be measured from *OX* to *OA*; S_1 and S_3 being "positive," and S_2 "negative."

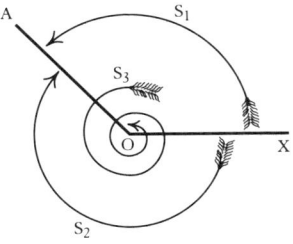

Figure 4.

Let us now consider how far the "goniometrical ratios" enable us to determine an angle under each of these senses.

It is evident that these ratios depend *only* on the *position* of *OA* with regard to *OX,* and are unaffected by the method in which we may choose to measure the angle; so that they are of no use in determining the "angle of *revolution*"; this can only be expressed by referring it to some angular unit.

How far, then, will they help us in determining the "angle of *position*"? We may see by the annexed figure [Fig. 5] that if the *sine* only of XOA_1 were given, we could not distinguish it from XOA_2; nor, by its *cosine* only, could we distinguish it from XOA_3; nor, by its *tangent* only, from XOA_4; and the same may be said of its *cosecant,* its *secant,* and its *cotangent,* which are the reciprocals of the former three. Hence *no one* of the goniometrical ratios is sufficient by itself to determine an "angle of position": and since from any one we may determine the *magnitudes* of *all* the rest, and the *magnitude* and *sign* of the *reciprocal* ratio, we may lay it down as a rule, that in order to determine an "angle of position" we require to know the *magnitude* of some *one* ratio, and the *signs* of *two,* and that these two must *not* be reciprocals of each other, (to which we may add, that they must *not* be the cosine and versed-sine).

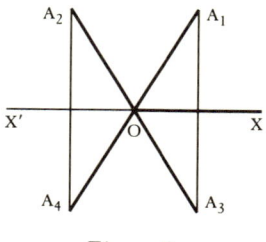

Figure 5.

But in the case of the "*geometrical* angle," (as, for instance, when we are treating of one of the angles of a triangle, which is always positive, and can never exceed two right angles,) can *this* never be determined unless *two* "goniometrical ratios" are given? In some cases it can: but not in all. If in the last figure we confine our attention to the *positive* angles only, we see that if the *sine* only of

XOA_1 were given, we could not distinguish it from XOA_2, (and the same may be said of its reciprocal, the cosecant); but that from any one of the remaining ratios the angle might be absolutely determined.

Hence the statement, which I quoted at the beginning of this dissertation, that "if any of the quantities, sin A, cos A, &c., be given, the angle A may be determined," ought to be read thus: "to determine the angle A as *a geometrical angle,* one of the 4 quantities, cos A, tan A, sec A, cot A, must be given (sin A and cosec A not being sufficient for this purpose): to determine it as *an angle of position,* the magnitude of *one* of these 6 quantities must be given, and the signs of *two* (which two must not be reciprocals): to determine it *as angle of revolution,* it must be referred to some angular unit."

N.B. The phrase "goniometrical ratio," besides being too long for constant repetition, conveys the false notion of its *measuring* the angle, (whereas the measure always varies as the thing measured,) instead of merely *indicating* its value: and does *not* convey the notion that *two* of these ratios are generally necessary to determine the angle; for this latter purpose some term analogous to "co-ordinate," (which, by the way should rather be "co-ordin*ant*,") is needed. I propose, then, to call them the "co-indicants" of the angle.

In the following Formulæ, the "data" are separated from the "quæsita," so that, by covering half of the page, the student may test for himself the accuracy of his recollection of them.

The following are the new symbols introduced:

Symbol.	Name.	Meaning.
⌒.	sin.	sine.
⌢.	cos.	cosine.
⌥.	sec.	secant.
⌯.	cosec.	cosecant.
⌃.	tan.	tangent.
⌣.	cot.	cotangent.
⌳.	versin.	versed-sine.

Part III.

Trigonometry.

It should be observed that the angles of a rectilinear Figure are considered as angles of *absolute magnitude only*, i.e. as "geometrical angles." Hence the angle A *can* be determined from cos A, or sec A, or tan A, or cot A; but it can *not* be determined from sin A, or cosec A: this gives rise to what is called the "ambiguous case" in the solution of Triangles.

Formulæ of Part I.

Given an angle expressed in one of the 3 measures, English, French, and radial; to express it in another.	
N.B. The number of English degrees in an angle is represented by "E"; the number of French grades by "F"; the number of radial angles by "Θ"; and the number of radial angles contained in two right angles, (i.e. 3.14159 &c,) by "π."	
(1) Formula connecting E and F,	$E:F::9:10.$
E and Θ,	$E:\Theta::180:\pi.$

Formulæ of Part II.

I. FORMULÆ CONCERNING *ONE* ANGLE ONLY.	
(α) *Given magnitude and sign of a co-indicant: to find magnitude and sign of one other.*	
(1) The pairs of reciprocals are,	⌒ and ⟩, ⌢ and ⟨, ⌐ and ⌙.
(β) *Given magnitude of a co-indicant: to find magnitudes of the rest.*	
(2) Formula connecting ⌒ and ⌢,	$⌒^2 + ⌢^2 = 1.$

THE FORMULÆ OF PLANE TRIGONOMETRY 133

(3) ⌢, ⌣, and ⌒, $\quad\quad ⌒ = \dfrac{⌢}{⌣}.$
(4) ∝ and ⌒, $\quad\quad ∝^2 = ⌒^2 + 1.$
(5) ⋈ and ⌣, $\quad\quad ⋈ = 1 - ⌣.$

(γ) *Given certain angles: to find their co-indicants.*

(6)   ⌢, ⌣, and ⌒ of $0° = $ | $0, 1, 0.$
$\quad\quad\quad\quad\quad\quad\quad 90° = $ | $1, 0, \dfrac{1}{0}.$
$\quad\quad\quad\quad\quad\quad\quad 180° = $ | $0, -1, 0.$
$\quad\quad\quad\quad\quad\quad\quad 270° = $ | $-1, 0, \dfrac{1}{0}.$
$\quad\quad\quad\quad\quad\quad\quad 45° = $ | $\dfrac{1}{\sqrt{2}}, \dfrac{1}{\sqrt{2}}, 1.$
$\quad\quad\quad\quad\quad\quad\quad 60° = $ | $\dfrac{\sqrt{3}}{2}, \dfrac{1}{2}, \sqrt{3}.$
$\quad\quad\quad\quad\quad\quad\quad 30° = $ | $\dfrac{1}{2}, \dfrac{\sqrt{3}}{2}, \dfrac{1}{\sqrt{3}}.$

II. FORMULÆ CONCERNING 2 OR MORE ANGLES.

(α) *Given the co-indicants of 2 angles: to find the co-indicants of their sum and difference.*

(7) $⌢\overline{A + B} = $ $\quad ⌢A.⌣B + ⌣A.⌢B.$
$\quad\quad ⌢\overline{A - B} = $ $\quad ⌢A.⌣B - ⌣A.⌢B.$
$\quad\quad ⌣\overline{A + B} = $ $\quad ⌣A.⌣B - ⌢A.⌢B.$
$\quad\quad ⌣\overline{A - B} = $ $\quad ⌣A.⌣B + ⌢A.⌢B.$

Hence $⌢2A = $ $\quad 2⌢A.⌣A.$
$\quad\quad ⌣2A$, in terms of $⌢A$ and $⌣A$, $= \quad ⌣^2A - ⌢^2A.$
$\quad\quad\quad\quad$ in terms of $⌢A$ only, $= \quad 1 - 2⌢^2A.$
$\quad\quad\quad\quad$ in terms of $⌣A$ only, $= \quad 2⌣^2 - 1.$

(8) $⌒\overline{A + B} = $ $\quad \dfrac{⌒A + ⌒B}{1 - ⌒A.⌒B}.$

$\quad\quad ⌒\overline{A - B} = $ $\quad \dfrac{⌒A - ⌒B}{1 + ⌒A.⌒B}.$

Hence $⌒2A = $ $\quad \dfrac{2⌒A}{1 - ⌒^2A}.$

(9) The formulæ of (7) may also be written thus, by taking $\tan^{-1} t$ to mean "the angle whose tangent is t."

$$\tan^{-1} t_1 + \tan^{-1} t_2 = \tan^{-1}\frac{t_1 + t_2}{1 - t_1 t_2}.$$

$$\tan^{-1} t_1 - \tan^{-1} t_2 = \tan^{-1}\frac{t_1 - t_2}{1 + t_1 t_2}.$$

$$2\tan^{-1} t = \tan^{-1}\frac{2t}{1 - t^2}.$$

(β) *Given the co-indicants of 3 angles: to find the co-indicants of their sum.*

(10) $\overline{\sin A + B + C} = \cos A . \cos B . \cos C . (\tan A + \tan B + \tan C - \tan A . \tan B . \tan C).$

$\overline{\cos A + B + C} = \cos A . \cos B . \cos C . (1 - \tan B . \tan C - \tan C . \tan A - \tan A . \tan B)$

Hence $\sin 3A = \sin A . (3 - 4\sin^2 A).$

$\cos 3A = \cos A . (4\cos^2 A - 3).$

(11) $\overline{\tan A + B + C} = \dfrac{\tan A + \tan B + \tan C - \tan A . \tan B . \tan C}{1 - (\tan B . \tan C + \tan C . \tan A + \tan A . \tan B)}.$

Hence $\tan 3A = \dfrac{3\tan A - \tan^3 A}{1 - 3\tan^2 A}.$

(12) $\tan^{-1} t_1 + \tan^{-1} t_2 + \tan^{-1} t_3 = \tan^{-1}\dfrac{t_1 + t_2 + t_3 - t_1 . t_2 . t_3}{1 - (t_2 t_3 + t_3 t_1 + t_1 t_2)}.$

Hence $3\tan^{-1} t = \tan^{-1}\dfrac{3t - t^3}{1 - 3t^2}.$

(γ) *Given 2 angles: to reduce the sum or difference of their corresponding co-indicants to one term (for logarithmic computation).*

(13) $\sin A + \sin B = 2\sin\dfrac{A + B}{2} . \cos\dfrac{A - B}{2}.$

$\sin A - \sin B = 2\cos\dfrac{A + B}{2} . \sin\dfrac{A - B}{2}.$

THE FORMULÆ OF PLANE TRIGONOMETRY

$$\smile A + \smile B = 2\smile\frac{A+B}{2} \cdot \smile\frac{A-B}{2}.$$

$$\smile A - \smile B = -2\frown\frac{A+B}{2} \cdot \frown\frac{A-B}{2}.$$

$$\frown A + \frown B = \frac{\frown(A+B)}{\smile A.\smile B}.$$

$$\frown A - \frown B = \frac{\frown(A-B)}{\smile A.\smile B}.$$

III. Formulæ concerning the powers of co-indicants.

(15) Demoivre's theorem, viz.:

$$(\smile\theta \pm \sqrt{-1}.\frown\theta)^n = \smile n\theta \pm \sqrt{-1}.\frown n\theta.$$

Hence, if $2\smile\theta = v + \dfrac{1}{v}$,

then $2\sqrt{-1}.\frown\theta = v - \dfrac{1}{v}.$

$2\smile n\theta = v^n + \dfrac{1}{v^n}.$

$2\sqrt{-1}.\frown n\theta = v^n - \dfrac{1}{v^n}.$

(16) Formula expressing $\smile^n\theta$ in terms of $\smile\theta, \smile 2\theta, \smile 3\theta$, &c.
Rule is let $(2\smile\theta)^n = \left(v + \dfrac{1}{v}\right)^n.$

(17) Formula expressing $\frown^n\theta$ in terms of $\frown\theta, \smile\theta, \frown 2\theta, \smile 2\theta$, &c.
Rule is let $(2\sqrt{-1}.\frown\theta)^n = \left(v - \dfrac{1}{v}\right)^n.$

(18) Formula expressing $\frown n\theta$ in terms of $\frown\theta, \frown^2\theta, \frown^3\theta$, &c.
Rule is in Demoivre's theorem, equate the possible and impossible parts on both sides, and divide one by the other.

IV. Summation of series of co-indicants.

(19) $\frown A + \frown(A+B) + \ldots + \frown(A + \overline{n-1}.B) = \dfrac{\frown\dfrac{nB}{2}}{\frown\dfrac{B}{2}} . \frown\left(A + \dfrac{(n-1).B}{2}\right).$

$\smile A + \smile(A+B) + \ldots + \smile(A + \overline{n-1}.B) = \dfrac{\frown\dfrac{nB}{2}}{\frown\dfrac{B}{2}} . \smile\left(A + \dfrac{(n-1).A}{2}\right).$

Hence $\frown A + \frown 2A + \ldots + \frown nA = \dfrac{\frown\dfrac{nA}{2}}{\frown\dfrac{A}{2}} . \frown\dfrac{(n+1).A}{2}.$

$\smile A + \smile 2A + \ldots + \smile nA = \dfrac{\frown\dfrac{nA}{2}}{\frown\dfrac{A}{2}} . \smile\dfrac{(n+1).A}{2}.$

V. Formulæ connecting the co-indicants of an angle with its radial measure.

N.B. The number $\left(1 + \dfrac{1}{\lfloor 1} + \dfrac{1}{\lfloor 2} + \ldots \text{ad inf.}\right)$, i.e. the number 2.71828 &c., is represented by "ϵ." Also, when the symbol θ is used without any co-indicant symbol, it represents the angle in *radial* measure.

(20) Formulæ expressing $\frown\theta$ and $\smile\theta$ in terms of θ, θ^2, θ^3, &c.

$$\smile\theta = 1 - \dfrac{\theta^2}{\lfloor 2} + \dfrac{\theta^4}{\lfloor 4} - \text{\&c.}$$

$$\frown\theta = \theta - \dfrac{\theta^3}{\lfloor 3} + \dfrac{\theta^5}{\lfloor 5} - \text{\&c.}$$

(21) Gregorie's series, expressing θ in terms of $\frown\theta$, $\frown^3\theta$, &c. (i.e. expressing $\frown^{-1}t$ in terms of t, t^3, &c.)

$$\theta = \dfrac{\frown\theta}{1} - \dfrac{\frown^3\theta}{3} + \dfrac{\frown^5\theta}{5} - \text{\&c.}$$

i.e. $\frown^{-1}t = \dfrac{t}{1} - \dfrac{t^3}{3} + \dfrac{t^5}{5} - \text{\&c.}$

Hence to find π.   Rule is   let $\theta = \tfrac{1}{2}$ a right angle.

THE FORMULÆ OF PLANE TRIGONOMETRY   137

(22) Euler's series to find π. Rule is | in the formula
$$\frown^{-1}t_1 - \frown^{-1}t_2 = \frown^{-1}\frac{t_1 - t_2}{1 + t_1 t_2},$$
let $t_1 = 1$, and $t_2 = \dfrac{1}{2}$, and use Gregorie's series.

(23) Machin's series to find π. Rule is | in the formula
$$2\frown^{-1}t = \frown^{-1}\frac{2t}{1 - t^2}, \text{ let } t = \frac{1}{5},$$
multiply both sides by 2, and reduce the right-hand side by the same formula; subtract the equation $\left(\dfrac{\pi}{4} = \frown^{-1}1\right)$ from the result, and use Gregorie's series.

(24) Formulæ expressing $\frown\theta$ and $\smile\theta$ in terms of θ and ϵ.
$$\frown\theta = \frac{1}{2} \cdot (\epsilon^{\theta\sqrt{-1}} + \epsilon^{-\theta\sqrt{-1}}).$$
$$\smile\theta = \frac{1}{2\sqrt{-1}} \cdot (\epsilon^{\theta\sqrt{-1}} - \epsilon^{-\theta\sqrt{-1}}).$$

Formulæ of Part III.

(α) *Given certain magnitudes concerning a triangle: to find certain other magnitudes.*

N.B. The 3 sides are represented by "*a, b, c*"; the opposite angles by "*A, B, C*"; and the quantity $\dfrac{a + b + c}{2}$ by "*S.*"

(1) "Formula of sines."
$$\frac{\frown A}{a} = \frac{\frown B}{b} = \frac{\frown C}{c}.$$

(2) "Formula of sides."
$$\smile A = \frac{b^2 + c^2 - a^2}{2bc}.$$

(3) "Formula of tangents."
$$\frac{a - b}{a + b} = \frown\frac{A - B}{2} \cdot \frown\frac{C}{2}.$$

(4) $\sin\frac{A}{2}$, in terms of the sides, $= \sqrt{\dfrac{S \cdot (S-a)}{bc}}$.

$\cos\frac{A}{2}$, in terms of the sides, $= \sqrt{\dfrac{(S-b) \cdot (S-c)}{bc}}$.

(5) $\sin A$, in terms of the sides, $= \dfrac{2}{bc}\sqrt{S \cdot (S-a) \cdot (S-b) \cdot (S-c)}$.

(6) Area of triangle, in terms of the sides, $= \sqrt{S \cdot (S-a) \cdot (S-b) \cdot (S-c)}$.

(7) Area of triangle, in terms of 2 sides and the included angle, $= \dfrac{bc}{2} \cdot \sin A$.

N.B. The radius of the circle inscribed in a triangle is represented by "r"; the radius of the circumscribed circle by "R"; and the radii of the 3 escribed circles, respectively touching the sides "$a, b, c,$" by "$R_a, R_b, R_c.$" Also the quantity $\sqrt{S \cdot (S-a) \cdot (S-b) \cdot (S-c)}$ is represented by M.

(8) r, in terms of the sides, $= \dfrac{M}{S}$.

(9) R, in terms of the sides, $= \dfrac{abc}{4M}$.

(10) $R_a, R_b,$ and R_c, in terms of the sides, $= \dfrac{M}{S-a}, \dfrac{M}{S-b},$ and $\dfrac{M}{S-c}$.

(β) *Given certain magnitudes connected with a quadrilateral figure, whose opposite angles are supplementary: to find certain other magnitudes.*

N.B. The 4 sides are represented by "a,b,c,d"; the quantity $\dfrac{a+b+c+d}{2}$ by "S"; and the quantity $\sqrt{(ab+cd) \cdot (ac+bd) \cdot (ad+bc)}$ by "D".

(11) Area of figure, in terms of the sides, $= \sqrt{(S-a) \cdot (S-b) \cdot (S-c) \cdot (S-}$

THE FORMULÆ OF PLANE TRIGONOMETRY   139

(12) Length of diagonal lying between the sides a, b, and the sides c, d, $= \dfrac{D}{ab + cd}$.

(13) Length of diagonal lying between the sides a, d, and the sides b, c, $= \dfrac{D}{ad + bc}$.

N.B. The radius of the circumscribed circle is represented by "R," and the quantity
$\sqrt{(S - a) \cdot (S - b) \cdot (S - c) \cdot (S - d)}$ by M.

(14) R, in terms of sides, $= \dfrac{D}{4M}$.

(γ) *Given certain magnitudes connected with a regular polygon: to find certain other magnitudes.*

N.B. The length of each side is represented by "a," and the number of the sides by "n."

(15) Area, in terms of a, $= \dfrac{na^2}{4} \cdot \cot \dfrac{180°}{n}$.

N.B. The radius of the inscribed circle is represented by "r," and of the circumscribed by "R."

(16) r, in terms of a, $= \dfrac{a}{2} \cdot \cot \dfrac{180°}{n}$.

(17) R, in terms of a, $= \dfrac{a}{2} \cdot \csc \dfrac{180°}{n}$.

(18) Area, in terms of r, $= nr^2 \cdot \tan \dfrac{180°}{n}$.

(19) Area, in terms of R, $= \dfrac{nr^2}{2} \cdot \sin \dfrac{360°}{n}$.

6. Formulæ (Group C)

[Undated: ANC 58: Brabant]

There is no mention of this rare pamphlet in the standard references. Although it is unsigned as well as undated, it is certainly by Dodgson. The handwriting is his, and on the third page he has made some corrections using purple ink. Charles Lovett writes that it was produced with an "electric pen"; 1877, the year Dodgson acquired the pen, is its earliest possible date. The holes made by this predecessor to the typewriter created a stencil from which multiple copies could be made.[1] Dodgson used purple ink from about 1871 to 1891, but it is likely that this item was produced at the same time he made the cyclostyled sheets "Algebra" (item 14) in 1877 and "Formulæ" (item 15) in 1878. At the top of the first sheet of the "Formulæ" Dodgson made it clear that his intended audience consisted of "pupils," a term often used to refer to undergraduates. The formulas correspond to the topics in sections G and L of *A Guide to the Mathematical Student in Reading, Reviewing, and Working Examples* (item 43).

N.B. The pupil need not commit to memory the formulæ marked thus "["; but he should be able to work them out readily.

Formula connecting E, F		$E : F :: 9 : 10$
	E, Θ	$E : \Theta :: 180 : \pi$
Approximate values of π		$\dfrac{22}{7}, \dfrac{355}{133}, 3.14159$ &c.
Reciprocal ratios		sin, cosec; cos, sec; tan, cot
Formula connecting sin, cos		$\sin^2 + \cos^2 = 1$
" tan, sin, cos		$\tan = \dfrac{\sin}{\cos}$
" sec, tan		$\sec^2 = \tan^2 + 1$
Sin, cos; tan, of	$0°$	$0, 1, 0$
	$90°$	$1, 0, \dfrac{1}{0}$

1. Charles Lovett, *Alice in North Carroll-ina: Exhibition Notes* (Winston-Salem: privately printed, 1989), item 58, 20.

FORMULÆ (GROUP C)

$180°$	$0, -1, 0$
$[270°$	$-1, 0, \dfrac{1}{0}$
$45°$	$\dfrac{1}{\sqrt{2}}, \dfrac{1}{\sqrt{2}}, 1$
$60°$	$\dfrac{\sqrt{3}}{2}, \dfrac{1}{2}, \sqrt{3}$
$30°$	$\dfrac{1}{2}, \dfrac{\sqrt{3}}{2}, \dfrac{1}{\sqrt{3}}$
Sin $(A + B)$	$\sin A \cdot \cos B + \cos A \cdot \sin B$
" $(A - B)$	" $-$ "
Cos $(A + B)$	$\cos A \cdot \cos B - \sin A \cdot \sin B$
" $(A - B)$	" $+$ "
Tan $(A + B)$	$\dfrac{\tan A + \tan B}{1 - \tan A \cdot \tan B}$
" $(A - B)$	$\dfrac{\tan A - \tan B}{1 + \tan A \cdot \tan B}$
Sin $2A$	$2 \cdot \sin A \cdot \cos A$
Cos $2A$	
in terms of cos, sin	$\cos^2 A - \sin^2 A$
of cos only	$2 \cos^2 A - 1$
of sin only	$1 - 2 \sin^2 A$
Tan $2A$	$\dfrac{2 \tan A}{1 - \tan^2 A}$
[Cos $\dfrac{A}{2}$, in terms of cos A	$\sqrt{\dfrac{1 + \cos A}{2}}$
[Sin $\dfrac{A}{2}$, "	$\sqrt{\dfrac{1 - \cos A}{2}}$
Tan$^{-1} t_1$ + tan$^{-1} t_2$	$\tan^{-1} \dfrac{t_1 + t_2}{1 - t_1 t_2}$
" $-$ "	$\tan^{-1} \dfrac{t_1 - t_2}{1 + t_1 t_2}$
[Hence $2 \tan^{-1} t$	$\tan^{-1} \dfrac{2t}{1 - t^2}$
Sin A + sin B	$2 \sin \dfrac{A + B}{2} \cdot \cos \dfrac{A - B}{2}$
" $-$ "	$2 \cos \dfrac{A + B}{2} \cdot \sin \dfrac{A - B}{2}$

141

Cos A + cos B	$2 \cos \dfrac{A+B}{2} \cdot \cos \dfrac{A-B}{2}$
" − "	$-2 \sin \dfrac{A+B}{2} \cdot \sin \dfrac{A-B}{2}$

Triangles

Formulæ of sines	$\dfrac{\sin A}{a} = \dfrac{\sin B}{b} = \dfrac{\sin C}{c}$
" sides	$\cos A = \dfrac{b^2 + c^2 - a^2}{2bc}$
" tangents	$\tan \dfrac{B-C}{2} = \dfrac{b-c}{b+c} \cdot \cot \dfrac{A}{2}$
$\cos \dfrac{A}{2}$, in terms of sides	$\sqrt{\dfrac{s \cdot (s-a)}{bc}}$
$\sin \dfrac{A}{2}$, "	$\sqrt{\dfrac{(\;-b) \cdot (s-c)}{bc}}$
$\tan \dfrac{A}{2}$, "	$\sqrt{\dfrac{(s-b) \cdot (s-c)}{s \cdot (s-a)}}$
[sin A, "	$\dfrac{2}{bc} \cdot \sqrt{s \cdot (s-a) \cdot (s-b) \cdot (s-c)}$
a, in terms of b, c, B, C	$b \cdot \cos C + c \cdot \cos B$
Area, in terms of two sides and included angle	$\dfrac{bc}{2} \cdot \sin A$
in terms of sides	$\sqrt{s \cdot (s-a) \cdot (s-b) \cdot (s-c)}$

If Area be denoted by "M," and radii of inscribed, circumscribed, and escribed circles by 'r, R, R_a, R_b, R_c;

r	$\dfrac{M}{s}$
R	$\dfrac{abc}{4M}$
[R_a, &c	$\dfrac{M}{s-a}$, &c

Polygons. (*n* sides)

[Each angle	$180° - \dfrac{360°}{n}$
[Formula connecting *r*, *a*	$\dfrac{a}{2r} = \tan \dfrac{180°}{n}$
[” ” *R*, *a*	$\dfrac{a}{2R} = \sin \dfrac{180°}{n}$
[Area, in terms of sides	$\dfrac{na^2}{4} \cdot \cot \dfrac{180°}{n}$
[” ” of *r*	$nr^2 \cdot \tan \dfrac{180°}{n}$
[” ” of *R*	$\dfrac{nR^2}{2} \cdot \sin \dfrac{360°}{n}$

Logarithms

If base be denoted by "*a*";

$\log a$	1
$\log 1$	0
$\log mn$	$\log m + \log n$
$\log \dfrac{m}{n}$	$\log m - \log n$
$\log m^n$	$n \cdot \log m$
$\log \sqrt[n]{m}$	$\dfrac{\log m}{n}.$

7. Simple Facts about Circle-Squaring

CHAPTER I

[1882: LCH 152a, 512: Princeton]

Although incomplete, this introductory chapter, together with item 8, two theorems, were clearly written to be part of a book Dodgson planned to publish on circle-squaring. They are included primarily because of their close connection with item 9, Question 11530. The introductory chapter is dated 20 April 1882; the galleys are not dated. Without these items, it would be difficult to understand both the motivation and importance of Question 11530, the most significant of Dodgson's publications on trigonometry. The Parrish Collection contains much material (Packet 5 of "The Mathematical Manuscripts of Charles L. Dodgson") meant for a treatise on circle-squaring. If it had been published, Dodgson could have referred erroneous amateurs to the treatise to refute their false arguments, without Dodgson having to examine the details of each one.

Dodgson referred to the treatise several times in his diary in 1881: on 24 July, 6 October, and again on 23 November. Each time he mentioned the work he used the title, "Limits of Circle-Squaring." But we know he considered several titles such as "The Elements of Circle-Squaring," "The Elements of Cyclometry," "The Cyclometer's Friend," "The Limits of Cyclometry," and "Circle-Squaring Simplified," as well as the subtitle, "Simple Facts About Circle-Squaring."

CHAPTER I (*Introductory*)

April 20, 1882.

Suppose that a controversy had arisen about the details of the battle of Waterloo, and that in a certain Debating Society the question had been raised as to the exact time when Bulow's Prussian

---

First printed in the *Lewis Carroll Centenary in London, Special Edition*, edited by Falconer Madan (London: J. & E. Bumpus, 1932), 122–25.

Corps appeared on the field of battle. Disputants, who supported the theory that it was a little before 6 p.m., or a little after, would no doubt be patiently listened to: but what would the Society say to a member who proposed to prove that it took place at 4 p.m. on the *nineteenth* of June? Would they not exclaim with one voice, 'If there is one fact in History more certain than another, it is that the battle was fought on the *eighteenth*. To go outside the limits of that day is simply absurd. We cannot waste our time in listening to any one who does not accept the ordinary *data* of the subject.'

Now this is precisely the position I propose to take with reference to the theories of the "Circle-Squarers," under which term I include all who have attempted to give an *exact* value to the area of a circle, expressed in terms of the square on its radius. The mathematical world are agreed that it is somewhere very near 3.14159 times that square—so near, indeed, that the above number is too small to express it, while 3.1416 is too great. Any one, then, who should suggest the theory that it was a little more or less than this number, say 3.14161 or 3.14158, might perhaps find listeners: but what would be said to a theorist who proposed to prove it to be 4 1/2? "My good sir," we should exclaim, "if there is one fact in Geometry more certain than another, it is that the area of a circle is less than its circumscribed square and greater than its inscribed square; and that these two squares are respectively four times, and twice, the square of the radius. To go outside these limits is simply absurd. We cannot waste our time in listening to any one who does not accept the ordinary *data* of the subject."

For any Circle-Squarer, then, whose theory is that the area of a circle is more than four times, or less than twice, the square on its radius, what has been already said would be amply sufficient, and this little book would not need to be written. But the numbers proposed are in no case so wide of the mark as this; and if an answer of this kind is to be given to their proposers, the limits fixed must be very much closer together than the numbers 4 and 2.

And this, it has occurred to me, it is possible to do, without using more than the very simplest facts in Mathematics—facts to dispute which would be much the same thing as to deny that two and two make four.¹ To measure the area of the circle itself is a complicated matter, and the processes, by which the value 3.14159 has been calculated, are long and abstruse: and any Circle-Squarer, if called upon to disprove the results so obtained, before he can expect a hearing for his rival theory, might very reasonably reply, "As a mere question of wasting time, it is much more reasonable that *you* should give a few minutes to examining my Theorem, and to disproving it if you can, than that *I* should spend months, or even years, in mastering these difficult calculations."

"Why not, then," it may be asked me, "content yourself with simply *disproving* the Theorem of each Circle-Squarer you meet with? Their proofs are usually short: they seldom go beyond the range of elementary Geometry: and being, as we know they must be, untrue, they no doubt contain palpable logical fallacies."

That is all very true: but, in the first place, the *disproof* of his pet Theorem is precisely the very last fact in the world that a Circle-Squarer can be got even to listen to with patience: long contemplation of the result of his labours has made him as sure of its truth as of his own existence: and, in the second place, this would require a fresh argument to be composed for every fresh Circle-Squarer, instead of having, what I hope this little book will furnish, an answer equally applicable to all comers.

The course I propose to take is briefly this:—first, to give a list of the elementary truths I shall afterwards have occasion to quote: then to prove, by very simple methods (in which I shall make no attempt at measuring the *circle* at all, but shall merely measure certain *rectilinear figures* drawn within it and outside it), that, whatever be the *exact* value of the area, it is at any rate less than 3.1417 times, and greater than 3.1413 times, the square on its radius.

Hence, for any Circle-Squarer, whose value for this area lies outside these two limits, this little book will I hope be a sufficient answer. He cannot plead that the proofs here offered are too long, or too abstruse, for him to understand them: and he may fairly be

---

1. In the original manuscript Dodgson wrote "five," according to a note added by Madan, who changed it to "four" as it appears here.

called upon to disprove the truth of the above-named limits, before he can expect a hearing for a theorem which contradicts the opinion of most of the world. If he should take exception to any one of the preliminary truths here quoted, he is of course out of court at once, as they stand on the same footing as the fact that two and two are four: further discussion would be absolute waste of time. If, however, he accepts these preliminary truths, he cannot well avoid being led on, by irresistible logic, to accept the truth of the above-named limits. The method, by which they are obtained, is one that he can easily carry further for himself, and find new pairs of limits, each pair closer together than the preceding pair: so that, even if he has adopted a value a little within the limits 3.1417 and 3.1413, he may still find limits which will exclude the possibility of his value being true.

The exact value of π (the name usually given to "the ratio which the area of a circle bears to the square on its radius") has in all ages proved an *ignis fatuus,* that has led hundreds, if not thousands, of hapless mathematicians to waste valuable years in the hope of immortalizing themselves by discovering what has been so long sought in vain. I cherish the hope that this little book may fall into the hands of some who have been dazzled by its mocking light, and may prove the means of saving to them much time and labour that would otherwise be wasted.

8. *Proof Sheets: Propositions I, II*
[Undated: LCAT 630: HRHRC]

See the discussion of item 7.

PROP. I. THEOREM.

The area of a Circle is less than four times, and greater than twice, the Square on its radius.

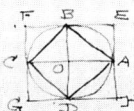

Let $ABCD$ be a Circle whose centre is O: and let AC, BD be two diameters at right angles to each other: and through A, B, C, D, let tangents be drawn to the Circle: and join AB, BC, CD, DA.

Then it is evident that OE is the Square on OA, and that $EFGH$ is 4 times this Square and $ABCD$ twice it.

Hence the area of the Circle, being $< EFGH$ and $> ABCD$, is < 4 times, and $>$ twice, the Square on its radius.

Q.E.D.

We will now prove this by another method, similar to what we shall use in the next Proposition.

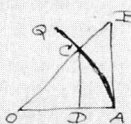

Take a Line OA, and from A draw AB at right angles to OA and equal to it. With centre O, at distance OA, describe the Circle ACQ: join OB, cutting this Circle at C: and from C draw CD at right angles to OA.

$$\begin{cases} \tan AOB = \dfrac{AB}{OA} = \dfrac{OA}{OA} = 1, \\ \text{and that is also} = \dfrac{CD}{OD}, \end{cases}$$

$$\therefore \dfrac{2D}{OD} = 1; \text{ i.e. } CD = OD.$$

also, $\because \tan AOB = 1$,

$\therefore$ angle $AOB = \lfloor tn \rfloor 1 = 45°$;

$\therefore$ the Sector $AOC = \frac{1}{8}$ of the Circle; i.e. the area of the Circle is 8 times that of the Sector.

Again, $\because$ the Triangle OAB has its altitude $AB =$ its base OA,

$\therefore$ it $= \frac{1}{2}$ of the Square on OA;

similarly, the Triangle $ODC = \frac{1}{2}$ of the Squares on OD.

Again, $\because$ the Square on $OC =$ the sum of the Squares on OD, DC, [Euc. I. 47.

$\therefore$ the Square on $OD = \frac{1}{2}$ the Square on OD;
$= \frac{1}{2}$ of the Square on OC;
$= \frac{1}{2}$ of the Square on OA;

$\therefore$ the Triangle $ODC = \frac{1}{2}$ of $(\frac{1}{2}$ of the Square on $OA)$;
i.e. it $= \frac{1}{4}$ of the Square on OA.

Now the area of the Sector AOC is evidently $<$ the Triangle OAB, and $>$ the Triangle ODC;

i.e. it is $< \frac{1}{2}$, and $> \frac{1}{4}$, of the Square on OA;

$\therefore$ the area of the Circle, being 8 times that of the Sector, is $< 8 \times \frac{1}{2}$, and $> 8 \times \frac{1}{4}$, of the Square on OA;

i.e. it is < 4 times, and $>$ twice, the Square on its radius.

Q.E.D.

PROP. II. THEOREM.

The area of a Circle is less than 3·4 times, and greater than 2·7 times, the Square on its radius.

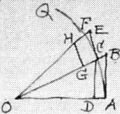

Take a Line OA, and from A draw AB at right angles to OA and equal to $\frac{1}{3}$ of it. With centre O, at distance OA, describe the Circle ACQ: join OB, cutting this Circle at C: and from C draw CD at right angles to OA. Also from C draw CE at right angles to OC and equal to $\frac{1}{3}$ of it.

Hence CE falls wholly outside the Circle. [Euc.

Join OE, cutting the Circle at F. Also from OC cut off $OG = OD$, and from G draw GH at right angles to OC.

Now, $\because$ $\begin{cases} \tan AOB = \dfrac{AB}{OA} = \dfrac{\frac{1}{3} \text{ of } OA}{OA} = \frac{1}{3}, \\ \text{and that is also} = \dfrac{CD}{OD}, \end{cases}$

$\therefore \dfrac{CD}{OD} = \frac{1}{3};$ i.e. $CD = \frac{1}{3}$ of OD;

similarly, $\tan COE = \frac{1}{3}$, and $HG = \frac{1}{3}$ of OG,
$= \frac{1}{3}$ of OD;

hence also, HG is not greater than CD,

Also, $\because$ in the right-angled Triangles ODC, OGH, $OG = OD$, and HG is not greater than CD,

$\therefore OH$ is not greater than OC;

$\therefore H$ is not outside the Circle;

$\therefore$ the Triangle OGH is within the Circle.

Now $\tan AOE = \tan (AOB + COE)$,

$= \dfrac{\tan AOB + \tan COE}{1 - \tan AOB . \tan COE},$

$= \dfrac{\frac{1}{3} + \frac{1}{3}}{1 - \frac{1}{3} \cdot \frac{1}{3}} \left(=\frac{3}{3}\right) = 1;$

$\therefore$ angle $AOE = \lfloor tn \rfloor 1 = 45°$;

$\therefore$ the Sector $AOF = \frac{1}{8}$ of the Circle; i.e. the area of the Circle $= 8$ times that of the Sector.

Again, $\because$ the Triangle OAB has its altitude $AB = \frac{1}{3}$ of its base OA,

$\therefore$ it $= \frac{1}{2}$ of ($\frac{1}{3}$ of the Square on OA);

similarly, the Triangle $OCE = \frac{1}{2}$ of ($\frac{1}{3}$ of the Square on OC)

$= \frac{1}{2}$ of ($\frac{1}{3}$ of the Square on OA);

$\therefore$ the sum of the Triangles OAB, OCE, $= \frac{1}{2}$ of $\{(\frac{1}{3} + \frac{1}{3})$ of the Square on $OA\}$.

Similarly, the sum of the Triangles ODC, OGH, $= \frac{1}{2}$ of $\{(\frac{1}{3} + \frac{1}{3})$ of the Square on $OD\}$.

Again, $\because$ the Square on $OC = $ the sum of the Squares on OD, DC,
 [Euc. I. 47.

$\therefore$ it $=$ the Square on OD with the Square on $\frac{1}{3}$ of OD;

i.e. it $= (1 + \frac{1}{9})$ of the Square on OD;

$= \frac{10}{9}$ of the Square on OD;

$\therefore$ the Square on $OD = \frac{9}{10}$ of the Square on OC;

$= \frac{9}{10}$ of the Square on OA;

$\therefore$ the sum of the Triangles ODC, OGH, $= \frac{1}{2}$ of $\{(\frac{1}{3} + \frac{1}{3})$ of the Square on $OA\} \times \frac{9}{10}$.

Now the area of the Sector AOF is evidently $<$ the sum of the Triangles OAB, OCE, and $>$ the sum of the Triangles ODC, OGH; i.e. it is $< \frac{1}{2}$ of $\{(\frac{1}{3} + \frac{1}{3})$ of the Square on $OA\}$, and $>$ the same $\times \frac{9}{10}$;

$\therefore$ the area of the Circle, being 8 times that of the Sector, is $< 8 \times \frac{1}{2}$ of $\{(\frac{1}{3} + \frac{1}{3})$ of the Square on $OA\}$, and $>$ the same $\times \frac{9}{10}$.

i.e. it is $< (2 + \frac{2}{3})$ of the Square on OA, and $> (2 + \frac{2}{3}) \times \frac{9}{10}$ of the same Square.

Now $2 = 2$.
$\frac{2}{3} = 1.3$, i.e. it is < 1.4, and > 1.3
$\therefore (2 + \frac{2}{3}) \qquad < 3.4$, and > 3.3;

$\therefore$ the area of the Circle, being $(2 + \frac{2}{3})$ of the Square on its radius, is *a fortiori* < 3.4 of this Square.

Now $\frac{9}{10} = 1 - \frac{1}{10}$;

$\therefore (2 + \frac{2}{3}) \times \frac{9}{10} = (2 + \frac{2}{3})(1 - \frac{1}{10})$,

$= (2 + \frac{2}{3}) - \dfrac{2 + \frac{2}{3}}{10};$

also $2 + \frac{2}{3} > 3 \cdot \frac{3}{10}$;

and $\dfrac{2 + \frac{2}{3}}{10} < \dfrac{3 \cdot \frac{4}{10}}{10}$; i.e. $< \cdot 6 \frac{8}{10}$; i.e. $< \cdot 7$;

$\therefore (2 + \frac{2}{3}) \times \frac{9}{10} > 3 \cdot \frac{3}{10} - \cdot 7$; i.e. $> 2 \cdot 7$;

$\therefore$ the area of the Circle, being $> (2 + \frac{2}{3}) \times \frac{9}{10}$ of the Square on its radius, is *a fortiori* $> 2 \cdot 7$ of this Square.

That is, the area of a Circle is $< 3 \cdot 4$ times and $> 2 \frac{2}{3}$ times the Square on its radius.
 Q.E.D.

We have thus found certain Limits, between which the area of Circle must lie: the Superior Limit being $3 \cdot 4$ times the Square on the radius, and the Inferior Limit being $2 \cdot 7$ times the same Square. Also we are now able to find fresh pairs of Limits, closer and closer together, *ad libitum*: for, by mere inspection of the preceding Theorem, we can obtain a rule for finding such pairs, whenever we can resolve the angle 45° into the sum of a series of angles whose tangents are of the form $\frac{1}{a}$, $\frac{1}{b}$, &c., where a is not greater than any other of the denominators.

The Limits are

$4 \times (\frac{1}{a} + \frac{1}{b} + \&c.)$, and $4 \times (\frac{1}{a} + \frac{1}{b} + \&c.) \times \dfrac{a^2}{a^2 + 1} \left(1 - \dfrac{1}{a^2 + 1}\right)$

of the Square on the radius. Hence, if we know that

$4 \times (\frac{1}{a} + \frac{1}{b} + \&c.) < x,$

we may take x times the Square on the radius as a Superior Limit for the area of a Circle.

We see also a simple method for finding a number less than

$4 \times (\frac{1}{a} + \frac{1}{b} + \&c.) \times \dfrac{a^2}{a^2 + 1}.$

for $\dfrac{a^2}{a^2 + 1} = 1 - \dfrac{1}{a^2 + 1};$

hence $4 \times (\frac{1}{a} + \frac{1}{b} + \&c.) \times \dfrac{a^2}{a^2 + 1}$

$= 4 \times (\frac{1}{a} + \frac{1}{b} + \&c.) \times (1 - \dfrac{1}{a^2 + 1}),$

$= 4 \times (\frac{1}{a} + \frac{1}{b} + \&c.) - \dfrac{4 \times (\frac{1}{a} + \frac{1}{b} + \&c.)}{a^2 + 1};$

and, if we know that

$4 \times (\frac{1}{a} + \frac{1}{b} + \&c.) < x,$ and $> y,$ so that

$\dfrac{4 \times (\frac{1}{a} + \frac{1}{b} + \&c.)}{a^2 + 1} < \dfrac{x}{a^2 + 1},$

we may conclude that their difference is $> y - \dfrac{x}{a^2 + 1}.$

Hence $(y - \dfrac{x}{a^2 + 1})$ times the Square on the radius may be taken as an Inferior Limit of the area of a Circle.

As an instance of this rule, let us resolve 45° into the sum of 3 angles.

We know already that

$45° = \lfloor tn \rfloor \frac{1}{2} + \lfloor tn \rfloor \frac{1}{3};$

now $\lfloor tn \rfloor \frac{1}{2} - \lfloor tn \rfloor \frac{1}{3} = \lfloor tn \rfloor \dfrac{\frac{1}{2} - \frac{1}{3}}{1 + \frac{1}{2 \cdot 3}}$

$= \lfloor tn \rfloor \frac{1}{7};$

$\therefore \lfloor tn \rfloor \frac{1}{2} = \lfloor tn \rfloor \frac{1}{3} + \lfloor tn \rfloor \frac{1}{7};$

$\therefore 45° = \lfloor tn \rfloor \frac{1}{3} + \lfloor tn \rfloor \frac{1}{3} + \lfloor tn \rfloor \frac{1}{7}.$

Hence, if $4 \times (\frac{1}{3} + \frac{1}{3} + \frac{1}{7}) < x,$ and $> y,$ we may take, as Limits, x times, and $(x - \dfrac{y}{3^2 + 1})$ times, the Square on the radius.

9. Question 11530

[1893: LCH 136: British]

The *Educational Times,* a monthly London periodical, began to publish several columns on mathematical problems and their solutions in 1862. These were then reprinted in two volumes per year as *Mathematical Questions and Solutions, from the "Educational Times," with Many Papers and Solutions in Addition to Those Published in the "Educational Times."* Mathematicians both major and minor contributed problems and solutions. Some of the more illustrious contributors were E. Beltrami of the University of Pisa, Arthur Cayley of Cambridge University, J. Hadamard and C. Hermite of Paris, B. Peirce of Harvard University, and James Sylvester of Oxford University. But there were many correspondents of less renown.

Dodgson contributed eleven items on various topics, all of which except one, on logic, are included in this volume.

Dodgson's contributions to the *Educational Times* are listed in the *Lewis Carroll Handbook,* p. 106, and in R. C. Archibald's "Bibliography of Lewis Carroll: Additions," *Notes and Queries,* vol. 179 (24 August 1940), 134–35. Both listings incorrectly give 1 May 1892, p. 274, as the reference for Question 11530 in the *Educational Times.* The correct citation is 1 March 1893, pp. 155–56. The *Lewis Carroll Handbook* omits "Response to 'Infinitesimal or Zero?'" (1886), included here as item 18. Question 14122, published after Dodgson's death, concerns a topic in logic and will be discussed in a subsequent volume in this series.

※

(Rev. C. L. Dodgson, M.A.)—Required a general investigation of the following trigonometrical formula, which is useful in calculating limits for the value of π. The problem which I set myself was to break up $\tan^{-1} 1/a$ into two angles of the same form. Let

Mathematical Questions and Solutions, vol. LIX, edited by W. J. C. Miller (London: Francis Hodgson, 1893), 71–73.

$$\tan^{-1}\frac{1}{a} = \tan^{-1}\frac{1}{a+x} + \tan^{-1}\frac{1}{a+y}$$
$$= \tan^{-1}\frac{2a+x+y}{a^2+a(x+y)+xy-1}.$$

Then, if $(xy - 1)$ were made equal to a^2, the denominator would become $a(2a + x + y)$; i.e., the fraction would become $1/a$. Hence we get the rule: Let $(a^2 + 1) = xy$; i.e., break up $(a^2 + 1)$ into any two factors, call them x and y, and use them in the formula with which we began. Thus, if $a = 3$, $a^2 + 1 = 10 = 2 \times 5$. Hence $\tan^{-1}\frac{1}{3} = \tan^{-1}\frac{1}{5} + \tan^{-1}\frac{1}{8}$. By the use of this formula, I have obtained 3.141597 and 3.141583 as limits for π.

Solution by H. J. Woodall, A.R.C.S.

If $\tan^{-1} 1/a = \tan^{-1} 1/(a+x) + \tan^{-1} 1/(a+y)$, or, as I prefer to write it (in Gauss's notation), A cot $a =$ A cot$(a + x)$ + A cot$(a + y)$, then

A cot $a =$ A cot$[\{a^2 + a(x+y) + xy - 1\}/\{2a + x + y\}]$;

and we get, with the assumption that the angles are the smallest possible,

$2a^2 + a(x+y) = a^2 + a(x+y) + xy - 1$;
whence $a^2 + 1 = xy$ ... (C₂).

This useful formula occurs in the 7th and 8th editions of the *Ency. Brit.*, Art. "Algebra" (Ed. 7, Vol. II., p. 497, Ed. 8, Vol. II., p. 557, both by Prof. Wallace). The series of relations which I obtained by the aid of $a^2 + 1 = xy$ reminds me of a remark by Prof. Chrystal in his *Algebra* (Vol. II., p. 309): "Gauss (*Werke*, Bd. II., p. 525) found, by means of the theory of numbers, two remarkable formulae of this kind, viz.:—

$$\tfrac{1}{4}\pi = 12 \tan^{-1} 1/18 + 8 \tan^{-1} 1/57 - 5 \tan^{-1} 1/239$$
$$= 12 \tan^{-1} 1/38 + 20 \tan^{-1} 1/57 + 7 \tan^{-1} 1/239$$
$$+ 24 \tan^{-1} 1/268 \ldots (1).$$

By means of this, π could be calculated with great rapidity should its value ever be required beyond the 707th place, which was reached by Mr. Shanks in 1873!"

Mr. Dodgson's formula (C_2) may be thus generalized:

If $\quad$ A cot a = A cot($a + x_1$) + A cot($a + x_2$) + $\ldots$
$$+ \text{A cot}(a + x_n),$$
then $\quad a = (_nS_n - {_nS_{n-2}} + {_nS_{n-4}} - \ldots)/$
$$({_nS_{n-1}} - {_nS_{n-3}} + {_nS_{n-5}} - \ldots) \ldots (C_n),$$

where $_nS_r$ denotes the sum of the products, r at a time, of the n quantities: $a + x_1, a + x_2, a + x_3, \ldots a + x_n$. ($_nS_0$ being taken $= 1$.)

The relation (C_n) may be expanded in various ways. (C_1) reduces to $a = a$, (C_2) reduces to $a^2 + 1 = xy$ (as already found), (C_3) reduces to $(a^2 + 1)\{2a + x_1 + x_2 + x_3\} = x_1 x_2 x_3$ (C_3), which includes C_2 if we make $x_3 = \infty$, and so on, the disadvantage of the use of any given formula varying with the suffix pertaining to that formula.

From the remarks at the end of the second volume of the *Werke*, it appears that Gauss had obtained a formula, of this class, which fully deserves the epithet "remarkable." We have

$$\text{A cot } a - \text{A cot}(2a + x_1) - \text{A cot}(2a - x_1)$$
$$= \text{A cot } a - \text{A cot}\{(4a^2 + x_1^2 - 1)/4a\}$$
$$= \text{A cot}[\{\tfrac{1}{4}(4a^2 - x_1^2 - 1) + 1\}/$$
$$\{\tfrac{1}{4}(4a^2 - x_1^2 - 1)/a - a\}]$$
$$= \text{A cot}\{-a(4a^2 - x_1^2 + 3)/(x_1^2 + 1)\}$$
$$= -\text{A cot } b \text{ say,}$$
then $\quad b = a\{4(a^2 + 1) - (x_1^2 + 1)\}/(x_1^2 + 1)$
$$= 4a(a^2 + 1)/(x_1^2 + 1) - a \ldots (2);$$

we have, in fact,

$$A \cot a + A \cot\{4a(a^2 + 1)/(x_1^2 + 1) - a\}$$
$$= A \cot(2a + x_1) + A \cot(2a - x_1).$$

Hence, in order that b shall be an integer, we must have $4a(a^2 + 1)/(x_1^2 + 1) - a$ an integer, that is to say, $4a(a^2 + 1)/(x_1^2 + 1)$ must be integral. Thus we give x_1 a value, and then, by means of a table of "prime factors of $aa + 1$," we can find suitable values for a, and hence for b.

Gauss's Editor, Schering, to whom we owe the abstraction and preservation of these manuscript remains, gives (*Werke*, Bd. II., p. 525) the following examples of the use of the formulæ (2):—

$a = 253,$	$x_1 = 6,$	$b = 1750507$
$a = 294,$	$x_1 = 11,$	$b = 832902$
$a = 119,$	$x_1 = 1,$	$b = 3370437$
$a = 57,$	$x_1 = 3,$	$b = 74043$
$a = 123,$	$x_1 = 9,$	$b = 90657.$

Another relation, which I have found and made use of, can be obtained from the equation $A \cot a = A \cot(a - x_1) - A \cot(a + x_2)$, whence

$$a(x_1 + x_2) = a^2 + a(x_2 - x_1) - x_1 x_2 + 1;$$
$$\therefore a^2 + 1 = x_1(2a + x_2) \ldots (3).$$

By the aid of the above-given formulæ, but principally of (C_2) and (3), we may very readily "expand" a relation in A cots; thus we have, for $\frac{1}{4}\pi$, the several values:—

$$A \cot 1 = A \cot 2 + A \cot 3 = 2A \cot 3 + A \cot 7,$$
$$2A \cot 5 + A \cot 7 + 2A \cot 8 = 3A \cot 7 + 2A \cot 18$$
$$+ 2A \cot 8,$$
$$5A \cot 8 + 3A \cot 57 + 2A \cot 18 = 5A \cot 13 + 5A \cot 21$$
$$+ 3A \cot 57 + 2A \cot 18,$$
$$5(A \cot 18 + A \cot 47 + A \cot 21) + 3A \cot 57 + 2A \cot 18,$$
$$7A \cot 18 + 5(A \cot 57 + A \cot 268 + A \cot 21) + 3A \cot 57,$$
$$7A \cot 18 + 8A \cot 57 + 5(A \cot 21 + A \cot 268) \ldots \ldots (A).$$

Now consider A cot 18 − A cot 21 − A cot 268 − A cot 239, with relation to the formula

$$(a^2 + 1)(2a + x_1 + x_2 + x_3) − x_1 x_2 x_3 = 0 \ldots \ldots (C_3).$$

Here $a = 18$, $x_1 = 3$, $x_2 = 250$, $x_3 = 221$; substituting in (C_3), we get

$$325(36 + 474) − 3 \times 250 \times 221 = 0 \text{ identically;}$$

therefore

$$\text{A cot } 18 − \text{A cot } 21 − \text{A cot } 268 − \text{A cot } 239 = 0.$$

Multiply by 5 and add to (A), when we get

$$\tfrac{1}{4}\pi = 12\text{A cot } 18 + 8\text{A cot } 57 − 5\text{A cot } 239,$$

which is Gauss I; in a similar way, but with use of (3), we can easily find

$$\tfrac{1}{4}\pi = 12\text{A cot } 38 + 20\text{A cot } 57 + 7\text{A cot } 239 + 24\text{A cot } 268,$$

which is Gauss's II.

Algebra

William Spottiswoode, F.R.S. (1825–1883), mathematician and physicist. President of the Royal Society, 1878–1883.

Introduction

Dodgson's algebraic writings consist of one book, *An Elementary Treatise on Determinants with Their Application to Simultaneous Linear Equations and Algebraical Geometry* (1867); Question 9995 in the *Educational Times* of 1 February 1889 (item 16); three related Pillow Problems, 8, 25, and 68; and seven pamphlets: *Notes on the First Part of Algebra* from 1861, unavailable but in the Hartley collection; *Condensation of Determinants* from 1866 (item 10); *Algebraical Formulæ* (item 11) and *Formulæ in Algebra* (item 12) from 1868 (?); *Algebraical Formulæ and Rules* (item 13) from 1870; *Algebra* from 1877 (item 14); and *Formulæ* from 1878 (item 15).

The contents of the pamphlet *Condensation of Determinants* is included (with some alteration, particularly in notation) in Appendices II, III, and IV to his book on determinants. In the preface to that book, Dodgson acknowledged several of his sources. He cited the German mathematician Heinrich Richard Baltzer's influential 1857 treatise on determinants, which had been translated into French in 1861, Isaac Todhunter's work on determinants in his theory of equations, and George Peacock's algebra text (presumably his 1830 *Treatise on Algebra*).

Dodgson's contributions to linear algebra constitute his most significant mathematical work. Thomas Muir, author of the authoritative five-volume survey, *The Theory of Determinants in the Historical Order of Development*, commented on the qualities of originality, logical consistency, and completeness in both the pamphlet and the book.

> Dodgson's condensation-process for the evaluation of determinants whose elements are arithmetical has a different basis from the theorems of Hermite (1849) and Chio (1853)....
>
> By applying the rule to a determinant with general elements we may obtain not only the necessary justification for its use but a fuller and better understanding of the whole process.[1]
>
> This is a text-book quite unlike all its predecessors. Professedly its main aim is logical exactitude. In pursuance thereof all definitions, conventions, axioms, propositions and corollaries are carefully formulated, labelled and numbered: every step in a train of reasoning is

---

1. Thomas Muir, *The Theory of Determinants in the Historical Order of Development*, vol. III (London: Macmillan, 1920; reprint, New York: Dover Publications, 1960), 17–18.

kept scrupulously separate from its antecedent and consequent, and in order to guard against possible contamination of the reasoning illustrative examples are relegated to the footnotes....

The most important chapters are iii, iv, v, vi, the subjects which they concern being treated in more detail than in any previous text-book.[2]

Dodgson used a similar format, separating the geometric and algebraic developments and using footnotes for additional illustrations and examples, in his 1868 pamphlet, *The Fifth Book of Euclid Treated Algebraically*.

In his article "Lewis Carroll as a Probabilist and Mathematician," Eugene Seneta has included an item of considerable importance for assessing Dodgson's contribution to linear algebra. It is a quotation from volume 1, chapter 2 of *Matematika XIX Veka*, a three-volume study of nineteenth-century mathematics edited by the distinguished Russian mathematicians A. N. Kolmogorov and A. P. Yushkevich. The authors of Chapter 2 are I. G. Basmakova and A. N. Rudakov. What follows is a precise translation of the quotation in Seneta's paper.

> The concept of rank of a matrix and the theorem of Kronecker-Capelli were discovered independently by several investigators. The first printed proof of this theorem is due to C. L. Dodgson (1832–1898), author of the splendid stories *Alice in Wonderland* and *Alice Through the Looking Glass*. The theorem was published in his *An Elementary Treatise on Determinants* (London, 1867) in the following formulation: "For a system of n inhomogeneous equations with m unknowns to be consistent, it is necessary and sufficient that the order of the largest minor different from zero be the same in the augmented and non-augmented matrix of the system."[3]

Dodgson's own version of this theorem, proposition II in Chapter IV of his book, illustrates one of the major shortcomings of his mathematical writing—awkward prose.

> If there be given n Equations, not all homogeneous, containing Variables: a test for their being consistent is that either, first, there is one of them such that, when it is taken along with each of the remaining

---

2. *Ibid.*, 24.

3. The quotation appears in Eugene Seneta, "Lewis Carroll as a Probabilist and Mathematician," *Math. Scientist,* 9 (1984), pp. 91–92. Seneta's free English translation is practically identical to the precise translation of the quotation rendered by Abe Shenitzer, which is used here.

Equations successively, each pair of Equations, so formed, has its B-Block evanescent; or secondly, there are m of them, where m is one of the numbers 2 . . . . . n, which contain at least m variables, and have their V-Block not evanescent, and are such that, when they are taken along with each of the remaining Equations successively, each set of Equations, so formed, has its B-Block evanescent.[4]

As Seneta explains, Dodgson is dealing with rectangular arrays, or matrices, in the context of a linear system of equations that we would write as $\mathbf{A}x = b$. $\mathbf{A}$ is the unaugmented matrix; $(\mathbf{A}, b)$ is the augmented matrix. If $\mathbf{B}$ is an $(m \times n)$ matrix whose rank is the smaller of (m, n) and the determinant of each of its $(r \times r)$ submatrices is zero, then b is evanescent.

Henry J. S. Smith, Savilian Professor of Geometry at Oxford and one of the best British mathematicians of the nineteenth century, was also interested in linear systems of equations but in connection with number theory. Contrast the statement of a theorem from one of his papers, "On Systems of Linear Indeterminate Equations and Congruences," with Dodgson's just from the point of view of written English.

If every determinant of the augmented matrix of a redundant system of linear equations is equal to zero, while the determinants of the unaugmented matrix are not all zero, the system admits of one solution, and only one.[5]

Dodgson's convoluted style certainly reflects how his mind dealt with complicated problems. He often made connections that more conventional scholars might not see, particularly when subtle differences of meaning were involved. Unfortunately, his style also hindered broader acceptance of his stronger ideas, like this one. His insistence on inventing new terms, such as *block* for *matrix*, also distracted from the mathematical issues at hand: "But surely the former word means rather the mould, or form, into which algebraical quantities may be introduced, than an actual assemblage of such quantities."[6]

Dodgson introduced another problem in his writing that made it difficult to get at the main points of his arguments: bizarre notation.

---

4. C. L. Dodgson, *An Elementary Treatise on Determinants* (London: Macmillan, 1867), 61.
5. *The Collected Mathematical Papers of Henry John Stephen Smith*, edited by J. W. L. Glaisher, vol. I (1894; reprint, New York: Chelsea Publishing Co., 1965), 369.
6. *An Elementary Treatise on Determinants*, iv–v.

50, Grosvenor Place, S.W.

8 Feb[ruary]

Letter from William Spottiswoode to Charles L. Dodgson concerning Dodgson's methods of handling determinants, dated 8 February [1867].

My dear Sir,

Pray send me the rest of your work, when it is in print; it will give me great pleasure to read it.

The method of breaking up a determinant such as

$$\begin{vmatrix} a & b & c & d \\ e & f & g & h \\ \alpha & \beta & \gamma & \delta \\ \epsilon & \zeta & \eta & \theta \end{vmatrix}$$

into $\sum \pm \begin{vmatrix} a & b \\ e & f \end{vmatrix} \begin{vmatrix} \gamma & \delta \\ \eta & \theta \end{vmatrix}$

is of great use in the algebraical development of determinants. There are it is true few, or almost no, general rules about it; but the exact mode of procedure will be suggested by each special case. E.g. If it were proposed to eliminate x, y, from

$$(a, b, \ldots)(x, y)^n = 0$$
$$(\alpha, \beta, \ldots)(x, y)^n = 0$$

we should have as the resultant

$$\begin{vmatrix} a & \alpha b & \cdots \\ \cdot & a & \cdots \\ \vdots & \vdots & \\ \alpha & \alpha\beta & \cdots \\ \cdot & \alpha & \cdots \\ \vdots & \vdots & \end{vmatrix} = 0$$

$$= \sum \begin{vmatrix} a & \alpha b & \cdots \\ \cdot & a & \cdots \\ \vdots & \vdots & \end{vmatrix} \begin{vmatrix} \alpha & \alpha\beta & \cdots \\ \cdot & \alpha & \cdots \\ \vdots & \vdots & \end{vmatrix}$$

v. to Block v. Matrix &
Complemental v. Complementary
&c; If the expressions to wh.
they refer were new, I am
not sure I but that I shd.
prefer yours; but it seems

to me undesirable to introduce new terms, — or to multiply terms, — unless some
considerable advantage
was to be obtained.

I am inclined to suggest
$V(=B)=0$, or even $V=B=0$,
when the V-Block & B-Block
are one & the same.

If you have read Sylvester's
papers you will realise my
horror of new terminology.
Believe me dear Sir
yrs sincerely
W. H. H. Hudson

His notation for a matrix element in row 1, column 2 is $\overline{1}2$. In selecting this notation, he successively argued against the notation $a_{1,2}$ introduced by one of the originators of calculus, Gottfried Leibnitz, as well as the notation used by Baltzer, (1,2). For the former he wrote, "But it seems a fatal objection to this system that most of the space is occupied by a number of a's, which are wholly superfluous, while the only important part of the notation is reduced to minute subscripts, alike difficult to the writer and the reader." For the latter he reasoned, "This system, though tedious for writing, might serve very well, were it not for its liability to be confused with the notation, common in Plain Algebraical Geometry, by which (1,1) denotes the Point x = 1, y = 1." He said of his own notation, "The symbol ... which I have ventured to suggest ... will be found, I have great hopes, sufficiently simple, distinct, and easy to be written. I have turned the symbol towards the left, in order to avoid all chance of confusion with $\int$, the symbol for integration."[7] Alas, Dodgson was not to meet with success in this venture. Leibnitz had the last word!

Of great interest is Dodgson's pamphlet on determinants (item 10), published a year earlier than his book and presented as a paper to the Royal Society. The paper had been looked over by his friend and mentor Bartholomew Price, and by William Spottiswoode, Oxford author of the first separately published elementary monograph on determinants and later president of the Royal Society. As Dodgson noted in his diary on 25 March 1866, "Heard from Mr. Spottiswoode to whom Price had sent my question as to the shortest way of computing Determinants arithmetically, saying that he knows of no short way and that he would be very glad to hear from me."[8]

Dodgson's condensation method is a very simple way to find the determinant of a higher order matrix; it avoids the extensive calculations required by the standard method of his time and ours, Laplace's method. As an example, consider the matrix

$$\mathbf{B} = \begin{bmatrix} -2 & -1 & -1 & -2 \\ -1 & -2 & -1 & -3 \\ -1 & -1 & 2 & 2 \\ 2 & 1 & -3 & -4 \end{bmatrix}.$$

---

7. *Ibid.,* iv.

8. *The Diaries of Lewis Carroll,* edited by Roger L. Green (New York: Oxford University Press, 1954), vol. I, 240.

Taking the determinant of each adjacent four terms and forming a new matrix with their entries produces

$$\mathbf{B}' = \begin{bmatrix} 3 & -1 & 1 \\ -1 & -5 & 4 \\ 1 & 1 & -2 \end{bmatrix}.$$

Again, taking the determinant of each four adjacent terms and forming a matrix, we arrive at

$$\begin{bmatrix} -16 & 1 \\ 4 & 6 \end{bmatrix}.$$

Now, dividing each of these terms by the corresponding entries in the interior of the original matrix (in the box), we obtain

$$\begin{bmatrix} 8 & -1 \\ -4 & 3 \end{bmatrix}.$$

Taking the determinant of this matrix and then dividing by the interior element of matrix $\mathbf{B}'$, -5, we obtain -4, the determinant of matrix $\mathbf{B}$.

An important feature of the condensation method is that nowhere does a determinant of a higher order than two have to be computed. Compared to the standard method based on Laplace's expansion theorem, which uses cofactors, Dodgson's is a model of computational simplicity. It implies that the determinant of a square matrix is a rational function of all minors formed from consecutive rows and columns of any two consecutive sizes. But his method has an obvious flaw: if a zero divisor should develop at any stage, the method fails. Dodgson was aware of this and recommended the use of row transposition to eliminate zeros, a technique that involves little additional computation. However, if more than two zeros occur in the interior of a matrix, row transposition will not eliminate them. Dodgson knew this, too. In the preface to his book on determinants he wrote,

> This process, though extremely convenient where no ciphers, or where one or two at most, occur in the interior of a block, nevertheless fails entirely, it must be admitted, where they occur in larger numbers: I therefore offer it merely as a fanciful addition to the pro-

cesses already in use, which may in some cases lessen the labour of computation.[9]

Dodgson wrote both the paper and the book on determinants during one of the most creative periods of his life, between the *Alice* books. We know that he first mentioned his work on determinants in his diary entry of 28 October 1865: "I have been at work for some days on an elementary pamphlet on Determinants which I think of printing."[10] On 27 February 1866, he wrote, "Discovered a process for evaluating Arithmetical Determinants, by a sort of condensation: and proved it up to 4^2 terms."[11]

He based his condensation method on a generalization of a theorem of Carl Jacobi. Jacobi's theorem, as Dodgson states it in item 10 (page 175), is: "If the determinant of a block = R, the determinant of any minor of the mth degree of the adjugate block is the product of R^{m-1} and the coefficient which, in R, multiplies the determinant of the corresponding minor."

Dodgson's theorem, Proposition VII from chapter II of his book is:

If there be a square Block of the nth degree, and if in it any Minor of the mth degree be selected: the Determinant of the corresponding Minor in the adjugate Block is equal, in absolute magnitude, to the product of the $\overline{m-1}$th power of the Determinant of the first Block, multiplied by the Determinant of the Minor complemental to the one selected.[12]

The condensation method is an application of this theorem for the case $m = 2$. Dodgson's approach is to rearrange the entries of the final 2×2

---

9. *An Elementary Treatise on Determinants,* v. Dodgson certainly would have appreciated a computer program designed to remove interior zeros easily. Mark Lotkin devised such a program after reading "an apparently new and interesting method for the evaluation of determinants . . . by [R. H.] Macmillan" ["A New Method for the Numerical Evaluation of Determinants," *Royal Aeronautical Society Journal* 59 (1955), 772]. Lotkin's article, "Note on the Method of Contractants," appeared in the *American Mathematical Monthly* 66 (1959), 476–79. R. H. Johnston points out that both Macmillan and Lotkin were really discussing Dodgson's condensation method. I am indebted to Martin Gardner for sending me Johnston's article, "On the Method of Contractants," *America Mathematical Monthly* 67 (1960), 865.

10. *Diaries,* vol. I, 236.

11. *Diaries,* vol. I, 240. Collingwood erroneously transcribed "4^2" as "42."

12. *An Elementary Treatise on Determinants,* 25. Baltzer had also extended Jacobi's theorem in his *Theorie und Anwendung der Determinanten* [Leipzig: S. Hirzel, 1857; trans. J. Hoüel (Paris: Mallet-Bachelier, 1861)]. Muir renders it as "Any m^{th} order minor of the adjugate of any determinant $\triangle$ is equal to the product obtained by multiplying the cofactor of the corresponding minor in $\triangle$ by $\triangle^{m-1}$" (Muir, vol. II, 98).

matrix so that they become the entries in the corresponding minor of the adjugate matrix. Applying his theorem results in the division of the determinant of this minor by the interior element(s) in the complementary minor, thus producing the determinant of the original matrix. His definition of an adjugate matrix is not the usual one. In Dodgson's definition, an adjugate is a matrix in which each entry is the determinant of the complemental minor of the corresponding term of the original square matrix. He defines the complemental minor to have only those entries not in any row or column common to the rows and columns of the entries in the selected minor.

Complemental minors, adjugate blocks, and row and column transpositions use many of the ideas that are involved in inversions and mirror images, notions that are so prevalent in the *Alice* books. Curiously, the condensation method uses a construction that might be termed a double mirror image. To see this, we return to Dodgson's example, matrix **B** above. The selected minor is

$$\begin{bmatrix} -2 & -2 \\ 2 & -4 \end{bmatrix}.$$

The complementary minor is the interior of **B**

$$\begin{bmatrix} -2 & -1 \\ -1 & 2 \end{bmatrix}.$$

The corresponding minor is

$$\begin{bmatrix} 3 & -4 \\ -1 & 8 \end{bmatrix}.$$

Hence,

$$\begin{bmatrix} 8 & -1 \\ -4 & 3 \end{bmatrix} = \det \mathbf{B} \times \begin{bmatrix} -2 & -1 \\ -1 & 2 \end{bmatrix},$$

so det **B** = −4.

Seneta notes that the determinant on the left is unaffected by transposition over both diagonals simultaneously; therefore it is in effect a double mirror image![13]

Dodgson's condensation method has appeared in texts on numerical analysis, including P. S. Dwyer's 1951 *Linear Computations;* recently it has been used in quite another context, that of enumerative combinatorics. In a remarkable series of papers, David P. Robbins and his coworkers have obtained two formulas expressing the determinant of a matrix using Dodgson's method (which they call connected minors) involving alternating sign matrices. (In an alternating sign matrix all the entries are 0, 1, and -1; its rows and columns sum to 1; omitting zeros, the 1s and -1s alternate in each row and column.) One of their formulas expresses the determinant as a sum in which alternating sign matrixes index the terms just as permutation matrices index the ordinary expansion of the determinant. Connecting the alternating sign matrices with descending plane partitions leads to a proof of an open problem in combinatorics known as the Macdonald Conjecture.[14]

Dodgson used techniques from linear algebra to remedy problems in other areas. Between 1873 and 1876 he published three papers on election procedures and ranking processes. He was deeply concerned with cyclical majorities and their effect on choosing the best candidate in a committee election. (A cyclical majority occurs when a is preferred to b and b is preferred to c, but c is preferred to a). He was critical both of plurality and of successive elimination as solutions to the problem and proposed instead an inversion approach that uses the idea of a set of n-truples of linear orders defined on the set of candidates. Peter Fishburn has analyzed this method, calling it Dodgson's function. Michael Dummett writes, "It is a matter for the deepest regret that Dodgson never completed the book that he planned to write on this subject. Such was his lucidity of exposition and his mastery of the topic that it seems possible that, had he ever published it, the political history of Britain would have been significantly different."[15]

---

13. Personal communication.

14. The papers most directly related to Dodgson's work are David P. Robbins and Howard Rumsey, Jr., "Determinants and Alternating Sign Matrices," *Advances in Mathematics* 62 (1986), 169–84, and D. P. Robbins, "The Story of 1, 2, 7, 42, 429, 7436 . . . ," *The Mathematical Intelligencer* 13 (1991), 12–19.

15. A complete discussion of Dodgson's ranking methods can be found in Francine Abeles, "Ranking by Inversion: A Note on C. L. Dodgson," *Historia Mathematica* 6 (1979), 310–17. Peter Fishburn's analysis of the inversion method appears in his book, *The Theory of Social Choice* (Princeton, N.J.: Princeton University Press, 1973), 150, 172–73. Michael Dummett's statements appear in *Voting Procedures* (Oxford: Clarendon Press, 1984), 5 n..

A decade later, Dodgson developed a theory of apportionment for proportional representation that was closely related to his theory of unrepresented electors. His quota method is the foundation of two solutions that he proposed as an election reform to allot seats to specific candidates from a list of candidates. In both solutions he takes up the problem of surplus votes and informally works with the notions of zero-sum and coalition games. What is extraordinary is that on an intuitive level Dodgson showed a grasp of ideas that were not made precise until the 1920s.[16]

Dodgson's interest in solving problems involving lack of fairness also included the unjust way in which prizes are awarded in tennis tournaments, which he discussed in a landmark essay. In his discussion of minimum-comparison selection, Donald Knuth acknowledges that "the history of this question [best possible procedures to select the tth largest of n elements] goes back to Rev. C. L. Dodgson's amusing (though serious) essay on lawn tennis tournaments which appeared in *St. James's Gazette*, August 1, 1883, pp. 5–6."[17] Dodgson published this essay under his pen name, Lewis Carroll, although he usually used his given name when writing about serious mathematical topics. Perhaps the setting for this problem, involving as it did a sport, caused him to publish it under his pen name.

The essay is really an extension of the earlier problem of choosing the winner from a ranked set of candidates in the presence of a cyclical majority. In lawn-tennis tournament language, Dodgson worked with a round-robin tournament, the proper setting for ranking all the candidates in a list. He was concerned with choosing the best three players in a tournament. He tried to determine the true best second and third players under a transitive ranking: If player A beats player B and player B beats

---

16. All of Dodgson's mathematical-political pamphlets are printed as an appendix to Duncan Black's book, *The Theory of Committees and Elections* (London: Cambridge University Press, 1958). Black's analysis of Dodgson's work appears on pages 201–33. For additional commentary by Black, see "Lewis Carroll and the Cambridge Mathematical School of P.R.; Arthur Cohen and Edith Denman," *Public Choice*, 8 (1970), 1–28; "The Central Argument in Lewis Carroll's *The Principles of Parliamentary Representation*," *Papers on Non-Market Decision Making*, 3 (1967), 1–17; "Lewis Carroll and the Theory of Games," *American Economic Review*, 59 (1969), 206–15; and "Evaluating Carroll's Theory of Parliamentary Representation," *Jabberwocky* 4 (1970), 19–21. A complete survey of Dodgson's work on elections and committees can be found in the essay by Francine Abeles, "The Mathematical-Political Papers of C. L. Dodgson," *Lewis Carroll: A Celebration*, edited by Edward Guiliano (New York: Clarkson N. Potter, 1982), 195–210.

17. Donald E. Knuth, *The Art of Computer Programming*, vol. III: *Sorting and Searching* (Reading, Mass.: Addison-Wesley, 1973), 209.

player C, then we can conclude that player A would beat player C, so he would not actually have to play him. (A cyclical majority is an antitransitive ranking because he would not be able to assume that A would beat C.)

The formal analysis of how players are ranked did not come into existence until 1929 in Ernst Zermelo's work. The general problem that Dodgson was dealing with was solved in 1932 by J. Schreier, but a correct proof of the procedure did not appear until 1964.[18]

Knuth remarks,

> It would be nice to report that Lewis Carroll's tournament turns out to be optimal, but unfortunately that is not the case. . . . His procedure makes more comparisons than necessary, and it is not formulated precisely enough to qualify as an algorithm. On the other hand, it has some rather interesting aspects from the standpoint of parallel computation. And it appears to be an excellent plan for a tennis tournament, because he built in some dramatic effects. . . . But tournament directors presumably thought the proposal was too logical, and so Carroll's system has apparently never been tried. Instead a method of "seeding" is used to keep the best players in different parts of the tree.[19]

It is not a simple matter to see the connections between Dodgson's idea for a fair tournament and his work on committees and elections. Knuth, Black, and Abeles develop this theme extensively. Dodgson himself established the link between voting and playing in the final section of his pamphlet *A Method of Taking Votes on More Than Two Issues:* "This principle of voting [voter manipulation] makes an election more of a game of skill than a real test of the wishes of the electors. . . . I think it desirable that all should know the rule by which the game can be won."[20] The more appropriate tournament for selecting just the top player (the winner) is the knockout, or single-elimination type. He understood the difference between this method and one that can properly select the top three: he named the second section of *Lawn Tennis Tournaments,* "A proof that the present method of assigning prizes is, except in the case of the first prize, entirely unmeaning." What Dodgson succeeded in doing in this pamphlet was to construct a method—in modern terminology, a triple-elimination tournament—that, although flawed, captured the essence of what was re-

---

18. *Ibid.,* 211–15, contains a complete discussion of the problem.
19. *Ibid.,* 211.
20. Dodgson published this pamphlet in 1876. The quotation is taken from its reproduction in Black, *Committees and Elections,* 232–33.

quired at minimum cost.[21] It is clear, however, that Dodgson believed in the prescriptive type of preference behavior—one that assumes a rational person will behave in a certain way and presupposes that his preferences will be transitive—rather than the descriptive type, in which preferences are rarely transitive. Dodgson's affinity was for the prescriptive theory, and he used it as the basis for his development of a scheme for proportional representation as well.[22]

Question 9995 and Pillow Problems 8, 25, and 68 reflect Dodgson's interest in a type of puzzle structure that can be connected to the extraction of disclosing information, a problem that arises when dealing with confidentiality issues in survey and census materials.[23] This question and Pillow Problems 8 and 68 were constructed in February and March 1889; number 25 is earlier, 1876.

---

21. In addition to *A Method of Taking Votes* and *Lawn Tennis Tournaments*, Dodgson's other pamphlets on elections are *A Discussion of the Various Methods of Procedure in Conducting Elections* (1873), and *Suggestions as to the Best Method of Taking Votes, Where More than Two Issues Are to Be Voted On* (1874). These pamphlets will be included in another volume in this series.

22. A discussion of Dodgson's scheme for proportional representation is presented in F. Abeles, "C. L. Dodgson and Apportionment for Proportional Representation," *Ganita Bharati*, 3 (1981), 71–82.

23. I am grateful to William Kruskal for pointing out Dodgson's interest in puzzle problems of this sort.

10. *Condensation of Determinants,*
BEING A NEW AND BRIEF METHOD FOR COMPUTING THEIR
ARITHMETICAL VALUES

[1866: LCH 52, LCAT 343, 584: Princeton]

This pamphlet is a reprint of the paper presented to the Royal Society on 17 May 1866, which appeared in the Society's *Proceedings,* no. 84, for that year. Dodgson completed the paper on 12 May but had been working on it earlier. He wrote in his diary on 7 May "Went to Price to talk over my paper on *Condensation of Determinants,* which he has been looking over, and has undertaken to forward to the Royal Society."[1]

We know from his diary entry of 27 February 1866 that he formulated the condensation method around then. He began an extensive correspondence with William Spottiswoode about this and other topics on 24 March 1866, which continued until 9 November 1867. In a letter to Dodgson dated 2 April 1866, Spottiswoode wrote:

> Your method of computation, condensation is very successful. The Theorem upon which it is founded is, as you are doubtless aware, known; but the application of it is, as far as I am aware, quite original. [Spottiswoode is referring to a theorem appearing in C. G. J. Jacobi, "De binis quibuslibet functionibus homogeneis secundi ordinis per substitutiones lineares in alias binas transformandis, quae solis quadratis variabilium constant; una cum variis theorematis de transformatione et determinatione integralium multiplicium," *Journal für die Reine und Angewandte Mathematik* 12 (1834), 1–69.] I congratulate you upon it. . . . I should be much obliged if you would, at your leisure, allow me to see your rule for solving simultaneous equations. I see the general character from your example, but am not sure that [I] am quite master of the details. It may prove to be of great service in astronomical computations.[2]

From the *Proceedings of the Royal Society,* 84 (1866), 150–55.

1. *The Diaries of Lewis Carroll,* edited by Roger L. Green (New York: Oxford University Press, 1954), vol. I, 243.

2. From an unpublished letter in the Morris L. Parrish Collection of Victorian Novelists, Princeton University Library.

CONDENSATION OF DETERMINANTS

On 29 May, Dodgson noted in his diary that he had received and corrected a proof of the condensation paper.

※

If it be proposed to solve a set of n simultaneous linear equations, not being all homogeneous, involving n unknowns, or to test their compatibility when all are homogeneous, by the method of determinants, in these, as well as in other cases of common occurrence, it is necessary to compute the arithmetical values of one or more determinants—such, for example, as

$$\begin{vmatrix} 1, 3, -2 \\ 2, 1, 4 \\ 3, 5, -1 \end{vmatrix}.$$

Now the only method, so far as I am aware, that has been hitherto employed for such a purpose, is that of multiplying each term of the first row or column by the determinant of its complemental minor, and affecting the products with the signs + and − alternately, the determinants required in the process being, in their turn, broken up in the same manner until determinants are finally arrived at sufficiently small for mental computation.

This process, in the above instance, would run thus:—

$$\begin{vmatrix} 1, 3, -2 \\ 2, 1, 4 \\ 3, 5, -1 \end{vmatrix} = 1 \times \begin{vmatrix} 1, 4 \\ 5, -1 \end{vmatrix} - 2 \times \begin{vmatrix} 3, -2 \\ 5, -1 \end{vmatrix}$$

$$+ 3 \times \begin{vmatrix} 3, -2 \\ 1, 4 \end{vmatrix} = -21 - 14 + 42 = 7.$$

But such a process, when the block consists of 16, 25, or more terms, is so tedious that the old method of elimination is much to be preferred for solving simultaneous equations; so that the new method, excepting for equations containing 2 or 3 unknowns, is practically useless.

The new method of computation, which I now proceed to explain, and for which "Condensation" appears to be an appropriate name, will be found, I believe, to be far shorter and simpler than any hitherto employed.

In the following remarks I shall use the word "Block" to denote any number of terms arranged in rows and columns, and "interior of a block" to denote the block which remains when the first and last rows and columns are erased.

The process of "Condensation" is exhibited in the following rules, in which the given block is supposed to consist of n rows and n columns:—

(1) Arrange the given block, if necessary, so that no ciphers occur in its interior, This may be done either by transposing rows or columns, or by adding to certain rows the several terms of other rows multipled by certain multipliers.

(2) Compute the determinant of every minor consisting of four adjacent terms. These values will constitute a second block, consisting of $n-1$ rows and $n-1$ columns.

(3) Condense this second block in the same manner, dividing each term, when found, by the corresponding term in the interior of the first block.

(4) Repeat this process as often as may be necessary (observing that in condensing any block of the series, the rth for example, the terms so found must be divided by the corresponding terms in the interior of the $r-1$th block), until the block is condensed to a single term, which will be the required value.

As an instance of the foregoing rules, let us take the block

$$\begin{vmatrix} -2 & -1 & -1 & -4 \\ -1 & -2 & -1 & -6 \\ -1 & -1 & 2 & 4 \\ 2 & 1 & -3 & -8 \end{vmatrix}.$$

By rule (2) this is condensed into $\begin{vmatrix} 3 & -1 & 2 \\ -1 & -5 & 8 \\ 1 & 1 & -4 \end{vmatrix}$; this, again, by rule (3), is condensed into $\begin{vmatrix} 8 & -2 \\ -4 & 6 \end{vmatrix}$; and this, by rule (4), into -8, which is the required value.

The simplest method of working this rule appears to be to arrange the series of blocks one under another, as here exhibited; it will then be found very easy to pick out the divisors required in rules (3) and (4).

$$\begin{vmatrix} -2 & -1 & -1 & -4 \\ -1 & -2 & -1 & -6 \\ -1 & -1 & 2 & 4 \\ 2 & 1 & -3 & -8 \end{vmatrix}$$

$$\begin{vmatrix} 3 & -1 & 2 \\ -1 & -5 & 8 \\ 1 & 1 & -4 \end{vmatrix}$$

$$\begin{vmatrix} 8 & -2 \\ -4 & 6 \end{vmatrix}$$

$$-8.$$

This process cannot be continued when ciphers occur in the interior of any one of the blocks, since infinite values would be introduced by employing them as divisors. When they occur in the given block itself, it may be rearranged as has been already mentioned; but this cannot be done when they occur in any one of the derived blocks; in such a case the given block must be rearranged as circumstances require, and the operation commenced anew.

The best way of doing this is as follows:—

Suppose a cipher to occur in the hth row and kth column of one of the derived blocks (reckoning both row and column from the *nearest* corner of the block); find the term in the hth row and

kth column of the given block (reckoning from the corresponding corner), and transpose rows or columns cyclically until it is left in an outside row or column. When the necessary alterations have been made in the derived blocks, it will be found that the cipher now occurs in an outside row or column, and therefore need no longer be used as a divisor.

The advantage of *cyclical* transposition is, that most of the terms in the new blocks will have been computed already, and need only be copied; in no case will it be necessary to compute more than *one* new row or column for each block of the series.

In the following instance it will be seen that in the first series of blocks a cipher occurs in the interior of the third. We therefore abandon the process at that point and begin again, rearranging the given block by transferring the top row to the bottom; and the cipher, when it occurs, is now found in an exterior row. It will be observed that in each block of the new series, there is only *one* new row to be computed; the other rows are simply copies from the work already done.

$$\begin{vmatrix} 2 & -1 & 2 & 1 & -3 \\ 1 & 2 & 1 & -1 & 2 \\ 1 & -1 & -2 & -1 & -1 \\ 2 & 1 & -1 & -2 & -1 \\ 1 & -2 & -1 & -1 & 2 \end{vmatrix} \quad \begin{vmatrix} 1 & 2 & 1 & -1 & 2 \\ 1 & -1 & -2 & -1 & -1 \\ 2 & 1 & -1 & -2 & -1 \\ 1 & -2 & -1 & -1 & 2 \\ 2 & -1 & 2 & 1 & -3 \end{vmatrix}$$

$$\begin{vmatrix} 5 & -5 & -3 & -1 \\ -3 & -3 & -3 & 3 \\ 3 & 3 & 3 & -1 \\ -5 & -3 & -1 & -5 \end{vmatrix} \quad \begin{vmatrix} -3 & -3 & -3 & 3 \\ 3 & 3 & 3 & -1 \\ -5 & -3 & -1 & -5 \\ 3 & -5 & 1 & 1 \end{vmatrix}$$

$$\begin{vmatrix} -30 & 6 & -12 \\ 0 & 0 & 6 \\ 6 & -6 & 8 \end{vmatrix} \quad \begin{vmatrix} 0 & 0 & 6 \\ 6 & -6 & 8 \\ -17 & 8 & -4 \end{vmatrix}$$

$$\begin{vmatrix} 0 & 12 \\ 18 & 40 \end{vmatrix}$$

36.

CONDENSATION OF DETERMINANTS

The fact that, whenever ciphers occur in the interior of a derived block, it is necessary to recommence the operation, may be thought a great obstacle to the use of this method; but I believe it will be found in practice that, even though this should occur several times in the course of one operation, the whole amount of labour will still be much less than that involved in the old process of computation.

I now proceed to give a proof of the validity of this process, deduced from a well-known theorem in determinants; and in doing so, I shall use the word "adjugate" in the following sense:—if there be a square block, and if a new block be formed, such that each of its terms is the determinant of the complementary minor of the corresponding term of the first block, the second block is said to be *adjugate* to the first.

The theorem referred to is the following:—

"If the determinant of a block = R, the determinant of any minor of the mth degree of the adjugate block is the product of R^{m-1} and the coefficient which, in R, multiplies the determinant of the corresponding minor."

Let us first take a block of 9 terms,

$$\begin{vmatrix} a_{1,1} & a_{1,2} & a_{1,3} \\ a_{2,1} & a_{2,2} & a_{2,3} \\ a_{3,1} & a_{3,2} & a_{3,3} \end{vmatrix} = R;$$

and let $\alpha_{1,1}$ represent the determinant of the complemental minor of $a_{1,1}$, and so on.

If we "condense" this, by the method already given, we get the block $\begin{Bmatrix} \alpha_{3,3} & \alpha_{3,1} \\ \alpha_{1,3} & \alpha_{1,1} \end{Bmatrix}$, and, by the theorem above cited, the determinant of this, viz.

$$\begin{vmatrix} \alpha_{3,3} & \alpha_{3,1} \\ \alpha_{1,3} & \alpha_{1,1} \end{vmatrix} = R \times a_{2,2}.$$

Hence

$$R = \frac{\begin{vmatrix} \alpha_{3,3} & \alpha_{3,1} \\ \alpha_{1,3} & \alpha_{1,1} \end{vmatrix}}{a_{2,2}},$$

which proves the rule.

Secondly, let us take a block of 16 terms:

$$\begin{vmatrix} a_{1,1} & \cdots & a_{1,4} \\ \vdots & & \vdots \\ a_{4,1} & \cdots & a_{4,4} \end{vmatrix} = R.$$

If we "condense" this, we get a block of 9 terms; let us denote it by

$$\begin{vmatrix} b_{1,1} & \cdots & b_{1,3} \\ \vdots & & \vdots \\ b_{3,1} & \cdots & b_{3,3} \end{vmatrix}, \text{ in which } b_{1,1} = \begin{vmatrix} a_{1,1} & a_{1,2} \\ a_{2,1} & a_{2,2} \end{vmatrix}, \&c.$$

If we "condense" this block again, we get a block of 4 terms, each of which, by the preceding paragraph, is the determinant of 9 terms of the original block; that is to say, we get the block $\left\{ \begin{matrix} \alpha_{4,4} & \alpha_{4,1} \\ \alpha_{1,4} & \alpha_{1,1} \end{matrix} \right\}$; but, by the theorem already quoted, $\begin{vmatrix} \alpha_{4,4} & \alpha_{4,1} \\ \alpha_{1,4} & \alpha_{1,1} \end{vmatrix}$

$= R \times b_{2,2}$; therefore $R = \dfrac{\begin{vmatrix} \alpha_{4,4} & \alpha_{4,1} \\ \alpha_{1,4} & \alpha_{1,1} \end{vmatrix}}{b_{2,2}}$; that is, R may be obtained

by "condensing" the block $\left\{ \begin{matrix} \alpha_{4,4} & \alpha_{4,1} \\ \alpha_{1,4} & \alpha_{1,1} \end{matrix} \right\}$.

CONDENSATION OF DETERMINANTS

This proves the rule for a block of 16 terms; and similar proofs might be given for larger blocks.

I shall conclude by showing how this process may be applied to the solution of simultaneous linear equations.

If we take a block consisting of n rows and $\overline{n+1}$ columns, and "condense" it, we reduce it at last to 2 terms, the first of which is the determinant of the first n columns, the other of the last n columns.

Hence, if we take the n simultaneous equations,

$$a_{1,1}x_1 + a_{1,2}x_2 + \ldots + a_{1,n}x_n + a_{1,n+1} = 0,$$
$$a_{n,1}x_1 + \ldots\ldots\ldots\ldots\ldots + a_{n,n+1} = 0;$$

and if we condense the whole block of coefficients and constants, viz.

$$\begin{matrix} a_{1,1} & \cdots & a_{1,n+1} \\ \cdot & & \cdot \\ \cdot & & \cdot \\ \cdot & & \cdot \\ \cdot & & \cdot \\ a_{n,1} & \cdots & a_{n,n+1} \end{matrix},$$

we reduce it at last to 2 terms: let us denote them by S, T, so that

$$S = \begin{vmatrix} a_{1,1} & \cdots & a_{1,n} \\ \cdot & & \cdot \\ \cdot & & \cdot \\ \cdot & & \cdot \\ a_{n,1} & \cdots & a_{n,n} \end{vmatrix}, \text{ and } T = \begin{vmatrix} a_{1,1} & \cdots & a_{1,n+1} \\ \cdot & & \cdot \\ \cdot & & \cdot \\ \cdot & & \cdot \\ a_{n,2} & \cdots & a_{n,n+1} \end{vmatrix}.$$

Now we know that $x_1 = (-)^n \dfrac{T}{S}$, which may be written in the form $(-)^n S \cdot x_1 = T$.

Hence the 2 terms obtained by the process of condensation may be converted into an equation for x_1, by multiplying the first of them by x_1, affected with $+$ or $-$, according as n is even or odd. The latter part of the rule may be simply expressed thus:—"place the signs $+$ and $-$ alternately over the several columns, beginning with the last, and the sign which occurs over the column containing x_1 is the sign with which x_1 is to be affected."

When the value of x_1 has been thus found, it may be substituted in the first $\overline{n-1}$ equations, and the same operation repeated on the new block, which will now consist of $\overline{n-1}$ rows and n columns. But in calculating the second series of blocks, it will be found that most of the work has been already done; in fact, of the 2 determinants required in the new block, one has been already computed correctly, and the other so nearly so that it only requires the *last* column in each of the derived blocks to be corrected.

In the example given opposite, after writing $+$ and $-$ alternately over the columns, beginning with the last, we first condense the whole block, and thus obtain the 2 terms 36 and -72. Observing that the x-column has the sign $-$ placed over it, we multiply the 36 by $-x$, and so form the equation $-36x = -72$, which gives $x = 2$.

Hence the x-terms in the first four equations become respectively 2, 2, 4, and 2; adding these values to the constant terms in the same equations, we obtain a block of which we need only write down the last two columns, viz.

$$\begin{vmatrix} 2 & 4 \\ -1 & -2 \\ -1 & -2 \\ 2 & 6 \end{vmatrix}.$$

We then condense these into the column $\begin{vmatrix} 0 \\ 0 \\ 2 \end{vmatrix}$, and, supplying from the second block of the first series the column $\begin{vmatrix} 3 \\ -1 \\ -5 \end{vmatrix}$, we

CONDENSATION OF DETERMINANTS

obtain $\begin{vmatrix} 3 & 0 \\ -1 & 0 \\ -5 & 2 \end{vmatrix}$ as the last two columns of the *second* block of the new series; and proceeding thus we ultimately obtain the two terms 12, 12. Observing that the y-column has the sign $+$ placed over it, we multiply the first 12 by $+y$, and so form the equation $12y = 12$, which gives $y = 1$. The values of z, u, and v are similarly found.

It will be seen that when once the given block has been successfully condensed, and the value of the first unknown obtained, there is no further danger of the operation being interrupted by the occurrence of ciphers.

$$\begin{array}{ccccccc} - & & + & & - & & + & & - & & + \\ x & + & 2y & + & z & - & u & + & 2v & + & 2 & = & 0 \\ x & - & y & - & 2z & - & u & & & - & v & - & 4 & = & 0 \\ 2x & + & y & - & z & - & 2u & & & - & v & - & 6 & = & 0 \\ x & - & 2y & - & z & - & u & + & 2v & + & 4 & = & 0 \\ 2x & - & y & + & 2z & + & u & - & 3v & - & 8 & = & 0 \end{array}$$

$\begin{vmatrix} 1 & 2 & 1 & -1 & 2 & 2 \\ 1 & -1 & -2 & -1 & -1 & -4 \\ 2 & 1 & -1 & -2 & -1 & -6 \\ 1 & -2 & -1 & -1 & 2 & 4 \\ 2 & -1 & 2 & 1 & -3 & -8 \end{vmatrix}$ $\begin{vmatrix} 2 & 4 \\ -1 & -2 \\ -1 & -2 \\ 2 & 6 \end{vmatrix}$ $\begin{vmatrix} 2 & 6 \\ -1 & -3 \\ -1 & -1 \end{vmatrix}$ $\begin{vmatrix} 2 & 5 \\ -1 & -1 \end{vmatrix}$ $\begin{vmatrix} 2 & 4 \end{vmatrix}$ $\therefore -2v = 4 \ldots v = -2$

$\begin{vmatrix} 3 & 3 \end{vmatrix}$

$\begin{vmatrix} 3 & 0 \end{vmatrix}$ $\therefore 3u = 3$ .... $u = 1$

$\begin{vmatrix} -3 & -3 & -3 & 3 & -6 \\ 3 & 3 & 3 & -1 & 2 \\ -5 & -3 & -1 & -5 & 8 \\ 3 & -5 & 1 & 1 & -4 \end{vmatrix}$ $\begin{vmatrix} 3 & 0 \\ -1 & 0 \\ -5 & -2 \\ 6 & 0 \\ 8 & -2 \end{vmatrix}$ $\begin{vmatrix} -1 & -2 \\ 6 & 6 \end{vmatrix}$ $\therefore -6z = 6$ .... $z = -1$

$\begin{vmatrix} 0 & 0 & 6 & 0 \\ 6 & -6 & 8 & -2 \\ -17 & 8 & -4 & 6 \end{vmatrix}$ $\begin{vmatrix} 12 & 12 \end{vmatrix}$ $\therefore 12y = 12$ .... $y = 1$

$\begin{vmatrix} 0 & 12 & 12 \\ 18 & 40 & -8 \end{vmatrix}$

$\begin{vmatrix} 36 & -72 \end{vmatrix}$

$\therefore -36x = -72$ .... $x = 2$

$$\begin{array}{cccc} - & + & - & + \\ 5x + 2y & - 3z & + 3 & = 0 \\ 3x - y & - 2z & + 7 & = 0 \\ 2x + 3y & + z & - 12 & = 0 \end{array}$$

$\begin{vmatrix} 5 & 2 & -3 & 3 \\ 3 & -1 & -2 & 7 \\ 2 & 3 & 1 & -12 \end{vmatrix}$ $\quad \begin{vmatrix} -3 & 8 \\ -2 & 10 \end{vmatrix} \quad \begin{vmatrix} -3 & 12 \\ -7 & -14 \end{vmatrix} \therefore 3z = 12 \dots\dots\dots\dots\dots\dots\dots\dots z = 4$

$\begin{vmatrix} -11 & -7 & -15 \\ 11 & 5 & 17 \end{vmatrix} \therefore -7y = -14 \dots\dots\dots\dots\dots\dots\dots\dots\dots\dots\dots\dots\dots\dots y = 2$

$\quad |-22 \quad 22 \,|$
$\therefore 22x = 22 \dots\dots\dots\dots\dots\dots\dots\dots\dots\dots\dots\dots\dots\dots\dots\dots\dots\dots x = 1$

11. Algebraical Formulæ for the Use of Candidates for Responsions.

[1868: LCH 65]

The citation for this rare pamphlet, a collection of formulas dealing with monomials and binomials, lists its length as four pages. This copy has only three pages. It is unsigned and undated. In the *Lewis Carroll Handbook* the bibliographic reference from p. 433 of Collingwood (*Algebraical Formulæ for Responsions.* Oxford: Printed at the University Press, 1868), is used to describe it. Dodgson's diary entry of 21 May 1868, "Took to the University Press the MS. for Formulæ in Algebra for Responsions," quoted in the LCH citation, refers to this pamphlet, not to *Formulae in Algebra* (item 12) as suggested.[1]

Meaning of indices		a^x	$= a \times a \times a \times$ &c. (x factors).
" "		a^1	$= a.$
Multiplication		$a^x \times a^y$	$= a^{x+y}.$
Division		$a^x \div a^y$	$= a^{x-y}.$
		hence a^0	$= 1.$
		a^{-x}	$= \dfrac{1}{a^x}.$
		a^{-1}	$= \dfrac{1}{a}.$
Involution (single term)		$(a^x)^y$	$= a^{xy}.$
Evolution "		$\sqrt[y]{a^x}$	$= a^{x/y}.$
" "		$\sqrt[y]{a}$	$= a^{1/y}.$
Involution (two terms)			
(1) $(a+b)^2$			$= a^2 + 2ab + b^2.$
(2) $(a-b)^2$			$= a^2 - 2ab + b^2.$
(3) $(a+b)^3$			$= a^3 + 3a^2b + 3ab^2 + b^3.$
(4) $(a-b)^3$			$= a^3 - 3a^2b + 3ab^2 - b^3.$

1. *The Diaries of Lewis Carroll,* edited by Roger L. Green (New York: Oxford University Press, 1954), vol. II, 269.

(5) $(a + b)^n$ $= a^n + na^{n-1}b + \&c.$, where the index of a continually decreases by unity, while that of b increases, and where the coefficient of each term is formed from the preceding by the rule "Multiply together the coefficient and the index of a, and divide by the place of the term."

(6) $(a - b)^n$ $= a^n - na^{n-1}b + \&c.$, where each term is formed as in the last case, and the signs are alternately $+$ and $-$.

Resolution into factors
(1) $a^2 + b^2$ has no factors.
(2) $a^2 - b^2$ $= (a + b) \cdot (a - b)$.
(3) $a^3 + b^3$ $= (a + b) \cdot (a^2 - ab + b^2)$.
     $a^5 + b^5$ $= (a + b) \cdot (a^4 - a^3b + a^2b^2 - ab^3 + b^4)$.
     and generally, if n be an odd prime,
     $a^n + b^n$ $= (a + b) \cdot (a^{n-1} - a^{n-2}b + \ldots)$, where the index of a continually decreases by unity, while that of b increases, and the signs are alternately $+$ and $-$.

(4) $a^3 - b^3$ $= (a - b) \cdot (a^2 + ab + b^2)$.
     $a^5 - b^5$ $= (a - b) \cdot (a^4 + a^3b + a^2b^2 + ab^3 + b^4)$.
     and generally, if n be an odd prime,
     $a^n - b^n$ $= (a - b) \cdot (a^{n-1} + a^{n-2}b + \ldots)$, where the indices are as in the former case, and the signs are all $+$.

If T be a power of 2,
(5) $a^T + b^T$ cannot be resolved.
(6) $a^T - b^T$ can be resolved by form (2).
If N contain odd prime factors only,
(7) $a^N + b^N$ can be resolved by form (3).

(8) $a^N - b^N$ "   "   (4).

If M be even and contain an odd prime factor,

(9) $a^M + b^M$ can be resolved by form (3).
(10) $a^M - b^M$ "   "   (2) or   (4).

If x, y be commensurable,

(11) $a^x + b^y$ $\Big\}$ can be put into one of the first four
(12) $a^x - b^y$ forms, according to the nature of the G.C.M.

12. Formulae in Algebra.
[1868?: LCH 65a]

A collection of formulas that includes the formulas in item 11 above as well as formulas for quadratic equations, simultaneous linear equations, arithmetic and geometric series and means, the binomial theorem, permutations and combinations, interest, logarithms, and exponential and logarithmic series. This rare pamphlet is unsigned and undated. In the *Lewis Carroll Handbook* it is described as having eight pages; this copy has seven. It is cited as a possible proof copy of *Algebraical Formulæ for the Use of Candidates for Responsions* (item 11), which it is not.

Involution.

$(a + b)^2 = a^2 + 2ab + b^2$.

$(a - b)^2 = a^2 - 2ab + b^2$.

$(a + b)^3 = a^2 + 3a^2b + 3ab^2 + b^3$.

$(a - b)^3 = a^3 - 3a^2b + 3ab^2 - b^3$.

$(a + b)^n = a^n + n \cdot a^{n-1}b + $ &c., where the index of a continually decreases by unity, and that of b increases, and where the coefficient of each term is formed from the preceding by the rule "Multiply together the coefficient and the index of a, and divide by the place of the term."

$(a - b)^n = a^n - n \cdot a^{n-1}b + $ &c., where each term is formed as in the former case, and the signs are alternately $+$ and $-$.

Resolution into rational factors.

(1) $a^2 + b^2$ cannot be resolved.

(2) $a^2 - b^2 = (a + b) \cdot (a - b)$.

If n be an odd prime,

(3) $a^n + b^n = (a + b) \cdot (a^{n-1} - a^{n-2}b + \ldots)$.

FORMULAE IN ALGEBRA

(4) $a^n - b^n$ $= (a - b) \cdot (a^{n-1} + a^{n-2}b + \ldots)$.

If T be a power of 2,
(5) $a^T + b^T$ cannot be resolved.
(6) $a^T - b^T$ can be resolved by form (2).

If N contain odd prime factors only,
(7) $a^N + b^N$ "   "   (3).
(8) $a^N - b^N$ "   "   (4).

If M be even and contain an odd prime factor,
(9) $a^M + b^M$ "   "   (3).
(10) $a^M + b^M$ "   "   (2) or (4).

If x, y be commensurable,
(11) $a^x + b^y$ }   can be put into one of the first four
(12) $a^x - b^y$ }   forms, according to the nature of the G.C.M.

Rationalising binomial surds.
If r be odd,
(1) $\sqrt[r]{a} + \sqrt[r]{b}$ is submultiple of $a + b$.
(2) $\sqrt[r]{a} - \sqrt[r]{b}$ is   " $a - b$.

If s be even,
(3) $\sqrt[s]{a} + \sqrt[s]{b}$ is submultiple of $a - b$.
(4) $\sqrt[s]{a} - \sqrt[s]{b}$ is   "   "

Quadratic equation (one Variable).
If $x^2 - px + q = 0$,
     the values of the coefficients, in terms of the roots, are

     $p = $ sum of roots.
     $q = $ product of roots.

If $Ax^2 + Bx + C = 0$,
     the roots are
     and the test for

$$x = \frac{-B \pm \sqrt{B^2 - 4AC}}{2A}.$$

(1) roots equal with opposite signs $B = 0$.

(2) roots real $B^2 - 4AC \not< 0$.
(3)   "   imaginary   " < 0.
(4)   "   identical   " $= 0$.
(5)   "   rational   " $\not< 0$, and is a square.

Simultaneous Equations (two Variables).
If $A_1x + B_1y + C_1 = 0$,
$A_2x + B_2y + C_2 = 0$,
the roots are

$$\frac{x}{\begin{vmatrix} B_1, & C_1 \\ B_2, & C_2 \end{vmatrix}} = \frac{-y}{\begin{vmatrix} A_1, & C_1 \\ A_2, & C_2 \end{vmatrix}} = \frac{1}{\begin{vmatrix} A_1, & B_1 \\ A_2, & B_2 \end{vmatrix}}.$$

and test for Equations being

(1) consistent

$$\begin{vmatrix} A_1 & B_1 \\ A_2 & B_2 \end{vmatrix} \neq 0.$$

(2) inconsistent

$$\begin{vmatrix} A_1 & B_1 \\ A_2 & B_2 \end{vmatrix} = 0, \text{ and either}$$

$$\begin{vmatrix} A_1 & C_1 \\ A_2 & C_2 \end{vmatrix} \text{ or } \begin{vmatrix} B_1 & C_1 \\ B_2 & C_2 \end{vmatrix} \neq 0.$$

(3) identical

$$\begin{Vmatrix} A_1 & B_1 & C_1 \\ A_2 & B_2 & C_2 \end{Vmatrix} = 0.$$

Summation of Series.

Arithmetical.
If a = first term,
b = common difference,
n = number of terms,
l = last term,
S = sum;
the formula connecting

(1) a, b, n, S

$$S = (2a + \overline{n-1} \cdot b) \cdot \frac{n}{2}.$$

(2) a, l, S

$$S = (a + l) \cdot \frac{n}{2}.$$

Geometrical.
If a = first term,
r = common ratio,
n = number of terms,
S = sum;
the formula connecting

FORMULAE IN ALGEBRA

(1) a, r, n, S $S = a \cdot \dfrac{r^n - 1}{r - 1}.$

(2) a, r, S (where n is infinite) $S = \dfrac{a}{1 - r}.$

Arithmetical Mean, &c.
  If α, β, be two terms having one between them,

  (1) the Arithmetical Mean $\dfrac{\alpha + \beta}{2}.$

  (2) the Geometrical Mean $\sqrt{\alpha\beta}.$

  (3) the Harmonical Mean $\dfrac{2\alpha\beta}{\alpha + \beta}.$

Binomial Theorem.
  $(a \pm b)^n$

$$= a^n \pm na^{n-1}b + \dfrac{n \cdot (n - 1)}{\underline{|2}} \cdot a^{n-2}b^2 \pm \dfrac{n \cdot (n - 1) \cdot (n - 2)}{\underline{|3}} \cdot a^{n-3}b^3 + \ldots$$

the sum of the coefficients $= 2^n.$

Permutations and Combinations.
  Permutations of n things
  (1) taken r together $n \cdot (n - 1) \ldots (n - r + 1).$
  (2) taken all together $\underline{|n}.$

Combinations of n things taken r together $\dfrac{n \cdot (n - 1) \ldots (n - r + 1)}{\underline{|r}}.$

Permutations of n things taken all together, when there are p of one kind, q of another kind, &c. $\dfrac{\underline{|n}}{\underline{|p} \cdot \underline{|q} \ldots}.$

Total number of combinations of n things $= 2^n - 1.$

Interest, Discount, &c.
  If P = Principal,
    r = interest of £1 for 1 year,
    n = number of years,
    I = total interest,

M = amount,
D = discount,
A = annuity.

Simple interest:
   formula connecting P, r, n, I $I = Pnr.$
       "   " P, r, n, M $M = P + Pnr.$
       "   " M, r, n, D $D = \dfrac{Mnr}{1 + nr}.$

Compound interest:
   formula connecting P, r, n, M $M = P(1 + r)^n.$

Terminable annuities.
   formula connecting A, r, n, M $M = A \cdot \dfrac{(1 + r)^n - 1}{r}.$

       "   " A, r, n, M $V = A \cdot \dfrac{(1 + r)^n - 1}{r \cdot (1 + r)^n}.$

Perpetual annuity.
   formula connecting A, r, V $V = \dfrac{A}{r}.$

Deferred terminable annuity.
   If d = number of years for which it is deferred,
   n = number of years for which it is to continue,
   formula connecting A, r, d, n, V $V = A \cdot \dfrac{(1 + r)^n - 1}{r \cdot (1 + r)^{d+n}}.$

Deferred perpetual annuity.
   formula connecting A, r, d, V $V = A \cdot \dfrac{1}{r \cdot (1 + r)^d}.$

Logarithms.
   If a, b, &c., = the bases used
   M, N, &c., = any numbers,
   (1) $\log 1$ $= 0.$
   (2) $\log a$ $= 1.$
   (3) $\log_a M \times N \times$ &c. $\log_a M + \log_a N +$ &c.
   (4) $\log_a \dfrac{M}{N}$ $= \log_a M - \log_a N.$
   (5) $\log_a M^x$ $x \cdot \log_a M.$
   (6) $\log_b M$ (in terms of logs to base a) $\dfrac{\log_a M}{\log_a b}.$

Exponential and Logarithmic Series.

e (the Napierian base)
$$= 1 + 1 + \frac{1}{\lfloor 2} + \frac{1}{\lfloor 3} + \ldots \text{ ad infin.}$$
$$= 2.71828, \&c.$$

a^x (in a series of ascending powers of x)
$$= 1 + (\log_e a) \cdot x + \frac{(\log_e a)^2}{\lfloor 2} \cdot x^2 + \ldots$$

hence e^x
$$= 1 + x + \frac{x^2}{\lfloor 2} + \ldots$$

$\log_e(1 + x)$
$$= x - \frac{x^2}{2} + \frac{x^3}{3} - \frac{x^4}{4} + \ldots$$

$\dfrac{1}{\log_e 10}$
$$= .43429, \&c.$$

$\log_{10}(n + 1)$
$$= \log_{10} n + \frac{2}{\log_e 10} \cdot \left(\frac{1}{2n + 1} + \frac{1}{3(2n + 1)^3} + \frac{1}{5(2n + 1)^5} + \ldots \right).$$

13. Algebraical Formulæ and Rules for the Use of Candidates for Responsions.
[1870: LCH 76]

A rare four-page pamphlet that appears to be an extended and improved version of item 11 above. Collingwood's bibliographic reference on p. 433 provides the 1870 publication date and the University Press at Oxford as the description of this piece in the LCH.

Meaning of a^x	$a \cdot a \cdot a \cdot$ &c. (x factors).
hence a^1 =	a
Addition, when terms containing same power of same letters are added:—	
(α) two terms only:—	
(1) with same sign	add coefficients, and repeat sign.
(2) with different signs	take difference of coefficients, with sign of greater.
(β) many terms	collect coefficients of $+$ terms into one, and those of $-$ terms into one, and proceed as before.
Subtraction, when a term is taken from another containing same powers of same letters:—	
(α) when minuend is greatest, and signs are same	subtract, and repeat sign.
(β) otherwise	change sign of subtrahend, and proceed as in addition.
Brackets, to put on or take off:—	
(α) when sign outside is "+"	keeps signs within unchanged.
(β) " is "$-$"	change signs within.

ALGEBRAICAL FORMULAE AND RULES

Multiplication:—
  (α) as to indices, when two or more powers of same letter are multiplied   add indices.
     e.g. $x^a \cdot x^b \cdot x^c \cdot \&c. =$ $x^{a+b+c+\&c.}$.
  (β) as to signs, when two terms are multiplied:—
    (1) signs like   sign of answer is "+".
    (2) signs unlike   "   is "−".

Division:—
  (α) as to indices, when a power of a letter is divided by another power of same letter   subtract index of divisor from index of dividend.
     e.g. $x^a \div x^b =$ x^{a-b}.
     hence $x^0 =$ 1.
     $x^{-a} =$ $\dfrac{1}{x^a}$.
     $x^{-1} =$ $\dfrac{1}{x}$.
  (β) as to signs, when one term is divided by another   same rules as in multiplication.

Involution:—
  (α) a monomial:—
    (1) as to indices   multiply each index by index of required power.
   e.g. $(x^a \cdot y^b \cdot z^c \cdot \&c.)^n =$ $x^{an} \cdot y^{bn} \cdot z^{cn} \cdot \&c.$
    (2) as to signs:—
      (a) when index of required power is even   sign "+".
      (b) when it is odd   sign same as given quantity.
  (β) a binomial:—
     1. $(a+b)^2 =$ $a^2 + 2ab + b^2$.
     2. $(a-b)^2 =$ $a^2 - 2ab + b^2$.
     3. $(a+b)^3 =$ $a^3 + 3a^2b + 3ab^2 + b^3$
     4. $(a-b)^3 =$ $a^3 - 3a^2b + 3ab^2 - b^3$
     5. $(a+b)^n =$ $a^n + n \cdot a^{n-1}b + \&c.$ Index of a decreases while that of b increases. Coefficient of each

6. $(a - b)^n =$ | term is formed from preceding by multiplying together coefficient and index of a and dividing by place of term. Signs all "+".
$a^n - n \cdot a^{n-1}b +$ &c., as before. Signs alternately "+" and "−".

(γ) a quantity of 3 or more terms | collect the terms into 2 brackets, and proceed as before.

Evolution of a mononomial:—

(α) as to indices | divide each index by index of required root.

e.g. $\sqrt[n]{x^a \cdot y^b \cdot z^c \cdot \&c.} =$ | $x^{\frac{a}{n}} \cdot y^{\frac{b}{n}} \cdot z^{\frac{c}{n}} \cdot$ &c.

hence $\sqrt[n]{x} =$ | $x^{\frac{1}{n}}$.

(β) as to signs:—
(1) when index of required root is even:—
  (*a*) sign "+"
  (*b*) sign "−"
(2) when it is odd

sign of answer is "+" or "−".
root does not really exist.
sign same as given quantity.

Resolution of binomials, &c., into factors:—

general rule | divide out all monomial factors, placing them outside a bracket. For the factor within the bracket try the following formulæ.

(α) a binomial:—
1. $a^2 + b^2$ | has no factors.
2. $a^2 - b^2 =$ | $(a + b) \cdot (a - b)$.
3. $a^3 + b^3 =$ | $(a + b) \cdot (a^2 - ab + b^2)$.
4. $a^3 - b^3 =$ | $(a - b) \cdot (a^2 + ab + b^2)$.
5. $a^n + b^n$, (where n is an odd prime) $=$ | $(a + b) \cdot (a^n - a^{n-1}b +$ &c.) Index of a decreases while that of b increases. Signs alternately "+" and "−".
6. $a^n - b^n$, (where n is an odd prime) $=$ | $(a - b) \cdot (a^n - a^{n-1}b +$ &c.), as before. Signs all "+".

ALGEBRAICAL FORMULAE AND RULES

(β) a trinomial:—
1. $a^2 + 2ab + b^2$ = $(a+b)^2$.
2. $a^2 - 2ab + b^2$ = $(a-b)^2$.
3. $Ax^2 + Bxy + Cy^2$, where $B^2 - 4AC$ is a positive square (call it K^2) =
$$A \cdot \left(x + \frac{B+K}{2A} \cdot y\right) \cdot \left(x + \frac{B-K}{2A} \cdot y\right).$$

G. C. M. and L. C. M. of any number of mononomials:—
(α) G. C. M.   take each factor that occurs in *all*, with *lowest* index it bears.

(β) L. C. M.   take each factor that occurs, with *highest* index it bears.

G. C. M. of binomials, &c.:—
(α) two quantities:—
  (1) general rule   arrange both in order of indices of some one letter, bracketing coefficients of any terms which contain the same power of it: then divide greater by less, and divisor by remainder, and so on till there is no remainder: the last divisor is the G. C. M.

  (2) particular rules:—
    (*a*) when a factor is observed in *one* of the quantities   divide it out.
    (*b*) when in *both*   divide it out, and multiply the answer by it.

    (*c*) when first term of divisor will not exactly divide that of dividend   find L. C. M. of their coefficients, and multiply dividend by such a number as will raise coefficient of first term to this L. C. M.: but first try whether this multiplier, or any factor of it, will divide divisor.

(β) three or more quantities	find G. C. M. of the first two; then G. C. M. of answer and third quantity, and so on.
L. C. M. of binomials, &c.:—	
(α) two quantities	product divided by G. C. M.
(β) three or more quantities	find L. C. M. of first two; then L. C. M. of answer and third quantity, and so on.

14. Algebra [13]

[1877: LCH 119a, LCAT 402: HRHRC]

The authors of the *Lewis Carroll Handbook* consider that this sheet, which was probably the first product of Dodgson's new "electric pen," was apparently intended for schoolchildren. It is undated but in Dodgson's hand. However, the problems, not quite as simple as reported, correspond to the topics in section A of *A Guide to the Mathematical Student in Reading, Reviewing, and Working Examples* (item 43).

1. Define *Factor* and *Index*.
2. Multiply $\frac{2x}{3} - \frac{1}{2}$ by $\frac{x^2}{2} - \frac{x}{3} + \frac{1}{4}$.
3. Solve $ax - b = a + \frac{b^2 x}{a}$.
4. What sum of money exceeds four-ninths of it by 4s., 9d?
5. Resolve into factors $2a^5 - 2ab^2$.
6. Find G.C.M. of $a^3bc^2d^4$, $a^5c^3d^2$, $a^4b^3d^3$. Also of $x^2 - y^2 + ax + ay$, and $x^2 + y^2 + ax - ay - 2xy$.
7. Find L.C.M. of x^5y, x^3y^4, a^2y^2. Also of $(x - 1)^2$, $x^2 - 1$, $(x + 1)^3$, $x^3 + 1$.
8. Simplify

$$\cfrac{1}{x + \cfrac{1}{x - \cfrac{x}{x+1}}} ;$$

and $\left\{\dfrac{1}{3} + \dfrac{2a}{3(1-a)}\right\} \times \left\{\dfrac{1}{4} - \dfrac{a}{2(1+a)}\right\}$.

9. Solve $3x + \dfrac{y}{2} = 7x - y = 13$.

10. A is twice as old as B, and 3 times as old as C. One year ago, B was twice as old as C. Find A's age.

11. Find $\left(\dfrac{1}{x^2} - \dfrac{3}{y^2}\right)^3$; and the square root of $\dfrac{a^2}{b^4} + \dfrac{b^4}{a^2} - 2$.

12. Simplify $\dfrac{x^{1-m}}{x^{1-n}}$; $\sqrt[3]{x^{9a^3} \cdot a^3}$; and $\dfrac{x^{-1}}{y^0}$.

15. Formulæ.

[1878: Carlson]

A cyclostyled sheet containing eighteen formulas corresponding to the topics in section L of *A Guide to the Mathematical Student in Reading, Reviewing, and Working Examples* (item 43). This piece is not listed in any of the standard bibliographic references of Dodgson's work. It is dated "Mar.19.1878" in the lower right corner. Although unsigned, it is clearly by Dodgson.

FORMULÆ.

e, as series	$1 + 1 + \frac{1}{\lfloor 2} + \frac{1}{\lfloor 3} + \cdots$
in decimals	$2.71828182 8 \ldots$
e^x, as series	$1 + x + \frac{x^2}{\lfloor 2} + \frac{x^3}{\lfloor 3} + \cdots$
$\log_e a$, do.	$(a-1) - \frac{(a-1)^2}{2} + \frac{(a-1)^3}{3} - \cdots$
a^x, do.	$1 + \log_e a \cdot x + \frac{(\log_e a)^2 \cdot x^2}{\lfloor 2} + \cdots$
$\log_e(a+1)$ do.	$a - \frac{a^2}{2} + \frac{a^3}{3} - \cdots$
$\log_e(a+1) - \log_e a$	$2 \left\{ \frac{1}{2a+1} + \frac{1}{3 \cdot (2a+1)^3} + \frac{1}{5 \cdot (2a+1)^5} + \cdots \right\}$
$\log_e 10$, in decimals	2.3025851
$\log_{10} e$, do	$.4342945$
$\cos \theta$, in terms of θ	$1 - \frac{\theta^2}{\lfloor 2} + \frac{\theta^4}{\lfloor 4} - \cdots$
$\sin \theta$, do	$\theta - \frac{\theta^3}{\lfloor 3} + \frac{\theta^5}{\lfloor 5} - \cdots$
$\cos \theta$, exponential value	$\frac{e^{\theta i} + e^{-\theta i}}{2}, \; [i = \sqrt{-1}]$
$\sin \theta$, do.	$\frac{e^{\theta i} - e^{-\theta i}}{2i}$
$\tan^{-1} x$, in terms of x	$x - \frac{x^3}{3} + \frac{x^5}{5} - \cdots$
π, approximate values	$\frac{22}{7}, \frac{355}{113}$
in decimals	3.1415927
$\frac{\pi}{180}$, do.	$.0174533$
$\frac{180}{\pi}$, do.	57.2957795

Mar.19.1878.

16. Question 9995

[1889: LCH 136: NY Public]

This is the only question on algebra that Dodgson submitted to the *Educational Times*. Of his eleven contributions to this publication, four had to do with arithmetic computation and theory, four with probability, one with trigonometry, and one with logic.

✥

(C. L. Dodgson, M.A.)—A certain school contains not less than 90 boys nor more than 130. Latin, Greek, and French are taught, but no other languages. For every boy learning Latin, at least two learn Greek, but not French; for every three learning Greek, at least one learns French, but not Latin; and, for every two learning French, at least three learn Latin, but not Greek. Exactly half the school learn no languages. Find how many boys are learning each language.

Solution by J. C. St. Clair, L. Wiener, *and others.*

Let f, g, l represent the number of boys learning French, Greek, and Latin respectively. From the data it follows that

$$2l, g, 3f \text{ not} > \text{an unknown portion of } g, 3f, 2l;$$

adding,

$$2l + g + 3f \text{ is not} > \text{a portion of } (2l + g + 3f),$$

therefore the unknown portion is the whole, no boy learns two languages, and $g = 2l = 3f$. Now $N = 2(g + l + f) = 11f$, and is even. Therefore N is divisible by 22 and $= 110$; hence $f = 10$, $g = 30, l = 15$.

Mathematical Questions and Solutions from the "Educational Times," W. J. C. Miller, ed. (London: Francis Hodgson, vol. LI, 1889), 98.

Probability

[Copy]

[Sent to Mr. Potts - Ap. 14/83]

An engine is moving along a railway at a uniform rate. A is informed that the distance it goes in an hour is not less than 20 miles & not more than 30, & that he shall receive gold at the rate of 1 g. for every 10 m. so traversed. B is informed that the time it takes to go a mile is not less than 2 minutes & not more than 3, & that he shall receive gold at the rate of 1 g. for every minute so spent. What is the value of expectation in each case?

Letter from Charles L. Dodgson to Mr. Potts (possibly Robert Potts, mathematician at Trinity College, Cambridge) stating a probability problem, dated 14 April 1883.

Introduction

Dodgson's work on probability appears in three quite separate contexts. He published thirteen problems, formulated between 1876 and 1890, as Pillow Problems. Between 1885 and 1889 he carried on an extended commentary on a problem in the "Mathematical Questions and Solutions" section of the *Educational Times,* Question 7695 posed by J. O'Regan. To this he added a question of his own, Number 9588. The last context is a set of three pamphlets on elections and committees that he wrote earlier, between 1873 and 1876, which will be discussed in a later volume in this series.

Dodgson's knowledge of probability most likely came from works by Todhunter, De Morgan, Whitworth, Boole, and Venn. In his personal library he had Todhunter's *History of Probability,* De Morgan's *Formal Logic,* Whitworth's *Choice and Chance,* Venn's *The Logic of Chance,* and *An Introduction to Boole's "The Laws of Thought"* by J. P. Hughlings.[1] Eugene Seneta has stated that certain chapters of De Morgan's essay, "On Inverse Probabilities," and "On Direct Probabilities," together with examples 134 to 139 in the "Accession of Knowledge" section, seem to have motivated Dodgson's probability problems, as may have several sections of Whitworth's book involving inverse probabilities, such as the section entitled "Credibility of Testimony."[2]

Dodgson's contributions to probability are uneven in quality. Several of the Pillow Problems are poorly formulated or incorrectly solved (nos. 38, 41, 45, 58, 72); but at least one, no. 50, shows a flash of brilliance. To some degree this unevenness can partly be attributed to the state of knowledge

---

Eugene Seneta, Professor of Mathematical Statistics at the University of Sydney, wrote the headnotes for items 17–20 and the introductory material here pertaining to them.—Ed.

1. Refer to *Lewis Carroll's Library,* edited by Jeffrey Stern, Carroll Studies No. 5, University Press of Virginia, 1981, pp. 25, 26, 29, 58, 65, 71, 83. An analysis of the pillow problems dealing with probability appears in "Lewis Carroll as a Probabilist and Mathematician" by E. Seneta in *The Mathematical Scientist,* 9, 1984, pp. 79–94. A treatment of Dodgson's reasoning about probability models in his pamphlets on elections and committees can be found in "Ranking by Inversion: A Note on C. L. Dodgson" by F. Abeles in *Historia Mathematica* 6, 1979, 310–317. Notes 2–4 and 6 that follow are from these two articles.

2. Seneta, 82.

about probability in Dodgson's time. As Seneta notes, "Until the appearance of 'Pillow Problems' in 1893, this branch of mathematics [probability] appears to have played a relatively minor role in England."[3] Despite this fact, in grappling with the problem of cyclical majorities in his political pamphlets, Dodgson developed a clever scheme that anticipated the solution to the problem of estimating the maximum-likelihood weak stochastic ranking from a sample of paired comparisons—an issue that was not completely solved until 1964.[4]

In both the Pillow Problems and the political pamphlets, Dodgson used a process of reasoning we would expect from the author of the *Alice* books, namely, reasoning that uses inversion. For example, to solve the problem of selecting the best candidate from a set of ordered lists, Dodgson invoked a criterion that uses the smallest number of inversions necessary to restrict the order relation on the candidates; in this way he would obtain an order for which there exists a strict simple majority choice. Taking an example used by Dodgson, there are four candidates and 23 electors who have ranked the candidates in the following order:[5]

Electoral choices	Preferences (by pairs)
Two electors: $a\,b\,d\,c$	a over b by 7 electors
Four electors: $a\,c\,b\,d$	b over a by 16 electors
One elector: $a\,d\,b\,c$	a over c by 15 electors
Six electors: $b\,d\,a\,c$	c over a by 8 electors
Five electors: $c\,b\,a\,d$	a over d by 12 electors
One elector: $c\,b\,d\,a$	d over a by 11 electors
Two electors: $d\,b\,a\,c$	
Two electors: $d\,c\,b\,a$	

First, Dodgson discusses two possible voting solutions, one of which, plurality, would select a, and the other, successive elimination, would pick c as the winner. Then he introduces his own method by considering the number of changes of votes each candidate requires to give him a majority over all other candidates. For a to win, he needs five votes to give him a majority over b. To win, b needs one vote; c needs six, while d needs eight. Dodgson claims b should be the winner by this inversion method. The

---

3. Seneta, 81.
4. Abeles, 316.
5. Duncan Black, *The Theory of Committees and Elections* (London: Cambridge University Press, 1958), 229–30.

ordered lists can be decomposed to form a preference matrix from which the ranking is computed and a probability matrix can be associated with it.[6]

On 14 March 1885 Dodgson wrote to the editor of the *Educational Times* about an erroneous solution to a problem, Question 7695 (item 17) that had appeared in the "Mathematical Questions and Solutions" (*MQS*) section of the paper. The problem had been posed by J. O'Regan and the solution was provided by D. Biddle and W. J. Greenstreet (and others):

> Two persons play for a stake, each throwing two dice. They throw in turn, A commencing. A wins if he throws 6, B if he throws 7: the game ceasing as soon as either event happens. Show that A's chance is to B's as 30 to 31.

The analysis that follows shows that Dodgson provided a correct argument for countably infinite sample space, but he introduced in his solution his own difficulty in understanding countability.

The "solution" by Biddle, Greenstreet, and "others" notes that on a single throw of the dice, the probability of a six is 5/36 and of a seven is 6/36; this is correct, so the chances of not throwing a six (respectively not throwing a seven) are 31/36 (30/36). The incorrect solution procedure then takes the odds to be 30/36:31/36 = 30:31.

The correct solution procedure, taking more generally the probability of A's being successful on a toss as a, and B's being successful as b, gives the probability of A's ultimately winning as

$$a + (1 - a)(1 - b)a + (1 - a)^2(1 - b)^2 a + \ldots$$
$$= a/\{1 - (1 - a)(1 - b)\}, \quad (1)$$

where we are looking at the probability of A's winning on the first, or the third, or the fifth, . . . tosses, respectively. Similar reasoning gives the probability of B's winning as

$$(1 - a)b + (1 - a)^2(1 - b)b + (1 - a)^3(1 - b)^2 b + \ldots$$
$$= (1 - a)b/\{1 - (1 - a)(1 - b)\}. \quad (2)$$

The odds sought are therefore $a/\{(1 - a)b\}$, which when $a = 5/36$ and $b = 6/36$, do become 30/31. All this argument is given by Dodgson in his "Note on Question 7695" (with a typographical error from the original printing in the *Educational Times* of May 2 corrected), who is astounded

---

6. Abeles, 312, 315.

that the incorrect solution method gives the correct answer. It is not clear whether J. O'Regan in posing this problem was aware of this peculiarity.

We may explore more generally when the coincidence of correct and incorrect answers may occur by examining the question

$$a/\{(1 - a)b\} = (1 - b)/(1 - a),$$

i.e., $a = b(1 - b)$.

We see that any value of b, $0 < b < 1$, will give a legitimate value of a, $0 < a < 1$, probabilistically speaking. However, in the dice game here where the possible result of the toss of two dice can be any integer between 2 and 12 inclusively, whose corresponding probabilities are 1/36, 2/36, 3/36, 4/36, 5/36, 6/36, 5/36, 4/36, 3/36, 2/36, 1/36, the only one of these numbers (taken as b) that produces an a that is also one of these numbers is $b = 6/36$.

The controversy stems from Dodgson's statement that the ratio "of A's expectation to B's, is approximately $a/\{(1 - a)b\}$." He follows this by saying,

> The ratio 30/31 is only approximative, the expectation of A and B being just less than the fractions 30/61, 31/61. If this were not so, the sum total of their expectations would equal 1, i.e. it would be absolutely certain that one or other of them would win—whereas there is clearly a chance, though an indefinitely small one, that the game might go on forever without either winning.

In modern probability theory, the (conceivable) sample point (of the random experiment which the game represents) that the game goes on forever must be allocated the mathematical probability zero, since the sum of the probabilities of the sample points that A or B wins (so that the game finishes in a finite time) is, from (1) and (2), unity. Moreover, nowadays one does not, as Dodgson does, equate the mathematical statement that an event has probability zero of occurring with the physical statement that an event with probability zero is impossible.

Following Dodgson's discussion is a contrary view by Thomas Charles Simmons on Dodgson's last paragraph particularly. Simmons allows that the game may go on forever without either winning, but says that Dodgson's concluding arguments are extremely unmathematical, A's expectation being precisely 30/61. Simmons's view is thus more correct.

The nature of the problem of what probability to assign to the event that trials within a random experiment go on forever was not in fact new in Dodgson's time. The classical problem discussed by Blaise Pascal

(1623–1662) and Chevalier de Méré calculates the minimum times necessary, k_0, to throw a pair of dice to ensure at least one double-six with a better than even chance.[7] Implicit in these calculations is the conclusion that such an outcome will occur in some finite number of trials with probability one.

In 1886 Simmons continued the discussion of Question 7695 by posing an additional problem, Question 8200, which is introduced by the editor of the *MQS* section under the title, "Infinitesimal or Zero?" (item 18). The question reads, "A random point being taken on a given line, what is the chance of its coinciding with a previously assigned point?"

Question 8200 is of a different order of difficulty because the sample space, the set of possible outcomes of the random experiment, is no longer countable (as in Question 7695); it now consists of all the points of a line segment, and thus, if the segment is [0, 1], of all the numbers between 0 and 1 inclusively. Discrete probability arguments, which merely assign a probability value to each of a countable set of sample points, cannot be used here. Instead, the allocation of probabilities to certain subsets of the sample space, including individual points, is made to accord with A. Kolmogorov's general probability axioms of 1933 and with the physical reality of the experiment. With the description of randomness given in Question 8200, the appropriate allocation to the line segment of length 1 is the uniform distribution, which allocates to a subinterval a probability equal to the length of the subinterval, irrespective of where in [0, 1] the subinterval is located. Since any specific point is contained in an arbitrarily short interval, the probability assigned to any such point is 0, although it is not physically impossible that this point occurs.

One of Simmons's arguments (number 4 of his five points) supporting his conclusion that the probability is zero states that if the specified point is in fact the number k, $0 \leq k \leq 1$, then the probability of points to the right is $(1 - k)$, so the only probability that can be assigned to point k is zero. This argument clearly parallels the earlier round in which the controversy arose, and the dispute is over the same technical issue. Simmons's argument is, again, correct and still opposed by Dodgson.

Dodgson argues that the answer zero cannot be correct since the same probability, zero, would obtain for any point on the line, but some point will be chosen. In fact, there is no paradox because the correct reasoning merely implies that for any point on the line specified before the performance of the random experiment, the *probability* that the result of the experiment will yield that point is zero.

7. See E. Seneta, "Pascal and Probability," in *Interactive Statistics,* edited by D. McNeil (Amsterdam: North Holland, 1979), 225–33.

Of the other participants in this discussion, MacColl makes some sensible remarks on distinguishing events that are of probability zero but not impossible, e.g., a coin landing on a horizontal surface and staying upright on its edge, and events that are impossible, e.g., a coin landing and staying at an inclination of between 0 and 90 degrees to the horizontal (the examples he uses are actually different). In both cases he is inclined to assign a probability of zero.

In his final reply to Simmons, Dodgson states: "I reaffirm, as absolutely axiomatic, that when an event is possible, its chance of happening is not zero."

Two years later, in an item published in the *Educational Times* of 1888 as "Something or Nothing" (item 19), Dodgson reopens the discussion of Question 8200, supporting his position that the answer is not absolute zero (Simmons) but "some sort of infinitesimal." He seeks the probability that a point selected at random from [0, 1] is (1) a rational number, or (2) an irrational number. Since we now know from the work of mathematicians like Georg Cantor, who was writing at about the same time as Dodgson, that the set of rational numbers is denumerable, the probability assigned to them by the uniform distribution is shown to be bounded above by an arbitrarily small and positive number and is therefore zero. Therefore the probability assigned to the (complementary) set of irrational numbers is one. However, Dodgson states that his opponents would reason that the probability of selecting a rational point is zero (there are infinitely many rationals and each has probability zero) and of selecting a specified irrational point is likewise zero, thus giving a contradiction. Dodgson's reasoning is based on the false premise of additivity, that even if an uncountable set of points has zero probability for each point, the same must be true of the probability of the whole set. That probability is at most countably additive is one of the axioms of modern probability theory.

Although in retrospect Dodgson is quite wrong, he touches here on a number of issues, mainly countability of rationals and countable additivity, that were central in the proper axiomatization of probability as a mathematical discipline by Kolmogorov, some 45 years later!

Dodgson ends this item with a new problem that is related to Question 8200 and the subsequent discussion in "Something or Nothing." It was reprinted in the *MQS* in 1889 as Question 9588 (item 19) with solutions provided by Simmons and a Professor Tanner. Dodgson proposed: "A random point being taken on a given line, find the chance of its dividing the line into two parts (1) commensurable, (2) incommensurable." Curiously, the problem immediately preceding Dodgson's, Question 8861 by J. Brill, is very similar: "A rod of given length is broken at random into two pieces; find the probability that the lengths may be commensurable."

It appears that Dodgson turned the discussion in "Something or Nothing" into Question 9588 in order to elicit some response from his opponents. Simmons takes up the challenge and arrives at the correct answers, but his reasoning is shaky. It is based on the generation of all the rational numbers in [0, 1] by dividing the interval first into halves, then thirds, then quarters, and so on; however, Simmons is unable to cope with the entire denumerably infinite set. Dodgson's response is incorrect in that he states that the same reasoning would lead to the answer zero for the irrationals by considering consecutive points a distance $1/\sqrt{2}$ apart, then $1\sqrt{3}$ apart, then $1\sqrt{4}$ apart, etc. The flaw is that these do not generate all the irrational numbers in [0, 1]. In returning to the argument surrounding the problem in "Infinitesimal or Zero?" Tanner begins by assuming that the probability of each point in [0, 1] is $p \geq 0$. (In assuming that the probability is independent of the point, he is grappling with the difficult problem of equiprobability for points in an interval, later formalized, as noted above, by the uniform distribution.) The assumption $p > 0$ leads to a contradiction once one adds the probabilities corresponding to more than $1/p$ points. However, Dodgson misunderstands the reasoning and gives an erroneous counterargument.

As long as Dodgson stayed in the realm of random experiments where only a finite number of outcomes can occur (a finite sample space), he was quite a powerful manipulator of probabilities. But when he ventured outside to grapple with a denumerably infinite sample space or a noncountable sample space, he was less successful. In the first situation he is successful up to the point where he encounters "possible" events that nevertheless must, for mathematical consistency, be given zero probability. This he will not do, preferring some undefinable "infinitesimal." In the second situation, where not only the zero probability problem plagues him, but also the much more difficult related problem of how to assign probabilities, he is unsuccessful. In fairness to Dodgson, problems concerning rational and irrational numbers are quite sophisticated and were in the process of being resolved by much better mathematicians than he at about the same time.[8]

But looking at Dodgson's response to Question 7695, we notice at the outset that there is an element of playfulness in it.

> The solution given to this question ... is one of the most curious instances I have met with of the pitfalls to be found in mathematics: the answer is right, but the method of solution, beautifully simple as it looks, is entirely wrong.

8. For a full discussion of these issues in Dodgson's time, see M. W. Crofton, "Probability," in the *Encyclopedia Brittanica*, 9th ed., vol. 19, 1885, 768–88.

Again, when he reopens the discussion of this question two years later in "Something or Nothing," he believes he has found a paradox.

> And yet *one or other* of these two events [dividing a line into two commensurable parts; dividing a line into two incommensurable parts] *must* happen!

Dodgson was intrigued by problems that appeared to involve fallacious reasoning and seemed to enjoy questioning ordinary assumptions. It is unfortunate that he was not capable of actually resolving the deep probabilistic issues in which he became enmeshed.

17. Note on Question 7695

[1885: LCH 136: NY Public]

This note was reprinted from the *Educational Times,* vol. XXXVIII from 1 May 1885, 183. It contains Dodgson's criticism of the solution to Question 7695 (posed by J. O'Regan, New Street, Limerick) given by D. Biddle, W. J. Greenstreet, and others, as well as Dodgson's own and correct solution procedure. To provide the setting for Dodgson's response and the sequelae, O'Regan's question and its solution in *MQS,* vol. XLII (1885), 75, are included with this item. The last few sentences of Dodgson's discussion are followed by a contrary view by Thomas Charles Simmons—on the last paragraph in particular. The entry in the *Lewis Carroll Handbook* contains a typographical error, "Notes" rather than "Note," in the title.

※

7695. (By J. O'Regan.)—Two persons play for a stake, each throwing two dice. They throw in turn, A commencing. A wins if he throws 6, B if he throws 7: the game ceasing as soon as either event happens. Show that A's chance is to B's as 30 to 31.

Solution by D. Biddle; W. J. Greenstreet, B.A.; *and others.*

Out of 36 ways of throwing two dice, 6 may be turned up in 5 ways, viz., $1+5, 2+4, 3+3, 4+2, 5+1$; and 7 may be turned up in 6 ways, viz., $1+6, 2+5, 3+4, 4+3, 5+2, 6+1$. There are therefore 31 chances against throwing 6, but only 30 against throwing 7. The probability that B will have a throw after A is accordingly $\frac{31}{36}$; but that A will throw again after B, only $\frac{30}{36}$.

Mathematical Questions and Solutions from the "Educational Times," W. J. C. Miller, ed., vol. XLIII (London: Francis Hodgson, 1885), 86–87.

Note on Question 7695; *by* C. L. Dodgson, M.A.

The solution given to this question on p. 75 of Vol. 42, is one of the most curious instances I have met with of the pitfalls to be found in Mathematics: the answer is right, but the method of solution, beautifully simple as it looks, is entirely wrong.

This can be most easily demonstrated by a *reductio ad absurdum*. Let the winning throw, for A and B alike, be 6. Then, by this method of solution, their chances are equal, since "the probability that B will have a throw after A is $\frac{31}{36}$"; which is also the probability "that A will throw again after B." Yet is it obvious that, as A begins, his "expectation" is better than B's.

The true solution will be best given, first, in the general form; and the formula, so obtained, can then be applied to the particular case.

Let A's chance of making his winning throw, each time he throws, be k; and similarly let B's chance be l.

Then A's chance of winning, in his first throw, is k; in his second, $(1 - k) \cdot (1 - l) \cdot k$; in his third, $(1 - k)^2 \cdot (1 - l)^2 \cdot k$; and so on for ever. Hence the limit, to which his "expectation" approaches, is the limit of

$$k \cdot [1 + (1 - k) \cdot (1 - l) + (1 - k)^2 \cdot (1 - l)^2 + \&c.];$$

i.e., $\quad k \cdot \dfrac{1}{1 - (1 - k) \cdot (1 - l)};\quad$ *i.e.*, $\quad \dfrac{k}{k + l - kl}.$

Similarly, B's chance of winning, in his first throw, is $(1 - k) \cdot l$; in his second, $(1 - k) \cdot (1 - l) \cdot (1 - k) \cdot l$; in this third, $(1 - k)^2 \cdot (1 - l)^2 \cdot (1 - k) \cdot l$; and so on for ever. Hence his "expectation" approaches the limit of

$$(1 - k) \cdot l \cdot [1 + (1 - k) \cdot (1 - l) + (1 - k)^2 \cdot (1 - l)^2 + \&c.];\quad \textit{i.e.,}\quad \dfrac{(1 - k) \cdot l}{k + l - kl}.$$

NOTE ON QUESTION 7695   211

Hence the ratio, of A's expectation to B's, is approximately
$$\frac{k}{(1-k).l}.$$
In the given case, $k = \frac{5}{36}$, $l = \frac{6}{36} = \frac{1}{6}$; hence the required ratio = $\frac{30}{31}$. By a mere accident this happens to be the same as $\frac{1-l}{1-k}$, which accident has misled all the solvers into adopting this as a true formula.

In my "*reductio ad absurdum*" case, $k = l = \frac{5}{36}$; hence the required ratio = $\frac{36}{31}$.

It is worth noting that the ratio, $\frac{30}{31}$, is only *approximative,* the expectations of A and B being just *less* than the fractions $\frac{30}{61}$, $\frac{31}{61}$. If this were not so, the sum total of their expectations would equal 1; *i.e.,* it would be absolutely certain that one or other of them would win—whereas there is clearly a chance, though an indefinitely small one, that the game might go on for ever without either winning.

[Mr. Simmons remarks that the last portion of the above Note is "extremely unmathematical. A's expectation is represented with *perfect accuracy* by the series $\frac{5}{36}[1 + \frac{155}{216} + (\frac{155}{216})^2 + (\frac{155}{216})^3 + \ldots]$, and it is erroneous to say that the sum of this series is only *approximately* equal to $\frac{30}{61}$. When we say that $a = b$ approximately, we mean that a and b differ by at least some *conceivable* quantity. Thus, we say rightly that the ratio of the circumference of a circle to its diameter is approximately equal to 3.14159265; but it would be wrong to say that it is approximately equal to $4(1 - \frac{1}{3} + \frac{1}{5} - \frac{1}{7} + \ldots)$. The game may go on for ever without either A or B winning. True, but this is taken into account, and allowed for, by the above series going on for ever without stopping. Mr. Dodgson's reasoning, if it were correct, might be applied equally to almost every probability question. For instance, we might say that it is "worth noting" that, in the case of a triangle whose vertices are taken at random on the circumference of a given circle, the chance of its being acute-angled is only *approximately* $\frac{1}{4}$, and that of its being obtuse-angled only *approximately* ¾, '*because there is clearly a chance, though an indefinitely small one, that the triangle may be right-angled!*' Has not Mr. Dodgson, in his anxiety to avoid one of the aforesaid mathematical pit-falls, walked straight into another?"]

18. Response to "Infinitesimal or Zero?"

[1886: Cohen, I, 583–84: NY Public]

The general background to the controversy arising out of Question 7695 is given in a footnote to a letter Dodgson wrote to William John Clarke Miller, editor of *MQS*, dated 21 June 1885, which is published in *The Letters of Lewis Carroll*.[1] Dodgson restates Question 7695 and cites "Note on Question 7695" (item 17) and this item in the footnote. The next two items, 19 and 20, arising out of this controversy are not included in the footnote.

"Response to 'Infinitesimal or Zero'" does not appear in the *Lewis Carroll Handbook*. However, it is cited by R. C. Archibald in his "Bibliography of Lewis Carroll: Additions."[2] He remarks that the Miller article may not have appeared first in the *Educational Times*, but refers the reader to p. 32 of volume XXXIX (1886) of that publication, specifically to the notes on Question 7695 by the Rev. T. C. Simmons, M.A.

The editor of *MQS* leads on from Simmons's criticism (see item 13) in the form of another related problem consequently proposed by Simmons, Question 8200: "A random point being taken on a given line, what is the chance of its coinciding with a previously assigned point?" The editorial viewpoint is followed by Mr. Dodgson's reply; Mr. Simmons' response to Mr. Dodgson (in 5 points); observations by Mr. Biddle, Mr. MacColl, and Mr. Knowles; and finally, Mr. Dodgson's comments (in 4 points) on Mr. Simmons's previous response to his solution.

Infinitesimal or Zero? *By the* Editor.

"A random point being taken on a given line, what is the chance of its coinciding with a previously assigned point?" That is the question herein discussed, which has arisen incidentally

Mathematical Questions and Solutions from the "Educational Times," W. J. C. Miller, ed., vol. XLIV (London: Francis Hodgson, 1886), 24–27.

1. Morton N. Cohen, ed., *The Letters of Lewis Carroll* (New York: Oxford University Press, 1979), vol. I, 583–84.
2. *Notes and Queries,* 24 August 1940, 134–35.

out of Mr. Dodgson's remarks on Question 7695, with criticism thereon by Mr. Simmons, given on p. 86 of Volume XLIII., and has been distinctly proposed by him for further discussion, as Question 8200.

I. Mr. Dodgson rejoins as follows:—

"It is surely too late, in A.D. 1885, to seriously discuss the question whether a converging series does or does not reach its limit—in other words, whether an infinitesimal is or is not equal to zero. If the ordinary text-books have not shown Mr. Simmons the difference between them, how can I hope to do it? I will, however, try a *reductio ad absurdum*. I present Mr. Simmons with a line AB, in which I have selected a certain point C; and I ask him to take a point at random in AB, and to estimate its chance of coinciding with C. He will reply, 'If its chance of falling on one side of C be k, its chance of falling on the other side is, *with perfect accuracy*, $1 - k$. Hence its chance of missing C is *absolutely* 1; and its chance of coinciding with it is *absolutely zero*.' But the very same thing is true of *any other* point I might select in AB. Hence the new point *has no chance of falling anywhere!* If Mr. Simmons is partial to pitfalls, let me recommend this one to his notice; it is nice soft falling, and not very deep."

II. Mr. Simmons, having considered the foregoing reasoning, thus states his objections thereto:—

1. "Mr. Dodgson's anxiety for my comfort is most tender and considerate; but unfortunately in his first sentence he wanders from the question, which (to take a simple case) was whether the difference between two such quantities as 1 and $.\dot{9}$ can be said only *approximately* to equal zero. This use of the italicised word I still maintain to be incorrect.

2. "Here the matter might have ended, but for the remarkable assumption contained in what Mr. Dodgson calls a *reductio ad absurdum*. The argument therein, if it is to have any force at all, plainly assumes that a line may be considered as *wholly made up of points which can all previously be assigned,* a most unmathematical conception! For, even if we take an infinite number of points (which for simplicity of conception, and without affecting the ar-

gument, we may consider to be equidistant) on a line, there will still, between every consecutive two, be an infinite number of points remaining. That is to say, the points which have *not* been assigned will always be infinite in number compared with those which *have* been assigned. So that this conception of a line is entirely misleading. It would evidently imply that, when a line is split up into a number of portions sufficiently infinitesimal, the portions either cease to be lines at all, or else they become lines so small that each can contain only a finite number of points; that is to say, the space covered by each bears a finite ratio to the space covered by the points at its extremities!

3. "Now there are two ways in which the chance in question can be proved to be zero. We are of course bound by the definitions given in the text-books of the two words 'point' and 'chance.' For the former, we cannot do better than consult Mr. Dodgson's own edition of Euclid. Are we to infer that he does not consider this to be an 'ordinary' book? For a point is there asserted to have 'position but *no* magnitude.' From the definition of 'chance,' it follows at once that the chance of a point falling on any assigned region or regions of a line is equal to the ratio of the space covered thereby to the space covered by the whole line. Hence, from the two definitions combined, it is clear that the chance of a point falling on a previously assigned point is zero.

4. "Another way of proving it is as follows. Consider AB to be of length unity, and divided into two separate portions at C. If now the length of one portion be k, the length of the other portion will be *with perfect accuracy* $1 - k$. Mr. Dodgson will not, I think, challenge this statement. Is he then prepared to deny that it follows, as a necessary consequence, that the two chances are represented quite accurately by k and $1 - k$? But, if this be granted, it follows immediately that the chance of coincidence with C is absolutely zero. How Mr. Dodgson, *from the standpoint of the definitions,* can conclude it to be infinitesimal, I am unable to comprehend. I am quite willing to learn, and 'be shown' anything on the subject, but, in the manner given above, he certainly *cannot* 'hope to do it.' As to the 'difference' between infinitesimal and zero, I

suppose it is always infinitesimal, while the distinction (is this what Mr. Dodgson really means?) may in certain cases be infinite, especially where ratios are concerned.

5. "The zero result is no doubt remarkable, on the face of it. For does it not appear self-evident, *à priori,* that the new point *must have some* chance of coinciding with C? The apparent contradiction is by no means easy to explain. Is it that our notion of coincidence implies 'filling the same space,' and that this is inconsistent with the notion of a point as filling no space at all? Or is it that our conception of the possibility of the two points coinciding arises from the possibility of making them coincide by a conscious effort directed to that end, and which may be inconsistent with the 'random' conception? The difficulty, if any exists, is perhaps metaphysical rather than mathematical. At any rate, there seems no escape from the conclusion that, in spite of all preconceived notions to the contrary, the only accurate quantitative expression for the chance of the coincidence of two points, both taken at random on a given line, is absolute and undisguised zero, and nothing but zero."

III. Mr. Biddle considers it unwarrantable to define by the term "*absolute* zero" the probability of any event which is not impossible, and draws attention to the essential distinction between such an infinitesimal quantity as $\frac{1}{\infty^{\infty}}$, and $\frac{0}{1}$. "Probability," he adds, is "clearly a matter of *relativity;* otherwise we should have no right, in the instance given above, to say that the respective chances of the random point falling in AC, BC, were $\frac{AC}{AB}$, $\frac{BC}{AB}$. For, speaking *absolutely,* the length of a line makes no difference to the number of points that can be taken in it. Nor is the case mended by considering how the random point is chosen, viz., by some sort of line cutting AB; for, although speaking *relatively,* the average angular relations of AC, BC to the whole cycle of positions from which such line could be drawn, would give a greater probability to its crossing the longer portion; yet, speaking *absolutely,* we cannot say that more lines can be drawn through the one, than through the other portion. Again, speaking *absolutely,* there are as many points or positions in a square inch as in a square mile; and on that ground we have no right to indicate probabilities by ratios between areas. In fact, the *absolute* theory would do away with Local Probability altogether. In the above discussion, the fact seems to have been lost sight of, that C is not distinct from AC and BC, but belongs to both. The probability in regard to it cannot in reason be treated as if it were even hypothetically additional."

IV. From Mr. MacColl we have received the following observations:—

"Whether an *infinitesimal* chance is strictly and logically *zero* depends, of course, upon the definitions we agree to give of the words in italics. But I think it would be convenient if we agreed to define them so as *not* to make them synonymous. Let us take an illustration. A point is taken at random in the circle A, what is the chance that it will also be in the circle B? If the circles touch so as to have one point and one only in common, I should call the chance *infinitesimal;* but if, on the other hand, they neither touch nor intersect, and so have no point in common, I should call the chance *zero.* When conceptions (as in the illustrations just adduced) are different, it is generally convenient to mark this difference by a corresponding difference of language. Whether the same symbol 0 should be used to denote both conceptions, is another question. Practical convenience and the general custom of mathematicians seem in favour of so using it, at least in Probability."

V. Mr. Knowles expresses his views on the subject thus:—

"A few years ago I was present at a meeting of the members of the College of Preceptors, when a paper was read by Mr. A. J. Ellis, F.R.S., on 'Incommensurable Quantities,' and one of the subsequent speakers—and Mr. Ellis appeared to agree with him—stated that an incommensurable quantity might take the form of a recurring decimal. I differed on that occasion, and maintained that a recurring decimal was only an equivalent expression for a vulgar fraction. I am still of that opinion. It is well known that the root of any number never recurs, and if, in $\pi = 3.1415\ldots$ the decimals would only recur, the celebrated problem of squaring the circle would be solved.

"I think, therefore, that Mr. Simmons is correct when he states that 1 and .9̇ are absolutely equal. These remarks are strictly limited to circulating decimals, but this is important as they crop up in the very elements. With regard to the general question I offer no opinion; it has been a bone of contention ever since the Differential Calculus was discovered, and was the foundation of Bishop Berkeley's remarks on the illogical methods of the mathematicians of his day."

VI. Lastly, to Mr. Simmons' objections, contained above in Section II., Mr. Dodgson sends the following reply:—

1. "I re-affirm that the question whether a converging series does or does not reach its limit *is*, in other words, whether an infinitesimal is or is not equal to zero. *E.g.*—The converging series 2^{-1}, 2^{-2}, &c., 2^{-n}, has, for its limit, unity. Also its sum is $1 - 2^{-n}$. Hence, if when n is infinite, the series reaches its limit, the infinitesimal 2^{-n} must be equal to zero.

2. "I never assumed that 'a line may be considered as wholly made up of points which can all previously be assigned,' nor of points of any kind. A point, having no magnitude, can form no portion of a line.

3. "I admit that, if the length of AC, one portion of a line AB, be k, the length of the other portion CB will *with perfect accuracy* be $1 - k$. And I *am* 'prepared to deny that the two chances (of a point falling *in* the two portions) are represented quite accurately by k and $1 - k$.' For this would omit the 3 chances of its falling at A, at B, and at C. Suppose that, when the point falls at C, it is reckoned as falling in AB, and not in BC. Then, to deal fairly with the two portions, we must exclude A, and make unity represent the chance of the point falling somewhere in the line AB, excluding A, but including B. Then k is the chance of its falling between A and C, or else at C; and $1 - k$ the chance of its falling between C and B, or else at B.

4. "I re-affirm, as absolutely axiomatic, that, when an event is *possible*, its chance of happening is *not* zero."

19. Something or Nothing?

[1888: LCH 136: NY Public]

In this item, which originally appeared in the *Educational Times,* vol. XLI, (1 June 1888), 245, Dodgson reopens the discussion on item 18 above with a "new" view of the difficulty.

※

In the years 1885, 1886, there appeared in regard to a Solution of Quest. 7695 (see Vol. XLIII., p. 86, and XLIV., p. 24) a discussion about a difficulty in the Theory of Chances, of which the following question was treated as a typical example:—"A random point being taken on a given line, what is the chance of its coinciding with a previously assigned point?" On one side it was maintained that the chance is *absolute zero:* on the other side it was maintained, by myself and others, that it is some sort of *infinitesimal,* and *not* absolute zero. The arguments on both sides were fully stated, and my only excuse, for re-opening the discussion, is that I have a *new* view of the difficulty to offer to the supporters of the "absolute zero" theory.

I assume that both sides accept the following axioms:—(1) that no aggregate, however infinitely numerous, of *absolute zeroes* can constitute a *magnitude,* however infinitely small; (2) (an example of the preceding) that no aggregate, however infinitely numerous, of *points* can constitute any portion, however infinitely short, of a *line;* and hence (3) that, if the chance of a random point on a line coinciding with *a single selected point* be absolute zero, so also is its chance of coinciding with one or other of *a selected aggregate of points,* however infinitely numerous.

I now propose two questions:—

I. "A random point being taken on a given line, what is the

Mathematical Questions and Solutions from the "Educational Times," vol. XLIX, W. J. C. Miller, ed. (London: Francis Hodgson, 1888), 101–2.

chance of its dividing the line into two *commensurable* parts?" It seems clear that we are here dealing with *a selected aggregate of points,* since it is impossible to mark off any portion of the *line,* and to say "Wherever, in this portion, the random point shall fall, it will divide the whole line into two commensurable parts." I assume, then, that my opponents would answer "It is *absolute zero.*"

II. "And what is its chance of dividing the line into two *incommensurable* parts?" Here again they must answer "It is *absolute zero.*"

And yet *one or other* of these two events *must* happen! Hence, the sum of the two chances must be mathematically represented by unity; that is, one or other (though we cannot say which) must be—not only "*something,*" not only a certain *infinitesimal,* of some inconceivably high order—but must actually reach, if not exceed, the *finite* value of *one-half!*

20. Question 9588

[1889: LCH 136: NY Public]

The question, the solutions to it, and Dodgson's response to those solutions were printed in the *Educational Times,* vol. XLI, p. 247 on 1 June 1888. In this question, Dodgson puts forward a new problem related to the discussion of items 18 and 19. A very similar problem, Question 8861 by J. Brill, M.A., precedes Dodgson's.

A solution to the first part of Question 9588 was given by the Rev. T. C. Simmons, M.A., who obviously had maintained an interest in the controversy surrounding Question 7695. Unlike his comment on Dodgson's "Note on Question 7695," which he found "extremely unmathematical," Simmons finds Question 9588 "interesting." Simmons indicates that his own solution also solves Question 8861. Professor Tanner, M.A., provides a solution to the second part of Dodgson's question. Tanner is really responding to the question in item 18.

The entry in LCH mistakenly gives Thomas Charles Simmons's name as J. C. Simmons.

※

8861. (J. Brill, M.A.)—A rod of given length is broken at random into two pieces; find the probability that their lengths may be commensurable.

9588. (Charles L. Dodgson, M.A.)—A random point being taken on a given line, find the chance of its dividing the line into two parts (1) commensurable, (2) incommensurable.

Solutions by (1) Rev. T. C. Simmons, M.A.;
(2) Prof. Tanner, M.A.

1. One method of dealing with the interesting Question 9588 is as follows:—It will be on all hands admitted that, if a random point P be taken on an undivided line λ, it is infinitely more likely

Mathematical Questions and Solutions from the "Educational Times," vol. L, W. J. C. Miller, ed. (London: Francis Hodgson, 1889), 34–35.

to fall *between* the extremities than *on* either assigned extremity. Let λ now be divided, first into halves, then into thirds, quarters, fifths, sixths, and so on. At any stage of the operation suppose that n marks of division (including one of the original extremities of λ) have been in all obtained. The chance of P's falling *between* any two given consecutive marks will, as before, be infinitely greater than the chance of its falling *on* an assigned one of those two marks; so that its chance of falling *between some consecutive two* of the n marks, will be infinitely greater than its chance of falling *on some one* of the n marks. The ratio of the one chance to the other, having both its terms multiplied by n, will be independent of n, holding of course equally when n is made infinite. But, by making n infinite, all the possible commensurable divisions of λ can be apparently in time exhausted. Therefore the chance that P does not coincide with some possible commensurable division of λ is infinitely greater than the chance that it does so coincide; and on these grounds I venture the *opinion* that the answers to the question should be (1) zero, (2) unity.

In the same way the probability here required in Quest. 8861 would appear to be zero.

2. When Mr. Dodgson [see Note entitled *Something or Nothing*, on pp. 101, 2, of Vol. XLIX.] explains how his two selected aggregates of points can make up the whole of a line consistently with his axiom (2), he will go far to help the "opposition" to explain how, notwithstanding axiom (1), an aggregate of absolute zeros may be unity.

Represent by δ the chance of a random point in a given line of unit length coinciding with an assigned point. Consider a range of points P, each distant δ/n from its neighbours. If δ is not absolute zero the points P are discrete. In a segment of length $1/n$ there will be $1/\delta$ of these points, so that the chance of the random point coinciding with one or other of the points P is unity, and the chance of the random point falling outside the segment, however short, is zero. To avoid the absurdity, we must take $\delta = 0$.

[In reference to the reasoning and arguments used in the above solutions, Mr. Dodgson sends the following remarks:—

(1) In reply to the Rev. T. C. Simmons, if, instead of dividing his line by 2, 3, &c., he will divide it by $\sqrt{2}$, $\sqrt{3}$, &c., and if, where he has written "commensurable," he will write "incommensurable," he will find his argument quite as sound as before, and, instead of proving the two chances to be "zero" and "unity," he will prove them to be "unity" and "zero." An argument that proves with equal case either of two contradictories, needs very cautious handling.

(2) In reply to Professor Tanner, I must respectfully decline to explain how a thing can happen which I say cannot happen at all! No "aggregate of points," as I believe, can ever "make up the whole of a line," or any portion of it: so I must refer him, for the explanation he desires, to the "opposition," who are so ready to explain how "an aggregate of absolute zeros may be unity." While their hand is in, they may as well do the other little job. In the latter part of his letter he asserts, unless I misunderstand him, that, if the chance of a random point coinciding with one assigned point be δ, then its chance of coinciding with one or other of $1/\delta$ such points is unity. I suppose he would say, taking 10 bags, each containing 1 white counter and 9 black, that, since the chance of drawing a white from one bag is $\frac{1}{10}$, the chance of drawing a white from one or other of the 10 bags is unity. Does he accept this as a fair instance of the theorem?]

Arithmetic Computation
and Theory

Test of the divisibility of a number, consisting of more than 3 digits, by 7.

$N = 1000 p_1 + q_1$
$ = 1001 p_1 - p_1 + q_1$
$ = 1001 p_1 - \overline{p_1 - q_1}$ or $= 1001 p_1 + \overline{q_1 - p_1}$

Now since $1001 p_1$ is divisible by 7, if $\overline{p_1 - q_1}$ is divisible by 7, so is N.

Again, if $\overline{p_1 - q_1}$ is a number consisting of more than 3 digits, then as before
$p_1 - q_1 = 1001 p_2 \pm \overline{p_2 - q_2}$.

And so on till we come to $\overline{p_m - q_m}$ consisting of not more than 3 digits — then, if $\overline{p_m - q_m}$ is divisible by 7, so is N.

Working rule

Let the number $N = 67710164647$

1. Bar off the number in periods of three, beginning from the units place — thus

   ~~syyhan~~ 67/710/164/647

2. Place the last period under the preceding one — thus

   67/710/164/647
   647

3. Subtract, and set the answer under the preceding period — thus

   67/710/164/647
   $517/647/$

4. Subtract this answer from the figures above it (taking care to "pay", in the usual way, if one has been borrowed in the last step of the previous subtraction) and set the answer under the next preceding period — and so on till the periods are exhausted.

5. Take the difference between the last answer and the period above it (on whichever side it may be) and this will be $\overline{p_m - q_m}$ — the number containing the test. thus

   $67/710/164/647$
   $2.06/517/647/$
   $\overline{p_m - q_m} = 133$

   which is $\div$ ble by 7 — ∴ N is $\div$ ble by 7.

First page of "Test of the divisibility of a number consisting of more than 3 digits, by 7" (undated) by Charles Dodgson, father of Charles L. Dodgson.

Introduction

Dodgson published a variety of items that can be classified as "arithmetic" in the sense that they deal either with arithmetic problems or with the theory of arithmetic. They include the pamphlet, *Arithmetical Formulæ and Rules for the Use of Candidates for Responsions* from 1870 (item 21), *Examples in Arithmetic* from 1874 (item 22), *Arithmetic I* from 1870–74? (item 23), *Arithmetic II* from 1870–74? (item 24), and the undated *Arithmetic* (item 25); a response to the proposer of a problem entitled "Divisibility by Seven" from 1884 (item 27), five Pillow Problems, numbers 1, 14, 29, 44, and 61, dated either 1881 or 1884; and four items in the *Educational Times:* a letter to the editor titled "Practical Hints on Teaching: Long Multiplication Worked with a Single Line of Figures" from 1879 (item 26), Question 9636 from 1888 (item 29), Question 12650 from 1895 (item 30), and Question 13614 from 1897 (item 32). Additionally, he published three articles in *Nature:* "To Find the Day of the Week for any Given Date" from 1887 (item 28), "Brief Method of Dividing a Given Number by 9 or 11" from 1897 (item 33), and "Abridged Long Division" from 1898 (item 36). Dodgson also worked on several arithmetical problems that were never published, including a number puzzle called "Number-Guessing" (item 31), produced in 1896; an unfinished manuscript Dodgson intended to be "Curiosa Mathematica, Part III," which contains three chapters on "Brief Methods of Performing Some Processes in Arithmetic"; and finally, a galley proof, "Rule for Finding Easter-Day for Any Date till A.D. 2499" (item 35), written for a projected volume on games and puzzles that never materialized. *The Lewis Carrol Picture Book* contains three short numerical items, two of which are puzzles. Stuart D. Collingwood discusses one of them, "The Monkey and Weight Problem." This problem dating from December 1893 elicited different solutions from many mathematicians. An unidentified writer, E.T.C., linked it to Newton's third Law of Motion in an article, "Lewis Carroll's Monkey and Weight Puzzle" that appeared in the *English Mechanic* on 18 January 1901.[1] The articles

---

1. *The Lewis Carroll Picture Book,* edited by Stuart Dodgson Collingwood (London: T. Fisher Unwin, 1899); reprint (as *The Unknown Lewis Carroll*, New York: Dover, 1961), 267–69. For a complete discussion of "The Monkey and Weight Problem," the reader should consult Stuart D. Collingwood, *The Life and Letters of Lewis Carroll* (London: T. Fisher Unwin, 1898), 317–18. The Weaver Collection in the HRHRC contains several solutions to the puzzle.

Archdeacon Charles Dodgson, c. 1856.

Dodgson published in *Nature* are included because it would be impossible to perceive the connections among the unpublished items without them.

The most interesting aspect of these writings is the role that divisibility rules play in them. The origins of the study of properties of numbers like divisibility, greatest common divisor, and least common multiple go back almost as far as counting and the basic arithmetic operations. Their appeal to both dilettantes and mathematicians lies in the simple way problems can be stated and the great variety of methods available for their solution.

One of the first of Dodgson's pieces on this topic is a reply to a Mr. Askew who had asked for a proof of a particular divisibility method in a letter to the editor of *Knowledge*. The reply, "Divisibility by Seven," contains examples and a proof of a method that Dodgson's father had taught him thirty years earlier. He goes on to show that his father's method also provides a test for divisibility by 11 and by 13. The elder Dodgson's rule tests the divisibility by seven of a number N consisting of more than three digits. It depends on expressing

$$\begin{aligned} N \text{ as } 1000p_1 + q_1 &= 1001p_1 - p_1 + q_1 \\ &= 1001p_1 - (p_1 - q_1) \\ &= 1001p_1 + (q_1 - p_1). \end{aligned}$$

Since $1001p_1$ is divisible by 7 if $(p_1 - q_1)$ is divisible by 7, so is N. Now if $(p_1 - q_1)$ is a number consisting of more than three digits, repeat the above process to obtain

$$(p_1 - q_1) = 1001p_2 \pm (p_2 - q_2),$$

and so on until we get $(p_n - q_n)$ consisting of not more than three digits. Then, if $(p_n - q_n)$ is divisible by 7, so is N.

Dodgson adds to his father's rule that if the periods, $(p_i - q_i)$ are single digits, we get a test for divisibility by 11; with periods of two digits we get a test for divisibility by 101; and so for four or more digits. He adds a postscript tying in divisibility by 9, 99, 999, etc., and goes on to say, "Probably similar rules may be made for most primes. I have myself made fairly simple rules for 17 and for 19; but such processes are rather curious than useful."

In the introduction to the first of the three articles that he published in *Nature*, "To Find the Day of the Week for Any Given Date," Dodgson suggests that his method may appeal to rapid mental computers. It relies heavily on division by seven.

Take the given date in 4 portions, viz. the number of centuries, the number of years over, the month, the day of the month.

Compute the following 4 items, adding each, when found, to the total of the previous items. When an item or total exceeds 7, divide by 7, and keep the remainder only.

Dodgson's Question 13614 in the *Educational Times*, which appeared on 1 September 1897, ten years after his article in *Nature*, deals with the same problem: to find the day of the week corresponding to any given date. However, here Dodgson is trying to understand a formula given by Christopher Zeller that Walter William Rouse Ball had discussed in his book, *Mathematical Recreations and Problems*, to construct an algebraical equation for the number of days in any month.[2] Dodgson claims that he has been successful for every month except February. The difference between the two items is striking. In the former, the method is dependent on a month table, cleverly disguised in Dodgson's prose, while in the latter Dodgson is grappling with a complete arithmetical algorithm that produces the result.

Dodgson went on to develop other divisibility methods, generalizing them wherever possible. In "Brief Method of Dividing a Given Number by 9 or 11" he writes,

Now, there are shorter processes, for obtaining the 9-remainder or the 11-remainder of a given number, than my subtraction-rule (the process for finding the 11-remainder is another discovery of mine). Adopting these, I brought my rule to completion on September 28, 1897 (I record the exact date, as it is pleasant to be the discoverer of a new and, as I hope, a practically useful, truth)....

(2) Rule for finding the quotient and remainder produced by dividing a given number by 11.

To find the 11-remainder, begin at the unit-end, and sum the 1st, 3rd, &c., digits, and also the 2nd, 4th, &c., digits; and find the 11-remainder of the difference of these sums....

To find the 11-quotient, draw a line under the given number, and put its 11-remainder under its unit-digit: then subtract, putting the remainder under the next digit, and so on....

2. In a note on page 212 of the second edition of this book published by Macmillan in 1892, and again in the third edition of 1896 on page 242, Ball quotes Zeller's calendar formula. Zeller's article, "Kalender-Formeln" appeared in *Acta Mathematica* (Stockholm), vol. IX, 1887, pp. 131–36. There is no mention of Zeller by name in the Contents of the 1886 and 1887 volume years of *MQS*.

These new rules have yet another advantage over the rule of actual division, viz. that the final subtraction supplies a *test* of the correctness of the result. . . .

Mathematicians will not need to be told that rules, analogous to the above, will necessarily hold good for the divisors 99, 101, 999, 1001, &c. The only modification needed would be to mark off the given number in periods of 2 or more digits.

If we apply his method to the problem of dividing 64372583 by 11, we obtain first the remainder $(3 + 5 + 7 + 4) - (8 + 2 + 3 + 6) = 0$, so the number is divisible by 11. To carry out the division, we write the sequence of remainders below the given number, recalling that each remainder is obtained by subtracting the result of every downward subtraction from the appropriate power of 10.

$$11/6437258\ 3$$
$$5852053/0$$

In "Divisibility by Seven," Dodgson had written in connection with his example for single-digit periods, "The rule is to set the last digit under the next, and subtract, setting the remainder under the next, and so on. In this instance the test-number $= 0$; hence the given number $= 11 \times 5852053$.

$$64372583$$
$$05852053\ ''$$

The connections between "Divisibility by Seven" and "Brief Method of Dividing a Given Number by 9 or 11" are apparent. The second paper extends the ideas presented in the first one, but does not break new theoretical ground. The extension is novel: it permits his subtraction method for carrying out the division to be applied to numbers that are not divisible by 9 or 11 in that there will be a remainder. Dodgson's approach, incorporating a backward procedure, requires that the remainder be found by an addition process first; the quotient is obtained as a second step, quite the opposite of the usual process.

The casting out of nines or of multiples of seven or eleven as a computational check, as well as the related topic of testing the divisibility of one number by another, are of ancient origin. Many papers on these topics have appeared over centuries by all manner of mathematicians. In the third century, Hippolytos was interested in the division of sums of digits by 7 or

9; Leonardo Pisano discussed tests for 7 and 11, and gave a proof of the test for 9 in his *Liber Abbaci* (1202).[3]

In 1801, Carl Friedrich Gauss published his classic *Disquisitiones Arithmeticae*, in which divisibility rules were unified with the notion of congruence: if the difference of two integers A and B is divisible by m, they are called congruent modulo m, which we will write as A ≡ B (mod m). Let

$$N = a_n, a^{n-1}, \ldots, a_1, a_0 = a_n 10^n + a_{n-1} 10^{n-1} + \ldots + a_1 10 + a_0$$

denote any integer, where the a_i can have values 0, ..., 9. Recasting the first four equations in "Divisibility by Seven," we obtain a result equivalent to the third equation.

$$N - M = b(r + 1) + c(r^2 - 1) + d(r^3 + 1) + \&c,$$
$$\text{namely, } N - M = 0(\text{mod}(r + 1)).$$

Hence, if $M = 0(\text{mod}(r + 1))$, then $N = 0(\text{mod}(r + 1))$, which is equivalent to the last equation,

$$r + 1 = 1001 = 7 \times 11 \times 13.$$

Writing the divisibility by seven rule in this form, one sees its full generality and its simplicity.

A week after Dodgson's death, "Abridged Long Division" appeared in *Nature*. Here Dodgson generalizes "Brief Method" to division by a number of the form ($h \times 10^n \pm k$): "With certain limitations of the values of h, k, and n, this Method will be found to be a shorter and safer process than that of ordinary Long Division."

Dodgson credits his nephew, Bertram J. Collingwood, with the discovery of the method for the case where $k > 1$, and for his method of handling divisors of the form $10^n - k$. It is Collingwood's second method that Dodgson improves upon in this article. Dodgson sets up the actual computations in a rigorous way, setting out all the steps. He then compares the three methods, ordinary division, Collingwood's method, and his own, with regard to the number of digits written, the number of additions and subtractions, and the number of multiplications for a particular example. Predictably, his is the shortest with regard to the total number of operations required, while needing the smallest number of digits

---

3. For a complete discussion, the reader should consult chapter 12, volume I of the authoritative *History of the Theory of Numbers*, by Leonard E. Dickson, reprinted by Chelsea Publishing Co., 1969. This three volume work was originally published in 1919–23 by the Carnegie Institution in Washington.

[written]. In his version of his nephew's method, he incorporates a check by stages on the computations, which permits the early discovery of error—an important consideration when performing computations like 86781592485703152764092 divided by 9993!

Dodgson planned to incorporate several of the pieces reproduced here in a projected third volume (never published) of his *Curiosa Mathematica*. He took the two division articles that he published in *Nature* and, with some changes and additions, made them sections 1 and 2 of Chapter II, Book II. For Chapter III, he went on to apply the method of "Abridged Long Division," adding no new ideas. This chapter is rather short and possibly incomplete; his main concern seems to be abridgments of the earlier methods. Chapter I of this projected volume is entitled "Long Multiplication," a topic that he made the subject of a letter he wrote to the editor of the *Educational Times* on 1 November 1879 entitled, "Practical Hints on Teaching. Long Multiplication Worked with a Single Line of Figures." His work on multiplication demonstrates his concern with saving time and space and avoiding the risk of making mistakes. His method is graphical—the single line of digits is appealing to the eye and is concise—eliminating the intermediate lines required in the ordinary algorithm.

> The principle of this Method occurred to me on the 19th of September, 1879. I had been thinking of the great inconvenience arising, in the ordinary process of Long Multiplication, from the distance which often separates the two digits that are to be multiplied together, and what an advantage it would be if the sum could be so arranged that they should be close together. Then came the lucky thought that, by writing the lesser Number *backwards,* and moving it along above the other number, we should have, at each stage of its progress, visible all at once, the set of pairs of digits, whose products have to be added together to make one column of working in the ordinary way.[4]

Here is one of Dodgson's multiplication examples, 3819 × 574, first as he gave it, then worked out arithmetically.

$$
\begin{array}{r}
574 \\
\underline{3819} \\
513476 \\
\underline{167863} \\
2192106
\end{array}
$$

4. C. L. Dodgson, Fragment of "Curiosa Mathematica, Part III," from *The Lewis Carroll Picture Book,* edited by Stuart Dodgson Collingwood (London: T. Fisher Unwin, 1899; reprint [as *The Unknown Lewis Carroll*], New York: Dover, 1961), 240.

$3(1000) + 8(100) + 1(10) + 9 \times 5(100) + 7(10) + 4$

$4 \times 9 = 36 = 3(10) + 6$	=	36
$[4 \times 1(10)] + [7(10) \times 9] = 6(100) + 7(10)$	=	670
$[4 \times 8(100)] + [7(10) \times 1(10)] + [5(100) \times 9]$ $= 8(1000) + 4(100)$	=	8400
$[4 \times 3(1000)] + [7(10) \times 8(100)] + [5(100) \times 1(10)]$ $= 7(10000) + 3(1000)$	=	73000
$[7(10) \times 3(1000)] + [5(100) \times 8(100)]$ $= 6(100000) + 1(10000)$	=	610000
$5(100) \times 3(1000) = 1(1000000) + 5(100000)$	=	1500000
Adding the right-most column, we obtain the product:		2192106

Working out the example using Dodgson's computational form requires just two lines compared with the three of ordinary multiplication. His algorithm also permits checking the accuracy of each digit in the answer separately, a very nice feature to have. What is even more unusual about the method is that, for all practical purposes, only two lines are needed for any computation. In fact, for a computation to require four lines, the smaller number to be multiplied would have to contain at least 13 digits! Dodgson's computational shortcuts are imaginative and unusual but, like many of his contributions to mathematics, not influential in his time.

21. *Arithmetical Formulæ and Rules for the Use of Candidates for Responsions.*

[1870: LCH 77]

A rare four-page pamphlet, unsigned and undated, whose format follows that of *Algebraical Formulæ and Rules for the Use of Candidates for Responsions* (item 13). Collingwood's bibliographic reference on p. 433 provides the 1870 publication date and the publisher, the University Press at Oxford, as the description of this item in the LCH.

G.C.M.:—
    (α) two numbers — divide greater by less, and divisor by remainder, and so on till there is no remainder. The last divisor is the G.C.M.
    (β) three or more numbers — find G.C.M. of first two; then G.C.M. of answer and third number; and so on.

L.C.M.:—
    (α) two numbers — product divided by G.C.M.
    (β) three or more numbers — divide out all primes which are factors of 2 or more of them; then multiply together remaining numbers and primes so divided out.

Vulgar Fractions:—
    (α) to multiply by integer — multiply numerator.
    (β) to divide by integer — multiply denominator.
    (γ) to multiply by fraction — multiply numerators for new numerator, and denominators for new denominator.
    (δ) to divide by fraction — invert divisor, and proceed as in multiplication.

Decimal Fractions:—
  (α) to multiply together:—
    (1) rule for multiplying   multiply as in whole numbers.
    (2) rule for pointing answer   mark off in answer so many decimal places as there are in all the factors together.

  (β) to divide one by another:—
    (1) rule for dividing   divide as in whole numbers, annexing ciphers to decimal part of dividend if necessary.

    (2) rule for pointing answer:—
      (a) if dividend has *more* decimal places than divisor   mark off in answer so many places as difference denotes.
      (b) if *as many*   answer is integer.
      (c) if *less*   answer is integer; annex so many ciphers as difference denotes.

Circulating Decimals, to reduce to vulgar Fractions   for numerator, take to end of first circulating period, subtracting from it non-circulating period; for denominator, take so many nines, as there are figures that circulate, and so many ciphers as there are figures that do not.

TABLES OF WEIGHTS, &C.
Avoirdupois Weight:—

		Dram (*dr.*)
16	Drams	= 1 Ounce (*oz.*)
16	Ounces	= 1 Pound (*lb.*)
28	Pounds	= 1 Quarter (*qr.*)
4	Quarters	= 1 Hundredweight (*cwt.*)
20	Hundredweights	= 1 Ton.

Troy Weight:—

	Grain (*gr.*)	
24	Grains	= 1 Pennyweight (*dwt.*)
20	Pennyweights	= 1 Ounce (*oz.*)
12	Ounces	= 1 Pound (*lb.*)

Apothecaries' Weight:—

	Grain (*gr.*)	
20	Grains	= 1 Scruple (*scr.*) = 1 __
3	Scruples	= 1 Dram (*dr.*) = 1 __
8	Drams	= 1 Ounce (*oz.*) = 1 __
12	Ounces	= 1 Pound (*lb.*)

Miscellaneous Weights:—

14	Pounds (Avoird.) = 1 Stone (*st.*)
7000	Grains (Troy) = 1 Pound (Avoird.)
gr., oz., lb.	are same in Troy and Apoth. Weight.

Length:—

	Barleycorn.	
3	Barleycorns	= 1 Inch (*in.*)
12	Inches	= 1 Foot (*ft.*)
3	Feet	= 1 Yard (*yd.*)
220	Yards	= 1 Furlong.
8	Furlongs, *or*	} = 1 Mile (*m.*)
1760	Yards	
3	Miles	= 1 League.

Surface:—

	Square Inch.	
12 × 12	Square Inches	= 1 Square Foot.
3 × 3	Square Feet	= 1 Square Yard.
4840	Square Yards	= 1 Acre.
640	Acres, or	
1760 × 1760	Square Yards	} = 1 Square Mile.

Solid Contents:—

	Cubic Inch.	
12 × 12	Cubic Inches	= 1 Cubic Foot.
3 × 3 × 3	Cubic Feet	= 1 Cubic Yard.

Miscellaneous Lengths:—

4	Inches	= 1 Hand.
2	Yards, or	
6	Feet	} = 1 Fathom.
5½	Yards	= 1 Rod, Pole, or Perch.
40	Poles	= 1 Furlong.

Proportion:—

if $a:b::c:d$,
values of a, b, c, d, each in terms of the other three, are

$$\frac{bc}{d}, \frac{ad}{c}, \frac{ad}{b}, \frac{bc}{a}.$$

Proportional parts: to divide a given number into parts which shall be proportional to certain other given numbers

divide by sum of given numbers and multiply by each separate

ARITHMETICAL FORMULÆ AND RULES

Simple Interest, and Discount:—

let P = Principal, or Present value of future debt,
T = Time (in years),
R = Rate per cent.,
D = Interest on principal, or Discount on future debt,
F = Future amount of Principal, or Future debt;
then,
(α) Interest on £100 = $T \times R$.
(β) Amount of £100 = £100 + $T \times R$.
(γ) formula connecting P, D, F $P + D = F$.
(δ) P, T, R, D as £100 : P :: $T \times R \times D$.
(ϵ) P, T, R, F as £100 : P :: £100 + $T \times R : F$.

Compound Interest:—

given Principal, Time and Rate:
(α) to find Amount   find Amount of £1 in one year: multiply Principal by this so many times as there are years.
(β) Interest   =   Amount minus Principal.

Stocks:—

let S = amount of Stock,
R = Rate per cent.,
Y = Yearly income,
M = Market price of £100 Stock,
C = Cash value of amount of Stock,
formula connecting these   as £100 : S :: $R : Y$ :: $M : C$.

22. Examples in Arithmetic
[1874: LCH 101, LCAT 373: Princeton]

This collection of elementary unpublished problems contains examples on a variety of arithmetic subjects. The copy from the Parrish Collection has comments, corrections, and many of the answers as margin notes in Dodgson's hand. How many of these problems are his is unknown, but probably few are. Perhaps he worked them in order to decide if they might be useful questions for Responsions examinations. For example, his comment on the problem of dividing .0002 by .0163 is "too long."

CONTENTS

I.	(1)	Greatest Common Measure.
	(2)	Least Common Multiple.
II.	(1)	Vulgar Fractions: additions &c. by integers.
	(2)	" " multiplication and division by fractions.
	(3)	" " complex and continuous.
III.	(1)	Decimals: addition, subtraction, and multiplication.
	(2)	" division.
IV.	(1)	Reduction: vulgar fraction of concrete to lower denomination.
	(2)	" concrete to vulgar fraction of concrete.
	(3)	" decimal of concrete to lower denomination.
	(4)	" concrete to decimal of concrete.
	(5)	" vulgar fraction of concrete to vulgar fraction of concrete.
V.	(1)	Practice: simple.
	(2)	" compound.
VI.	(1)	Mensuration: plane.

EXAMPLES IN ARITHMETIC 241

	(2)	" " to find cost.
VII.	(1)	" solid.
	(2)	" " to find cost.
VIII.	(1)	Square Root: integers and vulgar fractions.
	(2)	" decimals.
IX.	(1)	Cube Root: integers and vulgar fractions.
	(2)	" decimals.
X.	(1)	Proportion: single.
	(2)	" double.
	(3)	" single: to find 1st, 2nd, or 3rd term.
XI.	(1)	Proportional Parts: integers.
	(2)	" fractions.
XII.		Simple Interest.

        P = Principal, or Present Value (of given Amount).
        T = Time.
        R = Rate.
        D = Interest (on given Principal), or Discount (on given Amount).
        A = Amount.

(1) Given P, T, R: to find   D and A.
(2)   T, R, D: P.
(3)   A, T, R: P and D.
(4)   P, T, and D or A: to find R.
(5)   P, R, and D or A: T.

XIII.   Compound Interest.
XIV.   Stocks.

        S = Stock.
        R = Rate.
        Y = Yearly Income.
        M = Market Price.
        C = Cash.
        S', R', &c. = new Stock, new Rate, &c.

(1) Given S, R: to find   Y.
(2)   S, Y: R.
(3)   R, Y: S.
(4)   S, M: C.

	(5)	S, C:	M.
	(6)	M, C:	S.
	(7)	R, Y, M:	C.
	(8)	R, Y, C:	M.
	(9)	R, M, C:	Y.
	(10)	Y, M, C:	R.

(11) S, R, M, R', M': to find difference of Y and Y'.
(12)   M, C: to find real interest on investment.
(13) Miscellaneous.

XV. Profit and Loss.
XVI. (1) Tables.
   (2) Definitions.
   (3) Rules.
   (4) Proofs.
XVII. Miscellaneous.

ARITHMETIC

I (1).

Find Greatest Common Measure of

1. 49 and 119.
2. 110 and 473.
3. 56 and 296.
4. 117 and 387.
5. 118 and 322.
6. 169 and 559.
7. 192 and 444.
8. 143 and 748.
9. 273 and 611.
10. 154 and 814.
11. 119 and 391.
12. 182 and 602.
13. 256 and 592.
14. 133 and 437.
15. 441 and 987.
16. 208 and 689.
17. 462 and 1034.
18. 231 and 979.
19. 357 and 799.
20. 161 and 729.
21. 247 and 1292.
22. 288 and 666.
23. 259 and 567.
24. 221 and 754.
25. 464 and 1537.
26. 407 and 891.
27. 442 and 1508.
28. 323 and 1102.
29. 529 and 1219.
30. 629 and 901.
31. 1161 and 1431.
32. 561 and 1914.
33. 1081 and 3726.
34. 874 and 2014.
35. 2121 and 1313.
36. 2793 and 2660.
37. 7056 and 7392.
38. 5325 and 8307.
39. 6327 and 23997.
40. 12321 and 54345.

EXAMPLES IN ARITHMETIC

I (2).

Find Least Common Multiple of

1. 8, 18, 50, 63.
2. 12, 18, 30, 70.
3. 20, 27, 42, 98.
4. 20, 30, 63, 105.
5. 27, 45, 125, 225.
6. 30, 42, 70, 90.
7. 25, 63, 66, 242.
8. 27, 66, 70, 84.
9. 27, 50, 54, 81.
10. 30, 63, 84, 135.
11. 30, 98, 242, 243.
12. 42, 45, 84, 120.
13. 45, 63, 70, 90, 210.
14. 18, 30, 70, 105, 126.
15. 27, 66, 70, 84, 90.
16. 12, 42, 105, 126, 180.
17. 30, 50, 81, 225, 405.
18. 42, 63, 88, 90, 120.
19. 18, 50, 54, 56, 60.
20. 30, 45, 84, 108, 120.
21. 50, 63, 125, 129, 135.
22. 63, 66, 105, 135, 405.
23. 45, 98, 126, 180, 630.
24. 42, 66, 90, 210, 300.
25. 45, 63, 70, 84, 630.
26. 27, 70, 81, 126, 180.
27. 30, 63, 84, 225, 405.
28. 98, 105, 126, 225, 300.
29. 70, 125, 135, 210, 243.
30. 36, 90, 135, 150, 405.
31. 16, 40, 54, 162, 180.
32. 40, 84, 126, 300, 405.
33. 24, 40, 60, 108, 300.
34. 36, 54, 84, 180, 405.
35. 56, 84, 120, 405, 630.
36. 210, 225, 243, 630, 2310.
37. 407 and 891.
38. 221 and 754.
39. 529 and 1219.
40. 874 and 2014.

II (1).

Vulgar Fractions, to dividing by integers.

1. Add together $\frac{1}{2}$, $\frac{2}{5}$, $\frac{1}{10}$: subtract $3\frac{3}{4}$ from 5: multiply $\frac{3}{11}$ by 15: and divide $\frac{4}{11}$ by 13.

3. Add together $\frac{5}{12}$, $\frac{1}{8}$, $\frac{7}{24}$: subtract $3\frac{3}{8}$ from $4\frac{1}{4}$: multiply $1\frac{2}{9}$ by 12: and divide $\frac{6}{11}$ by 16.

5. Add together $\frac{2}{3}$, $\frac{5}{6}$, $\frac{7}{12}$: subtract $2\frac{6}{7}$ from 5: multiply $4\frac{5}{6}$ by 8: and divide $\frac{15}{16}$ by 25.

7. Add together $\frac{4}{5}$, $\frac{7}{10}$, $\frac{4}{7}$, $\frac{2}{21}$: subtract $1\frac{1}{3}$ from $2\frac{3}{5}$: multiply $8\frac{3}{11}$ by 44: and divide $1\frac{5}{6}$ by 22.

9. Add together $\frac{7}{5}$, $\frac{4}{3}$, $\frac{1}{6}$, $\frac{11}{30}$: subtract $3\frac{4}{25}$ from 9: multiply $11\frac{11}{12}$ by 9: and divide $2\frac{2}{9}$ by 25.

11. Add together $\frac{1}{2}$, $\frac{5}{6}$, $\frac{3}{21}$: subtract $\frac{11}{60}$ from $10\frac{3}{5}$: multiply $5\frac{3}{20}$ by 25: and divide $3\frac{3}{8}$ by 15.

244   ITEM 22

13. Add together $\frac{1}{2}, \frac{1}{3}, \frac{1}{4}, \frac{1}{5}$: subtract $\frac{11}{12}$ from $\frac{17}{18}$: multiply $6\frac{11}{30}$ by 36: and divide $7\frac{1}{7}$ by 30.

15. Add together $\frac{3}{6}, 2\frac{1}{7}, 13\frac{3}{10}$: subtract $\frac{3}{11}$ from $\frac{4}{9}$: multiply $2\frac{10}{63}$ by 14: and divide $7\frac{7}{9}$ by 100.

17. Add together $\frac{5}{7}, \frac{6}{45}, \frac{9}{35}$: subtract $47\frac{1}{24}$ from $50\frac{1}{16}$: multiply $3\frac{13}{66}$ by 11: and divide $8\frac{1}{10}$ by 45.

19. Add together $\frac{2}{5}, \frac{3}{8}, \frac{7}{12}$: subtract $\frac{61}{126}$ from $3\frac{2}{9}$: multiply $8\frac{5}{9}$ by 15: and divide $3\frac{7}{11}$ by 16.

21. Add together $\frac{3}{5}, \frac{2}{7}, \frac{1}{3}$: subtract $\frac{3}{4}$ from $1\frac{4}{25}$: multiply $10\frac{4}{25}$ by 50: and divide $6\frac{2}{9}$ by 21.

23. Add together $\frac{3}{4}, \frac{4}{7}, \frac{7}{9}$: subtract $3\frac{1}{16}$ from $4\frac{1}{24}$: multiply $7\frac{6}{13}$ by 8: and divide $9\frac{7}{9}$ by 55.

25. Add together $\frac{10}{14}, \frac{2}{15}, \frac{18}{70}$: subtract $3\frac{8}{15}$ from $13\frac{2}{75}$: multiply $9\frac{3}{20}$ by 32: and divide $5\frac{1}{7}$ by 99.

27. Add together $\frac{11}{10}, \frac{11}{100}, \frac{11}{1000}, \frac{11}{10000}$: subtract $33\frac{5}{24}$ from $37\frac{4}{15}$: multiply $12\frac{8}{15}$ by 25: and divide $12\frac{6}{7}$ by 80.

29. Add together $\frac{1}{8}, \frac{7}{12}, \frac{5}{9}, \frac{9}{20}$: subtract $\frac{4}{21}$ from $17\frac{1}{35}$: multiply $12\frac{9}{16}$ by 24: and divide $8\frac{2}{5}$ by 18.

31. Add together $1\frac{15}{16}, 2\frac{23}{24}, 3\frac{24}{25}, 4\frac{29}{30}$: subtract $9\frac{7}{13}$ from $13\frac{5}{12}$: multiply $4\frac{5}{18}$ by 45: and divide $19\frac{3}{5}$ by 21.

33. Add together $2\frac{2}{3}, 3\frac{3}{4}, 4\frac{4}{5}, 5\frac{5}{6}, 6\frac{6}{7}$: subtract $\frac{4}{7}$ from $1\frac{8}{25}$: multiply $8\frac{2}{45}$ by 27: and divide $5\frac{8}{11}$ by 12.

35. Add together $\frac{3}{8}, 3\frac{14}{25}, 10\frac{2}{5}, \frac{9}{22}$: subtract $1\frac{16}{243}$ from $15\frac{32}{729}$: multiply $10\frac{3}{50}$ by 60: and divide $4\frac{4}{13}$ by 49.

37. Add together $\frac{11}{12}, \frac{17}{18}, \frac{29}{30}, \frac{47}{48}, \frac{59}{60}$: subtract $21\frac{1}{21}$ from $24\frac{1}{24}$: multiply $11\frac{7}{90}$ by 72: and divide $20\frac{5}{11}$ by 27.

39. Add together $\frac{11}{12}, 1\frac{2}{15}, \frac{7}{16}, 2\frac{11}{18}, \frac{1}{20}$: subtract $25\frac{12}{125}$ from $90\frac{10}{111}$: multiply $5\frac{5}{98}$ by 147: and divide $12\frac{14}{19}$ by 55.

II (2).

Vulgar Fractions, multiplying and dividing by fractions.

1. Add together, $\frac{1}{7}, \frac{2}{9}$, and $\frac{1}{3}$ of $\frac{2}{7}$: and simplify the expression
$$\left\{\frac{3\frac{1}{3}}{7} + \frac{2}{10\frac{1}{2}} - \frac{5}{18} \text{ of } \frac{4}{7}\right\} \times 1\frac{3}{4}.$$

EXAMPLES IN ARITHMETIC

2.* Find the sum of $\frac{5}{8}$, $\frac{3}{5}$ of $\frac{4}{7}$, and $\frac{2}{9}$ of 3: and simplify $\frac{113}{8} + \frac{3}{5}$ of $\frac{10}{9} - \frac{11}{15}$ of $6\frac{1}{4}$.
3. Add together $\frac{1}{3}$, $\frac{2}{5}$, and $\frac{1}{4}$ of $1\frac{1}{2}$: * and reduce to its simplest form the expression
$$\tfrac{1}{3} + 6\tfrac{1}{2} + \frac{3\frac{4}{9}}{7} - \tfrac{1}{3} \times \tfrac{2}{5}.$$
4.* Find the value of $\frac{5}{8}$ of $\frac{3}{7}$ of $\frac{1}{2}$ of $\frac{14}{15}$: and simplify $\frac{2}{3}$ of $\frac{5}{7} + \frac{3}{5}$ of $\frac{4}{9}$.
5. Multiply $\dfrac{5\frac{3}{8}}{7\frac{1}{9}}$ by $\dfrac{1}{2\frac{41}{44}}$; and * find the value of $24 \times \frac{8}{9}$ of $7 + \frac{5}{6}$ of $3 \times \frac{4}{15}$.
6.* Find the value of $2\frac{1}{5} \times 1\frac{5}{6}$ of $1\frac{2}{13} \times 3\frac{1}{4}$ of $1\frac{5}{11}$: and simplify
$$\frac{\frac{1}{3} + \frac{1}{5} + \frac{1}{7} + \frac{1}{9}}{\frac{2}{3} + \frac{2}{5} + \frac{2}{7} + \frac{2}{9}}.$$
7. Multiply $\frac{4}{17}$ by 5, then by $\frac{5}{3}$, then by $\dfrac{\frac{5}{3}}{7}$, then by $\dfrac{5}{\frac{3}{7}}$.
8. Add together ($\frac{4}{11}$ of $\frac{2}{3}$ of $\frac{1}{2}$), ($\frac{7}{8}$ of $\frac{4}{5}$ of $\frac{10}{33}$), and ($\frac{2}{5}$ of $1\frac{2}{3}$).
9.* Multiply $\frac{3}{7}$ of $\frac{9}{11}$ by $\frac{2}{15}$, and subtract the product from $\frac{8}{99}$.[1]
10. Find the simple fraction which is equal to the difference of $\frac{1}{3}$ of $3\frac{7}{8}$ and $\frac{1}{4}$ of $5\frac{1}{4}$.
11.* Divide $\frac{2}{7}$ by $\frac{7}{3}$, and $5\frac{1}{4}$ by $\frac{2}{5}$: and reduce the compound fractions $\frac{3}{11}$ of $1\frac{2}{9}$ of $5\frac{3}{5}$ of $\frac{1}{4}$, and $\frac{3}{8}$ of $6\frac{1}{2}$ of $\frac{13}{14}$ of $\frac{7}{26}$.
12.* Add together $\frac{2}{5}$, $\frac{1}{6}$, $\frac{7}{3}$: subtract $\frac{1}{2}$ of $\frac{1}{3}$ from $\frac{3}{4}$ of $\frac{4}{5}$: and divide $\frac{1}{2}$ by $\frac{1}{3}$, $\frac{4}{15}$ by $\frac{2}{5}$.
13. Add together $3\frac{1}{2}$, $4\frac{1}{3}$, $5\frac{1}{4}$, $\frac{3}{4}$ of $\frac{7}{8}$, and $\frac{1}{2}$ of $\frac{1}{3}$ of $\frac{5}{8}$.
14.* What is the sum, difference, product and quotient of $3\frac{7}{8}$ and $2\frac{1}{9}$?
15.* Add together $\frac{1}{4}$, $\frac{1}{3}$, $\frac{3}{20}$; and divide the sum by $\frac{2}{5}$.
16. Find the product of $\frac{32}{51}$, $\frac{85}{112}$, $\frac{189}{209}$, and $\frac{23}{36}$.
17. Find the product of $\dfrac{2\frac{1}{2}}{3\frac{2}{3}}$ and $\dfrac{2\frac{1}{3}}{3\frac{1}{2}}$: and simplify
$\frac{18}{17} \times (1 - \frac{64}{81}) + \frac{8}{11} \times \frac{1}{6} \times (\frac{1}{2} + \frac{5}{12})$.

---

1. Dodgson changed the original statement of the exercise which read, "subtract $\frac{8}{99}$ from the product." He gives no explanation for the asterisks on problems in sections II and III.

18.* Divide $5\frac{5}{18} - \frac{43}{36}$ by $1\frac{2}{7}$ of $8\frac{1}{6}$: and multiply $9 - 3\frac{6}{25}$ by $16\frac{2}{3}$ of $1\frac{1}{12}$.

19.* Find the value of $\frac{1}{2} - \frac{4}{9} + \frac{5}{12} - \frac{7}{16}$; and of $3\frac{1}{5}$ of $6\frac{1}{4} \div \dfrac{3}{1\frac{1}{5}}$.

20. Simplify $(\frac{2}{3}$ of $\frac{3}{4}$ of $\frac{7}{8}) + (\frac{1}{5}$ of $\frac{2}{3}$ of $1\frac{7}{8}) + \frac{1}{4}$.

21.* Divide $20\frac{5}{14}$ by $17\frac{1}{12}$: and find the value of $\dfrac{\frac{1}{6} + \frac{1}{8}}{1\frac{1}{2} + \frac{3}{14}}$.

22.* Find the value of $2\frac{5}{63} + (\frac{1}{7} \times \frac{4}{9}) - 2\frac{1}{2}$ of $(\frac{2}{3} \div \frac{7}{9})$.

23.* Multiply $\frac{2}{15}$ of $5\frac{5}{8}$ by six times $\frac{1}{5}$ of $2\frac{1}{2}$: and divide $\frac{3}{4}$ of $\frac{5}{6}$ by $\frac{9}{11}$ of $\frac{3}{4}$.

24.* Simplify the following expressions:
$$\dfrac{9\frac{7}{9}}{2\frac{1}{27}}, \dfrac{5\frac{3}{11}}{2\frac{1}{11}}, \dfrac{8\frac{3}{4}}{5\frac{5}{8}}, \dfrac{15\frac{3}{5}}{7\frac{4}{5}}.$$

25. Divide $\frac{3}{5}$ by 9, then by $\frac{7}{9}$, then by $\dfrac{7}{\frac{9}{11}}$, then by $\dfrac{\frac{7}{9}}{11}$.

26. Divide $\dfrac{2+3}{4+5}$ by $\dfrac{4+3\frac{1}{2}}{5+5\frac{1}{2}}$.

27. Multiply together $\frac{3}{4}, \frac{5}{7}, 7\frac{1}{5}, (\frac{2}{3}$ of $3\frac{8}{9})$: and divide $(\frac{4}{7}$ of $\frac{2}{8})$ by $(\frac{8}{9}$ of $\frac{1}{7})$.

28.* Divide $\frac{2}{3}$ of $\frac{1}{3}$ by $\frac{5}{7}$ of $\frac{3}{5}$: and find the value of $1\frac{28}{36} + (\frac{1}{6} \times \frac{7}{12}) - \frac{1}{2}$ of $(\frac{3}{4} \div \frac{2}{3})$.

29.* Reduce to simple fractions in lowest terms
$\frac{9193}{20677}$ and $(2\frac{1}{2} + \frac{1}{6}) \div (3\frac{2}{3} - \frac{1}{8})$:
and simplify $7\frac{1}{3}$ of $2\frac{5}{11} \div (\frac{9}{19}$ of $16\frac{5}{8})$.

30. Find the sum of $1\frac{1}{2}, 2\frac{2}{3}, 3\frac{5}{6}$: and divide the result by $\frac{4}{5}$ of $\frac{6}{8}$ of $\frac{21}{9}$.

31.* Find the value of $(2\frac{1}{2} + \frac{1}{6}) \div (2\frac{2}{3} - \frac{1}{8})$: and simplify
$$\dfrac{2\frac{1}{8} + 7\frac{1}{3}}{3\frac{1}{6} \times 9\frac{4}{5}}.$$

32.* Divide $\frac{9}{11}$ of $\frac{19}{81}$ by $\frac{33}{153}$ of $\frac{114}{121}$: and find the value of
$$\dfrac{3\frac{3}{7} \text{ of } 2\frac{11}{12}}{\frac{1}{33} \text{ of } 8\frac{9}{14}}.$$

33.* Simplify each of the following expressions:
$$\dfrac{4\frac{1}{2} \text{ of } \frac{5}{27}}{5\frac{3}{5} \text{ of } 1\frac{3}{7}}, \dfrac{3\frac{3}{4}}{5}, \dfrac{\frac{11}{12}}{7\frac{17}{18}}, \dfrac{23}{2\frac{2}{3} + \frac{2}{5}}.$$

EXAMPLES IN ARITHMETIC 247

34.* Divide $1 - (\frac{1}{2} + \frac{1}{3} + \frac{1}{24})$ by $1 - (\frac{1}{2}$ of $\frac{1}{3}$ of $\frac{1}{24})$:

35.* Reduce the following: $\dfrac{1\frac{1}{4} \text{ of } 2\frac{2}{15}}{\frac{2}{17} + \frac{11}{51}}, \dfrac{\{\frac{2}{9} \times \frac{2}{8}\} \div 2}{\frac{1}{6}\{\frac{4}{5} - \frac{1}{16}\}}$.

36.* Simplify the following:

(1) $\dfrac{10 \times \frac{1}{7}}{2 + \frac{9}{7}}$,   (2) $1\frac{1}{3} - (1\frac{1}{3}$ of $\frac{1}{2})$.

37. Which is the greater, $\frac{2}{5}$ of $\frac{7}{9}$ or $\frac{3}{8}$ of $\frac{5}{6}$?

38.* Divide $2\frac{6}{7}$ of $1\frac{2}{5}$ by $3\frac{1}{3}$ of $\frac{3}{5}$ of $\frac{1}{4}$: and find the value of

$$\frac{1\frac{1}{2}}{1\frac{1}{4}} \div \left(\frac{3\frac{7}{9}}{2\frac{4}{5}} \times \frac{5}{6}\right) + \frac{7}{16}.$$

39. Find the difference between $\frac{1}{3}$ of $3\frac{7}{8}$ and $\frac{1}{4}$ of $5\frac{1}{4}$.

40.* Divide $(\frac{1}{4}$ of $\frac{1}{5}$ of $4\frac{2}{7}$ of $2\frac{1}{3})$ by $(\frac{1}{9}$ of $4\frac{1}{5}$ of $3\frac{3}{4}$ of $1\frac{3}{7})$.

II (3).

Vulgar Fractions, complex and continued.

1.* Simplify $\dfrac{2\frac{1}{4} - 1\frac{3}{5}}{8\frac{1}{2} + 1\frac{9}{10}}$.

2. Simplify $\frac{1}{2} \times \dfrac{\frac{2}{3} - \frac{1}{2}}{2} \times \dfrac{\frac{3}{4} - \frac{2}{3}}{3}$.

3.* Simplify $\dfrac{1}{2 + \dfrac{3}{4 + \frac{5}{6}}}$.

4.* Simplify $\dfrac{\frac{1}{2} + \frac{4}{5} - \frac{2}{3}}{1\frac{1}{2} \times \frac{8}{15} \div 2\frac{1}{4}}$.

5.* Simplify (1) $\dfrac{1}{3 + \dfrac{3}{4\frac{1}{2}}}$,   (2) $\dfrac{2\frac{1}{4} - 1\frac{3}{5}}{8\frac{1}{2} + 1\frac{9}{10}}$.

6.* Reduce to its simplest form $\dfrac{\frac{3}{5}}{\frac{8}{9}} + \dfrac{\frac{5}{8}}{2\frac{1}{2} \times 1\frac{1}{3}} \times \frac{1}{80}$.

7.* Find the value of $\dfrac{3\frac{1}{3} \text{ of } 2\frac{2}{5}}{1\frac{3}{4} \text{ of } 1\frac{5}{7}} \div 1\frac{1}{2}$.

8. Simplify $\left(\dfrac{3\frac{1}{3}}{7} + \dfrac{2}{10\frac{1}{2}} - \frac{5}{18} \text{ of } \frac{4}{7}\right) \div \frac{4}{7}$.

ITEM 22

9. What is the value of $\left\{2\frac{2}{4} + \frac{5}{2} \text{ of } \dfrac{7}{3\frac{4}{5}} - \dfrac{1\frac{2}{3}}{2\frac{1}{2}}\right\} \div 1\frac{77}{228}$?

10.* Simplify $\dfrac{1 - \frac{1}{3}(\frac{1}{2} + \frac{1}{7})}{1 - \dfrac{1}{2 - \frac{1}{4}}}$.

11.* Find the value of $\dfrac{1}{2 + \dfrac{2}{3 + \frac{4}{5}}}$.

12.* Multiply $\dfrac{\frac{3}{7} \text{ of } \frac{5}{6}}{\frac{1}{4} \text{ of } \frac{9}{11}}$ by $\dfrac{\frac{5}{8} \text{ of } \frac{9}{111}}{\frac{3}{7} \text{ of } \frac{5}{6}}$.

13.* Simplify $\dfrac{5}{5 - \dfrac{5}{3 - \frac{1}{2}}} + \dfrac{6}{6 - \dfrac{6}{3 - \frac{2}{3}}}$.

14. What fraction multiplied by $\frac{2}{3}$ of $\frac{4}{5}$ of $3\frac{1}{2}$ gives $\frac{7}{9}$ as the result?

15. Find the value of $\dfrac{(2 + \frac{1}{5}) \div (3 + \frac{1}{7})}{(\frac{1}{2} - \frac{1}{3}) \times (4 - 3\frac{3}{7})}$.

16. Simplify $\dfrac{4\frac{4}{15} \text{ of } 2\frac{5}{8}}{5\frac{1}{5} - 4\frac{1}{2}}$ and $\frac{5}{7} \times \left(100 - \dfrac{200}{3} + \dfrac{7\frac{1}{3}}{2\frac{1}{4}}\right)$.

17.* Simplify (1) $\dfrac{1}{3 \times \dfrac{3}{4\frac{1}{2}}}$ and (2) $\dfrac{1}{2 + \dfrac{1}{2 + \frac{1}{2}}}$.

18. What fraction multiplied into the sum of $\frac{3}{8}$, $1\frac{13}{24}$, and $\frac{17}{36}$ will make the product 3?

19.* Find the value of $\dfrac{2\frac{3}{4} \text{ of } 3\frac{2}{3}}{2\frac{5}{6} + \frac{1}{2} + \frac{1}{3}}$ multiplied by $\frac{2}{33}$ of 6.

20. What is the continued product of $(\frac{3}{4} + \frac{4}{5} + \frac{7}{10})$, $(\frac{3}{4} \text{ of } \frac{4}{5} \text{ of } \frac{7}{10})$, $\dfrac{\frac{7}{12} + \frac{5}{9}}{\frac{7}{12} - \frac{5}{9}}$, and $\dfrac{1}{17\frac{2}{11}}$?

21. Simplify $\dfrac{1\frac{3}{4} - (\frac{7}{6} \text{ of } \frac{18}{28})}{(\frac{5}{6} \text{ of } \frac{12}{20}) + 5\frac{1}{2}} \div \frac{1}{6}$.

22. Multiply the sum of $(\frac{4}{7} \text{ of } \frac{1}{2})$ and $1\frac{1}{9}$ by $1\frac{3}{4}$ of the difference between $\frac{4}{11}$ and $\frac{1}{5}$.

EXAMPLES IN ARITHMETIC

23. Find the value of $\frac{3}{7}$ of $\dfrac{\frac{1}{5} + \dfrac{3\frac{1}{5}}{7} \div 2\frac{1}{3}}{\dfrac{3}{4\frac{2}{7}}}$.

24.* Simplify $\dfrac{\frac{1}{7} \text{ of } \frac{3}{4}}{1 - \dfrac{1}{1 + \frac{3}{25}}}$, and $\dfrac{3\frac{3}{4} \times 2\frac{1}{5}}{1\frac{15}{16} - 1\frac{1}{4}}$.

25.* Simplify $\dfrac{3}{\dfrac{2\frac{1}{4}}{1 + 3\frac{1}{4}} - \frac{3}{16} \text{ of } 1\frac{1}{3}}$, and $\dfrac{2\frac{3}{4} + 3\frac{1}{5}}{7\frac{7}{8} - 1\frac{7}{44}}$.

26. Shew that $\dfrac{2 + 4 + 6}{3 + 5 + 7}$ lies between the greatest and the least of the fractions $\frac{2}{3}, \frac{4}{5}, \frac{6}{7}$.

27.* Reduce to their simplest forms the following fractions:
(1) $\dfrac{2}{3 + \dfrac{4}{5 + \frac{6}{7}}}$, (2) $\dfrac{6\frac{5}{7} - (6 \times \frac{2}{7})}{5}$.

28.* Find the value of
$\dfrac{\frac{7}{10} - \frac{2}{3}}{\frac{9}{12} + \frac{1}{5}} \div \dfrac{\frac{1}{3}}{9\frac{1}{2}}$ and of $\frac{1}{2} + \frac{2}{9} - \frac{4}{15} + \frac{5}{18}$.

29.* Simplify $\dfrac{2\frac{3}{4} + 2\frac{1}{2} \text{ of } \dfrac{7}{3\frac{4}{5}} - \dfrac{1\frac{2}{3}}{2\frac{1}{2}}}{2 - \frac{151}{228}}$.

30.* Simplify the following fractions:
$\dfrac{2\frac{13}{16} \times \frac{1}{3}}{\frac{3}{5} \text{ of } \frac{5}{18} \div 5\frac{1}{6}} \div (\frac{1}{5} \text{ of } 1\frac{15}{16}.)$

31. Simplify $\dfrac{2\frac{4}{5} - 1\frac{1}{2} + 9\frac{1}{11}}{4\frac{1}{5} - 2\frac{1}{4} + 13\frac{7}{11}}$.

32. Simplify $\dfrac{(\frac{1}{2} + 1\frac{3}{7} + \frac{5}{6}) \times (\frac{4}{15} - \frac{3}{20})}{\frac{11}{18} \text{ of } 1\frac{14}{15}}$.

33. Simplify
$(\frac{1}{2} + \frac{1}{3} + \frac{1}{4}) \div \left(\dfrac{1}{2\frac{1}{2}} + \dfrac{1}{3\frac{1}{2}} + \dfrac{1}{4\frac{1}{2}}\right) - \frac{13}{24} \text{ of } \frac{576}{264}$.

ITEM 22

34. Simplify the fractions:
$$\frac{2\frac{1}{4} - \frac{2}{3} \text{ of } 1\frac{5}{6}}{\frac{1}{5} \text{ of } 3\frac{1}{3} + \frac{13}{36}}; \text{ and } \frac{1\frac{3}{4} - \frac{7}{6} \text{ of } \frac{18}{28}}{\frac{5}{6} \text{ of } \frac{12}{20} + 5\frac{1}{2}}.$$

35. Find the difference between
$$\left(\frac{\frac{1}{2} - \frac{1}{3}}{\frac{1}{2} + \frac{1}{3}} + \frac{\frac{1}{3} - \frac{1}{4}}{\frac{1}{3} + \frac{1}{4}}\right) \text{ and } \left(\frac{\frac{1}{4} - \frac{1}{6}}{\frac{1}{4} + \frac{1}{6}} - \frac{\frac{1}{6} - \frac{1}{8}}{\frac{1}{6} + \frac{1}{8}}\right).$$

36.* Simplify $\left\{\dfrac{6}{5\frac{1}{2}} - \dfrac{2}{2 + \dfrac{2}{2 + \frac{2}{3}}}\right\} \div (1\frac{5}{6} \times \frac{3}{2}).$

37.* Simplify $\dfrac{1}{2 + \dfrac{1}{1 + \dfrac{1}{2 + \dfrac{1}{1 + \frac{1}{2}}}}}.$

38.* Simplify $2 + \dfrac{1}{2 + \dfrac{1}{2 + \dfrac{1}{2 + \dfrac{1}{2 + \frac{1}{2}}}}}.$

39.* Reduce to its lowest terms—
$$\left\{1 + \dfrac{1}{3 + \dfrac{1}{3 + \frac{1}{3}}}\right\} \div (\tfrac{1}{3} + \tfrac{1}{11} \text{ of } 10\tfrac{2}{3}).$$

40. Simplify $\dfrac{(\frac{3}{5} \text{ of } \frac{2}{9}) - (\frac{4}{7} \text{ of } \frac{1}{8})}{(\frac{4}{15} \text{ of } \frac{1}{28}) + (\frac{3}{10} \text{ of } \frac{7}{9})}.$

III (2).

Find the value of
1. $15.625 \div 2.5$; and $.015625 \div 25$.
2. $1562.5 \div .00025$; and $1.5625 \div 25000$.
3. $181.3 \div .00037$; and $171.99 \div 27.3$.
4. $9.065 \div .049$; and $.03 \div .001$.

5. 8 ÷ .002; and 37.5 ÷ 7.68.
6. 15 ÷ 6.25; and 17.28 ÷ .0144.
7. .00128 ÷ 8.192; and 1708.4592 ÷ .00024.
8. .0002 ÷ .0163;[2] and 4 ÷ .00255.
9. 11.1 ÷ 32.76; and .0123 ÷ 3.21.
10. 2.117 ÷ .0073; and .032 ÷ 2.137.[3]

IV (3).

Find the value of
1. .45 of £1; .68125 of £1; and 2.325 of £1.
2. 32.5 of 5s.; 1.85 of 3s. 4d.; and 2.375 of 13s. 4d.
3. .13125 of £5; and .001953125 of £40.
4. 3.45 of 5 guineas; and .325 of 1½ ton.
5. 2.74 of 12s. 6d.; and 22.25 of £2 2s. 6d.
6. 3.225 of 2½ guineas; and 22.75 of £5 10s. 6d.
7. 3.03 of 10s. 5d.; and .0474609375 of £10 13s. 4d.
8. .2775 of 1 sq. yd. 3 ft. 72 in.; and 32.156 of 3 m. 330 yds.
9. .634375£ + .025 of 25s. + .325 of 30s.
10. 8.71875 of 8d. + 1.146875 of 6s. 8d. − .0625 of a guinea.
11. .375 of a guinea + .1875 of a crown + .3 of 7s. 6d. − .875 of 2d.
12. 3.83 of 4s.; and 6.15 of 2s. 9¾d.
13. .397916 of £1; and .40972 of a guinea.
14. .571428 of a qr.; and .285714 of a cwt.

IV (4).

Reduce
1. 9s. 6d. to the dec. of £1; and 2s. 2¼d. to the dec. of £5.
2. 5s. to the dec. of 13s. 4d.; and 17s. 3d. to the dec. of 10s.
3. £1 2s. 6d. to the dec. of £1; and 2s. 7½d. to the dec. of 10s.
4. 3s. 3¾d. to the dec. of £1 6s. 6d.; and £3 4s. 2d. to the dec. of 2s. 4d.

---

2. Dodgson indicated that this problem was "too long."
3. Dodgson wrote "reverse" under this problem, supposedly meaning to transpose the last two numbers.

5. 6s. 6¾d. to the dec. of a guinea.; and 7s. to 10½d. to the dec. of £2.
6. 9 oz. 2 dr. to the dec. of a lb.; and 3 fur. 33 yds. to the dec. of a mile.
7. 2 m. 1100 yds. to the dec. of a league; and 12 h. 55′ 21″ to the dec. of a day.
8. 15s. 6¾d. to the dec. of £4; and 1 cwt. 3 qrs. 7 lbs. to the dec. of 2½ tons.
9. 3¾ guineas to the dec. of £100; and 4½ lbs. to the dec. of 3 qrs. 12 lbs.
10. 13s. 4d. to the dec. of a crown; and 2 tons 4½ cwt. to the dec. of 1 ton 11¼ cwt.
11. 2R. 4P. to the dec. of 1R. 5P.; and £2 11s. 6¾d. to the dec. of £3.
12. 8 sq. ft. 20 in. to the dec. of 12 sq. in.; and 7s. 6½d. to the dec. of £1.
13. 2 w. 6¼d. to the dec. of 4 d. 3 hrs.; and £6 12s. 6¾d. to the dec. of 1½ guinea.
14. 3 hrs. 3′ 2¼″ to the dec. of a day; and £24 12s. 6¼d. to the dec. of £4.

V (1).

	£	s.	d.		£	s.	d.
1. 135 at	2	19	3½	7. 217 at	4	17	7¾
2. 273 at	3	18	4¾	8. 322 at	7	14	5½
3. 289 at		8	8½	9. 373 at		9	7¾
4. 431 at	5	17	11½	10. 397 at	6	15	10
5. 511 at		7	10¾	11. 623 at		11	9¼
6. 271 at	6	15	10¾	12. 333 at	5	18	11½

V (2).

1. 3 lbs. 5 oz. 14 dwts. 12 grs. at 17s. 6d. per oz.
2. 5 mo. 3 w. 4 d. at 17s. 6d. per week.
3. 9 mo. 1 w. 6 d. at £1 2s. 9d. per week.

VI (1).

1. What is the area of a court, 10 yds. 2 ft. long, and 5 yds. 1 ft. broad?
2. How many square yards of carpet will it take for a room 26 ft. by 32 ft.?
3. What is the surface of a marble slab, whose length is 5 ft. 7 in., and breadth 1 ft. 10 in.?
4. Find the area of a square building, whose side is 46 ft. 8 in.
5. How many square yards of paper will be required for a room 17 ft. long, 12 ft. 7 in. wide, and 8 ft. 5 in. high?
6. How much wainscoting is there in a square room, 18 ft. 3 in. long, and 8 ft. 6 in. high?
7. What is the length of a room, whose breadth is 11 ft. 11 in., and which it takes 17 sq. yds. 2 ft. 131 in. of drugget to cover?
8. How many yards of carpeting, 2 ft. 4 in. broad, will it take to cover a room whose dimensions are 26 ft. by 35 ft.?
9. It is found that 288 yds. of paper, 2 ft. 8 in. wide, will cover the walls of a room; how many would be required of paper 2 ft. 3 in. wide?
10. How many yards of matting, 2 ft. 3 in. wide, will be required for a square room, whose side is 18 ft. 9 in.?
11. If 45 bricks will pave a square yard, how many will be wanted for a space 34 ft. long and 14 ft. wide, allowing for a path, 2 ft. wide, all round?

VI (2).

1. A floor, 24 ft. 4 in. broad and 96 ft. 6 in. long, is to be laid at $1\frac{1}{2}d.$ per square foot; find the cost.
2. At $9\frac{3}{4}d.$ per square yard, what is the cost of painting a room which is 24 yds. round, and 10 ft. 4 in. in height?
3. What will be the expense of glazing a hall-window containing 60 squares, each 1 ft. 3 in. long, and $11\frac{1}{4}$ in. wide, at $5s.$ $4d.$ per sq. ft.
4. What is the cost of papering a room 15 ft. long, 12 ft. wide, and 10 ft. high, with paper 30 in. broad, at $7\frac{1}{2}d.$ per yard?

VII (1).

1. How many cubic feet of water can be contained in a vessel with square base, whose side is 3 ft. and height 2 ft. 10 in.?
2. What quantity of timber is there in a beam, whose length is 20 ft., breadth 2 ft., and thickness 2 ft. 6 in.?
3. Find the solid content of a cube, whose side is 7 ft. 5 in.
4. In making a square pond, whose side was 12 yds., there were taken out 336 cub. yds. of earth; how deep was it made?
5. What must be the length of a trench, 5 ft. 6 in. deep, and 10 ft. 8 in. wide, that it may contain 7040 cubic feet?
6. A straight plank is $3\frac{1}{2}$ in. thick, and $6\frac{1}{4}$ in. broad; what length must be cut off so as to contain $6\frac{1}{4}$ cubic feet of timber?
7. If a beam which is 10 in. wide, 8 in. deep, and 5 ft. 6 in. long, weigh 8 cwt. 1 qr., find the length of another beam, the end of which is a square foot, which shall weigh a ton.

VIII (1).

Find the Square Root of

1. 52441
2. 114244
3. 160801
4. 409600
5. 654481
6. $7\frac{9}{16}$
7. $9\frac{61}{100}$
8. $20\frac{61}{81}$
9. $1547\frac{1}{9}$
10. $1\frac{120}{841}$
11. 1042441
12. 1234321
13. 998001
14. 1522756
15. 52012944
16. $1\frac{40}{81} \times 4\frac{41}{100}$
17. $34\frac{4}{9} + 12\frac{1}{4}$
18. $15\frac{5}{8} \times 13\frac{11}{18}$

X (1).

1. A bankrupt owes £4726 10s., and his effects are worth £1181 12s. 6d.; how much will he be able to pay in the £?

2. If $3\frac{3}{4}$ shares in a speculation are worth £27 10s., what will $4\frac{5}{8}$ shares be worth?
3. If 39 cwt. 1 qr. 11 lbs. cost £59 6s. 6d., what will 13 cwt. cost at the same rate?
4. What weight of sugar may be bought for £374 8s., when the cost of 6 cwt. 2 qrs. is £27 14s. 8d.?
5. If the tax on £335 7s. 6d. amount to £58 13s. $9\frac{3}{4}d.$, what is that in the £?
6. How many gals. of wine, at the rate of £31 16s. 4d. for 46 gals., may be bought for £117 11s. 8d.?
7. If 17 cwt. 3 qrs. 14 lbs. of barley cost £8 18s. 9d., how much may be bought for £5 12s. 6d. at the same rate?
8. If $1\frac{2}{3}$ yds. of cotton cost 2s. 6d., what will be the cost of $24\frac{1}{2}$ yds.?
9. A wedge of gold, weighing 14 lbs. 3 oz. 8 dwt., is valued at £514 4s.; what is the value of an oz.?
10. If $2\frac{1}{4}$ yds. of cotton cost 3s. 9d., what will be the cost of $13\frac{5}{8}$ yds.?
11. What is the value of $\frac{3}{7}$ of $\frac{3}{4}$ of a ship, when $\frac{5}{8}$ of the whole is worth £525?
12. If £15 12s. pay 16 labourers for 18 days, how many labourers will £35 2s. pay for 24 days?
13. A bankrupt pays $3\frac{1}{2}d.$ in the pound, and the total of his payments amounts to £154; what was his debt?

X (2).

1. If 25 men do a piece of work in 24 days, working 8 hours a day, in how many days would 30 men do the same piece of work, working 10 hours a day?
2. If a tradesman with a capital of £2000 gain £50 in 3 months, what sum will he gain with a capital of £3000 in 7 months?
3. If 858 men in 6 months consume 234 quarters of wheat, how many quarters will be required for the consumption of 979 men for 3 months and a half?

4. If a man travels 90 miles in 3 days, by walking 8 hours a day, in what time will he travel 540 miles by walking 6 hours a day?
5. If the wages of 8 persons for 21 weeks be £92, what will be the wages of 14 persons for 33 weeks?
6. If 6 men can reap 34 acres of corn in 5 days, how many men will be required to reap $95\frac{1}{5}$ acres in 6 days?
7. If 3000 copies of a book of 11 sheets require 66 reams of paper, how much paper will be required for 5000 copies of a book of $12\frac{1}{2}$ sheets?
8. If 16 horses eat 3 bushels in 10 days, how long will 63 bushels keep 56 horses?
9. If £15 12s. pay 16 labourers for 18 days, how many labourers will £35 2s. pay for 24 days?
10. If 40 bushels of corn serve 12 horses 37 days, how many days would 195 bushels of corn serve 9 horses?
11. If £100 in 2 years gain £12 interest, what principal will gain £6 15s. in $4\frac{1}{2}$ months?
12. If 8 men earn £9 wages for 5 days' work, how much would 32 men earn for 24 days' work at the same rate?
13. If 12 men can complete a piece of work in 15 days, working 6 hours a day, how many can do it in $85\frac{1}{2}$ days, working $12\frac{12}{19}$ hours a day?
14. If 3 persons are boarded 4 weeks for £7, how long should 14 persons be boarded for £112?
15. If 7 men earn £9 10s. 6d. in $10\frac{1}{2}$ days, what sum will 28 men earn in $31\frac{1}{2}$ days?
16. If 30 cwt. are carried 15 miles for £5 8s. 9d., how far ought 8 cwt. to be carried for £29?
17. If £100 will pay the expenses of 5 persons for 22 w. 6 d., how long would 12 persons be supported by £150 under similar circumstances?
18. If 20 men can perform a piece of work in 12 days, required the number of men who could perform another piece of work 3 times as great in $\frac{1}{5}$th part of the time?
19. If 2 men in a tour of 3 months spent £140, how much at

the same rate would it cost a party of five persons, travelling 11 months?
20. If 7 masons can erect a certain piece of wall in $20\frac{5}{8}$ days of $9\frac{3}{5}$ hours each, how long would it take 3 masons to do $2\frac{3}{4}$ of the same work, reckoning 12 hours to the day?
21. If 12 persons spend £160 in 4 months, how many will £853 6s. 8d. last for 8 months?
22. If the 8d. loaf weighs 48 oz. when wheat is at 54s. per quarter, what should be the price of wheat when the 6d. loaf weighs 32 oz. 8 dwt.?
23. If 6 horses in 2 days, of 12 hours each, plough 17 acres, how much will 2 horses plough in 16 days, of 8 hours each?
24. If 12 men build 24 roods of wall in 30 days, working 8 hours a day, how many hours per day must 18 men work to build 72 roods in 40 days?
25. If 18 men eat 16s. worth of bread in 3 days, when wheat is at 54s., what value of bread will 45 men eat in 27 days, when wheat is at 45s.?
26. If 7 men can mow 84 acres in 12 days of $8\frac{1}{4}$ hours each, how many can be mowed by 20 men in 11 days of $7\frac{4}{5}$ hours each?
27. If 3 men can mow 7 acres of grass in 5 days of 9 hours each, in how many days of 8 hours each will 5 men mow $17\frac{1}{2}$ acres?
28. If 18 men can dig a trench, 30 yards long, in 24 days, by working 8 hours a day, how many will dig a trench, 60 yards long, in 64 days, working 6 hours a day?
29. If 6 men will dig a trench, 15 yards long and 4 broad, in 3 days of 12 hours each, in how many days of 8 hours each will 8 men dig a trench 20 yards long and 8 broad?
30. If 6 iron bars, 4 ft. long, 3 in. broad, and 2 in. thick, weigh 288 lbs., how much will 15 weigh, each $6\frac{1}{2}$ ft. long, 4 in. broad, and 3 in. thick?
31. If 8 men can dig a trench 100 ft. long, 3 ft. broad, and 4 ft. 6 in. deep, in 9 days, how many will be required to dig a trench 80 ft. long, 5 ft. broad, and 2 ft. deep, in $5\frac{1}{3}$ days?

32. If 5 men can reap a field, in length 800 ft. and breadth 700 ft., in 3½ days of 14 hours each, in how many days of 12 hours each will 7 men reap a field of 1800 ft. by 960 ft.?
33. A beam 16 feet long, 2¼ broad, and 8 inches thick, weighs 1280 lbs.: what must be the length of another beam of the same material, whose breadth is 3¼ feet, thickness 7½ inches, and weight 2028 lbs.?
34. If 100 men, in 6 days of 10 hours each, can dig a trench 200 yards long, 3 wide, and 2 deep, in how many days of 8 hours long will 180 men dig a trench of 360 yards long, 4 wide, and 3 deep?
35. If the wages of 25 men amount to £76 13s. 4d. in 16 days, how many men must work 24 days to receive £103 10s., the daily wages of the latter being one-half those of the former?
36. If 14 men, working 9 hours a day, dig a trench 140 yards long, 2 wide, and 2 deep, in 4 days, in how many days will 36 men, working 8 hours a day, dig a trench 280 yards long, 4 wide, and 2 deep?
37. The value of the wall-paper required for a room, supposing it ¾ yard wide, and 4½d. a yard, is £2 3s. 1½d.: what would it come to, if it were 2 feet wide, and 4d. a yard?
38. If a certain number of workmen can do a piece of work in 25 days, in what time will 1⅔ of that number of men do a piece of work twice as great, supposing 2 of the first set can do as much work in an hour as 3 of the second set can do in 1½ hours, and that the second set work half as long per day as the first set?
39. If 120 bricks, 16 inches long, 10 wide, and 8 thick, cost 6s. 8d., what length must bricks be, when 160 of them, 12 inches wide, and 7 thick, cost 8s. 9d.?
40. If three gold-diggers, working 6 hours a day, for 10 days, earn £80, when the nuggets occur, on the average, 2 in every cubic yard, how many so occur, when two diggers, working 5 hours a day, for 9 days, earn £90; the second party being only half as skilful as the first, but working three times as hard, the average size of the new nuggets

being double that of the old, and the price of gold having risen 50 per cent?

XII (1).

Find the Simple Interest, and the Amount, on the following Principals, put out for a certain time, at a certain rate per cent.

	Principal.			Time.		Rate.
	£	s.	d.	Years	Months	
1.	180	0	0	15	0	3
2.	65	0	0	10	0	$2\frac{1}{2}$
3.	375	0	0	18	0	$3\frac{1}{3}$
4.	15	0	0	2	6	4
5.	2000	0	0	7	0	$2\frac{1}{4}$
6.	1115	0	0		6	6
7.	80	0	0	12	0	4
8.	1660	0	0	2	0	$\frac{1}{4}$
9.	145	0	0	4	0	$3\frac{1}{2}$
10.	3295	0	0	8	0	$2\frac{1}{2}$
11.	45	10	0	3	0	5
12.	160	10	0	10	0	$1\frac{3}{4}$
13.	1065	5	0	5	0	4
14.	1000	5	0		5	8
15.		15	0	4	6	$2\frac{1}{2}$
16.	325	10	0	20	0	$2\frac{1}{4}$
17.	245	15	0	8	0	$8\frac{1}{3}$
18.	211	2	6	2	8	10
19.	40	1	4	2	6	15
20.	805	11	8	12	0	5
21.	105	2	6	8	0	$6\frac{1}{4}$
22.	84	3	4	4	0	$3\frac{1}{2}$
23.	650	8	4	5	4	$2\frac{1}{4}$
24.	72	1	8		3	$2\frac{1}{2}$
25.	320	8	4	8	0	$3\frac{1}{4}$
26.	450	16	3	15	0	$1\frac{1}{3}$
27.	5	5	5		9	10
28.	280	5	10	6	3	4
29.	1565	3	4	5	0	$1\frac{1}{4}$
30.	2050	7	6	10	0	$2\frac{1}{2}$
31.	165	10	10	2	6	5
32.	85	18	4	8	3	5
33.	1001	0	10	2	8	$4\frac{1}{2}$

	Principal.			Time.		Rate.
	£	s.	d.	Years	Months	
34.	5	6	$5\frac{1}{2}$	2	6	8
35.	48	3	$6\frac{1}{2}$		10	9
36.	525	8	$10\frac{1}{4}$	2	8	$4\frac{1}{2}$
37.	120	9	$4\frac{1}{2}$	5	5	24
38.	116	12	2	3	4	$6\frac{3}{7}$
39.	210	1	$0\frac{1}{2}$	3	6	$1\frac{1}{7}$
40.	3050	19	$3\frac{1}{4}$	2	11	$1\frac{5}{7}$

XII (3).

Find the Present Value, and Discount, of the following debts.

	Debt.			Time it has to run.		Rate of Interest
	£	s.	d.	Years	Months	Per cent.
1.	546	5	0	5	0	3
2.	734	8	0	9	0	4
3.	150	8	0	4	0	5
4.	918	0	0	10	0	2
5.	676	13	4		6	3
6.	63	6	8	4	0	$\frac{1}{3}$
7.	300	16	0	20	0	1
8.	2030	0	0		3	6
9.	1011	9	9	2	0	4
10.	723	13	4	6	0	5
11.	1022	19	6	1	0	8
12.	254	1	2	3	0	$2\frac{1}{2}$
13.	3215	16	8	4	6	3
14.	508	2	4	2	6	3
15.	423	2	6	3	8	$3\frac{1}{2}$
16.	2204[4]	4	0	12	0	3
17.	616	14	0	4	6	$2\frac{1}{4}$
18.	813	9	0	1	4	$4\frac{3}{4}$
19.	1102	2	0	24	0	$1\frac{1}{2}$
20.	227	5	9	17	0	$3\frac{1}{2}$
21.	254	1	2	3	0	$2\frac{1}{2}$
22.	1015	0	0	3	0	$\frac{1}{2}$
23.	1607	18	4	9	0	$1\frac{1}{2}$
24.	1626	18	0		8	$9\frac{1}{2}$
25.	1233	8	0	2	3	$4\frac{1}{2}$

4. Dodsgon crossed out 4 and wrote in 3.

| | Debt. | | | Time it has to run. | | Rate of Interest |
	£	s.	d.	Years	Months	Per cent.
26.	1549	11	0	5	8	$3\frac{4}{5}$
27.	3034	9	3	4	0	2
28.	1638	15	0	2	6	6
29.	1270	5	10	5	0	$1\frac{1}{2}$
30.	224	16	3	7	0[5]	$3\frac{3}{4}$
31.	211	11	3	3	6	$3\frac{2}{3}$
32.	519	1	$1\frac{1}{2}$	5	0	$3\frac{1}{4}$
33.	774	15	6	19	0	$1\frac{2}{15}$
34.	105	6	$0\frac{1}{2}$	4	3	$3\frac{1}{2}$
35.	374	6	$0\frac{1}{4}$	3	6	$4\frac{1}{4}$
36.	1038	2	3	3	3	5
37.	635	2	11	3	0	$2\frac{1}{2}$
38.	1157	7	$4\frac{1}{2}$	4	10	$4\frac{7}{8}$
39.	819	7	6	10	0	18
40.	387	7	$7\frac{1}{5}$	20	0	$\frac{3}{5}$

XIV (4).

When the ( ) per cents are selling at (M), what must be given for (S) Stock?[6]

| | M. | S. | | |
		£	s.	d.
1.	90	750	0	0
3.	88	180	0	0
5.	92	350	0	0
7.	105	700	0	0
9.	90	220	0	0
11.	86	850	0	0
13.	105	450	0	0
15.	86	750	0	0
17.	70	770	0	0
19.	99	660	0	0
21.	88	2460	0	0
23.	$52\frac{1}{2}$	900	0	0
25.	99	106	5	0

5. Dodgson crossed out 7 and wrote in 0.
6. Dodgson wrote the rule $\frac{MS}{100}$ in the margin.

	M.	S.		
		£	s.	d.
27.	$94\frac{3}{5}$	1112	10	0
29.	$95\frac{1}{3}$	2767	10	0
31.	$91\frac{7}{8}$	350	0	0
33.	$102\frac{2}{3}$	2767	10	0
35.	$78\frac{3}{4}$	1012	10	0
37.	$96\frac{3}{4}$	871	10	0
39.	$95\frac{1}{3}$	2818	18	4

XIV (9).

What is the annual income obtained by investing (C) in the (R) per cents, selling at (M)?[7]

	C.			R.	M.
	£	s.	d.		
1.	1008	0	0	3	84
3.	5580	0	0	4	93
5.	945	0	0	$4\frac{1}{2}$	105
7.	792	0	0	3	88
9.	1188	0	0	3	81
11.	3500	0	0	$3\frac{1}{2}$	98
13.	735	0	0	$3\frac{1}{4}$	98
15.	1638	0	0	$4\frac{1}{2}$	$93\frac{3}{5}$
17.	637	10	0	$5\frac{1}{2}$	85
19.	201	12	0	3	84
21.	787	10	0	4	105
23.	3000	0	0	3	$84\frac{3}{8}$
25.	3003	0	0	$2\frac{1}{2}$	$49\frac{1}{2}$
27.	2214	0	0	4	90
29.	4788	0	0	$3\frac{1}{2}$	105
31.	567	4	0	3	86
33.	2164	16	0	$3\frac{1}{2}$	88
35.	3500	0	0	$3\frac{1}{2}$	96
37.	3500	0	0	3	$82\frac{1}{2}$
39.	2000	0	0	3	$88\frac{1}{2}$

7. Dodgson wrote the rule $\dfrac{CR}{M}$ in the margin.

XIV (11).

What is the change in income produced by transferring (S) Stock, in the (R) per cents, selling at (M), into the (R') per cents, selling at (M')?[8]

	S.			R.	M.	R'.	M'.
	£.	s.	d.				
1.	5000	0	0	3	72	4	90
3.	2500	0	0	3	92	4	115
5.	1800	0	0	3	88	4	99
7.	4000	0	0	4	96	3	80
9.	1000	0	0	4	90	3	72
11.	750	0	0	7	102	5	90
13.	20000	0	0	2	46	3	92
15.	2300	0	0	$3\frac{1}{3}$	96	$3\frac{1}{2}$	$103\frac{1}{2}$
17.	11000	0	0	4	92	5	110
19.	2950	0	0	3	81	5	108
21.	750	0	0	$3\frac{1}{2}$	98	4	105
23.	850	0	0	5	90	$6\frac{1}{2}$	102
25.	1500	0	0	3	84	$4\frac{1}{2}$	120
27.	2000	0	0	3	90	$3\frac{1}{2}$	96
29.	2400	0	0	$3\frac{1}{2}$	$85\frac{1}{2}$	4	96
31.	2200	0	0	3	$97\frac{1}{2}$	$4\frac{3}{4}$	95
33.	29000	0	0	$3\frac{1}{2}$	99	3	$90\frac{5}{8}$
35.	1066	13	4	4	87	5	96
37.	240	13	4	$3\frac{1}{2}$	$86\frac{1}{2}$	5	95
39.	3000	0	0	3	$89\frac{3}{8}$	$3\frac{1}{2}$	$98\frac{1}{4}$

XVII.

1. If a person accepts £247 1*s.* 8*d.* as present payment of £252 0*s.* 6*d.* due 4 months hence, at what rate per cent. does he allow discount?
2. If £100 in 2 years gain £12 interest, what principal will gain £6 15*s.* in $4\frac{1}{2}$ months?
3. Find the present worth and discount on £226 1*s.* 11*d.*, due 7 months hence, at $4\frac{3}{4}$ per cent.

---

8. Dodgson wrote the rule $\dfrac{S(R'M - RM')}{100M'}$ in the margin.

4. What is the present worth of £131 12s. 6d., payable in $\frac{1}{4}$ of a year, at 5 per cent?
5. A person holding 50 shares in the London and North-Western Railway, sells out at 170; what income would he have by buying into the $3\frac{1}{2}$ per cents. at $93\frac{1}{3}$?
6. A person has £2950 in the Danish 3 per cents., at $75\frac{1}{4}$, which he transfers to the Russian 5 per cents., at $110\frac{5}{8}$; required the alteration in his income.
7. Two persons buy respectively, with the same sums, into the 3 and $3\frac{1}{2}$ per cents., and get the same amount of interest; the 3 per cents. being at 75, at what are the $3\frac{1}{2}$ per cents.?
8. The 3 per cent. stock is at $98\frac{3}{8}$, and the $3\frac{1}{2}$ per cents. at $106\frac{1}{4}$; into which is it most advantageous to buy?
9. How must nutmegs, which cost 18s. 9d. per lb., be sold, so as to gain 16 per cent.?
10. A plumber sold 96 cwt. of lead for £109 2s. 6d., and gained at the rate of $12\frac{1}{2}$ per cent.; what did it cost him per cwt.?
11. If cheese, which was bought at £1 2s. 11d. per cwt., be sold at £1 5s. 8d., what is the gain per cent.?
12. If eggs be bought at the rate of 5 a penny, how many should be sold for 7d., to gain 40 per cent.?
13. Divide 1065 into parts, which shall be to each other in the ratio of 3, 5, 7; and also into parts which shall be in the ratio of $\frac{1}{3}, \frac{1}{5}, \frac{1}{7}$.
14. If 3 lbs. of tea be worth 4 lbs. of coffee, and 6 lbs. of coffee be worth 20 lbs. of sugar; how many lbs. of sugar can be had for 9 lbs. of tea?
15. If I lose $1\frac{1}{4}d$. in 3s. 4d., how much do I lose per cent.?
16. A general levied a contribution of £870 on four villages, containing 250, 300, 400, and 500 inhabitants respectively; what must they each pay?
17. Divide £16 0s. 10d. among 4 persons in the proportion of the fractions $\frac{1}{2}, \frac{1}{3}, \frac{1}{4}, \frac{1}{5}$.
18. If 7 oxen are worth 42 sheep, and 3 sheep cost £10, what must be given for 100 oxen?
19. In the centigrade thermometer the freezing point is zero, and the boiling point is 100°; in Fahrenheit's the freezing

point is 32°, and the boiling point is 212°; what degree C, corresponds to 68 F.?

20. How much water must be added to a cask, containing 40 gallons of spirits at 13s. 8d., to reduce the price to 10s. 6d.?
21. A can do a piece of work in 10 days, which B could do in 13; in what time would they do it together?
22. If A can do a piece of work in 10 days, and A and B can do it together in 7, in what time would B alone do it?
23. A cistern has two pipes, by one of which it may be filled in 40 min., and by the other in 50 min.; it has also a discharging pipe, by which it may be emptied in 25 min. If all these three were opened together, in what time would the cistern be filled?
24. If 3 men, 5 women, or 8 children, could do a quantity of work in $26\frac{1}{2}$ hours, in what time will 2 men, 3 women, and 4 children complete it?
25. How much stock must be bought at 88 per cent., in order that, by selling out when the stocks are at 90, 20 guineas may be gained?

23. Arithmetic. I.

[1870–74?: Carlson]

This four-page pamphlet, headed "Arithmetic, I," consists of templates for thirty-three examination questions, divided into five sections A through E. The problems are stated in skeleton form so that the numerical parts requiring calculation can be inserted easily and different ones substituted to yield new problems. This piece, unsigned and undated, does not appear in the standard bibliographic references of Dodgson's work. The question types follow closely those in *Examples in Arithmetic* (item 22).

※

[*You are requested to answer   of the   questions marked* ✢, *and to return this paper with your answers.*]

A. INTEGRAL NUMBERS.

1. Multiply   by
2. Divide   by
3. Find the Greatest Common Measure of
4. Find the Least Common Multiple of

B. VULGAR FRACTIONS.

1. Add together
2. Subtract   from
3. Multiply together
4. Divide   by
   and   by
5. Reduce to their lowest terms
6. Simplify the expressions

ARITHMETIC. I.

C. DECIMAL FRACTIONS.

1. Add together
2. Subtract   from   ; and   from
3. Multiply together
4. Divide   by
5. Reduce the following vulgar fractions to decimals:

Reduce the following decimals to vulgar fractions in their lowest terms:

D. CONCRETE NUMBERS.

1. Reduce   to   and   to
2. Multiply   by   ; and   by
3. Divide   by   ; and   by
4. Reduce   to the fraction of   ; and   to the fraction of
5. Reduce   to the decimal of   ; and   to the decimal of
6. Find the value of   of   ; and of   of
7. Find the value of   of   and of   of
8. Find by Practice the cost of   at £   each.
9. What will it cost to carpet a room   feet long, and   feet wide, with carpet   feet wide, at £   per

E. RULE OF THREE, &c..

1. If   cost £   : what will   cost?
2. If   cost £   : how much can be bought for £
3. A bankrupt's debts amount to £   and his assets to £   : how much can he pay in the pound?
4. If   men can mow   acres in   days: how many acres can   men mow in   days?

5. If   men can mow   acres in   days: how long will   men be in mowing   acres?
6. If £   are put out to simple interest for   years, at   per cent: what will be the interest, and the amount?
7. If £   are due   years hence, interest (simple) being reckoned at   per cent: what is the present value, and the discount?
8. Find the square root of

24. Arithmetic. II.

[1870–74?: Carlson]

These three pages, headed "Arithmetic. II," consist of twenty templates for examination questions divided into three sections D through F. This newly discovered piece is not in the *Lewis Carroll Handbook* and is unsigned and undated. Since the numbering of the questions is continuous with that of sections D and E of "Arithmetic. I" (item 23) and the format is the same, it extends the contents of that pamphlet.

❦

[*You are requested to answer*   *of the*   *questions marked* ✠, *and to return this paper with your answers.*]

D. CONCRETE NUMBERS.

10. What will it cost to paper a room   feet long,   feet wide, and   feet high, with paper   feet wide, at £   per
11. What will it cost to dig a trench   feet long,   feet wide, and   feet deep, at £   per cubic
12. If   apples are worth   pears, and   pears are worth   oranges, and   oranges are worth   lemons, and   lemons are worth   nuts, and if nuts are sold at   a penny, what is the price of apples?

E. RULE OF THREE, &C.

[N.B. *In the following questions it is required to find the value of the quantity denoted by x.*]

9. A bankrupt's debts amount to £   , and his assets to £   : he can pay   in the pound.

10. If   men can dig a trench   feet long,   feet wide, and   feet deep, in   days, working   hours per day: then-   men can dig one   feet long,   feet wide, and   feet deep, in   days, working   hours per day.

11. If a grocer mixes   lbs. of sugar at £   per lb. with   lbs. at £   per lb: the mixture is worth £   per lb.

12. If £   are put out for   years, at   per cent, simple interest: the interest will be £   and the amount £

13. If £   are put out for   years, at   per cent, compound interest: the interest will be £   and the amount £

14. If £   are due   years hence, interest being reckoned at   per cent: the present value is £   and the discount £

15. If £   be invested in the   per cents, selling at   : the quantity of stock obtained will be £   and the yearly income £

16. If £   stock in the   per cents, selling at   , be sold out: the money obtained will be £

17. Find the square root of   the last to   places of decimals.

F. DEFINITIONS, &C.

1. Explain the terms
2. Give the rules for finding G.C.M. and L.C.M.
3. Give the rules for multiplication and division of decimals.
4. Give the rules for reducing circulating decimals to vulgar fractions.
5. Give the rule for Simple Interest.
6. Give the rule for Compound Interest.
7. Give the rule for Discount.
8. Give the tests for a number being divisible by

25. *Arithmetic.*

[Undated: Carlson]

These two pages headed "Arithmetic" consist of templates for eleven examination questions that are similar to but simpler than those in "Arithmetic. I." This newly discovered item, unsigned and undated, is not in the standard bibliographic references of Dodgson's work. [Matriculation, Ch. Ch.] appears at the bottom of the second page.

I.

Find the G.C.M. of   and   ; and the L.C.M. of   ,   ,   , and   .

II.

Add together   ; subtract   from   ; multiply   by   ; and divide   by   .

III.

Find the value of   of   ; and reduce   to the fraction of

IV.

Multiply together   ; and divide   by   by, , and by

V.

If   cost   , what will   cost?

VI.

Find by practice the cost of   articles, at   each.

VII.

What will it cost to carpet a room   feet long, and   feet wide, with carpet   feet wide, at   a yard?

VIII.

Find the value of   of £   , and reduce   to the decimal of £

IX.

What will   amount to, in   years, at £   per cent, simple interest?

X.

Find the square root of   , of   , and of

XI.

Reduce to vulgar fractions, in their lowest terms,

[*Matriculation, Ch. Ch.*]

26. *Practical Hints on Teaching*
LONG MULTIPLICATION WORKED WITH A SINGLE LINE OF FIGURES
[1879: LCH 101: Colindale]

In this letter to the editor, Dodgson provides an example to illustrate his new method of multiplication. In a fragment of "Curiosa Mathematica, Part III," published in *The Lewis Carroll Picture Book,* Dodgson wrote at length about this method, the idea for it having occurred to him for the first time on 19 September 1879.[1]

TO THE EDITOR OF THE EDUCATIONAL TIMES.

SIR,—If the following brief method of working Long Multiplication should prove to be new, I hope you may think it worth publishing:—

Suppose we wish to multiply 56248 by 3726. We set the sum in the usual way, thus:—

$$\frac{56248}{3726}$$

We then write out the upper line, *backwards,* on the lower edge of a separate slip of paper, placing a mark over the unit-digit, as a guide to the eye: with this slip we cover the upper line of the given sum, bringing the marked digit over the unit of the lower line, thus:—

$$\frac{\overline{8}4265}{3726}$$

Educational Times, vol. XXXII, (1 November 1879), 307–8.
1. Stuart Dodgson Collingwood, *The Lewis Carroll Picture Book* (London: T. Fisher Unwin, 1899; reprint [as The *Unknown Lewis Carroll*], New York: Dover, 1961), 240–46.

We then take the product of the digits which are in the same vertical line (viz., 8, 6); this gives us 48; we write the unit of this (viz., 8) vertically under the scored digit, and "carry" the 4, thus:—

$$\frac{\overline{8}4265}{8}$$
$$3726$$

We then shift the slip one place to the left, thus:—

$$\frac{\overline{8}4265}{8}$$
$$3726$$

We then add together the carried digit and the products of the digits which are in the same vertical lines, and write the result as before. The mental process being, "4 + 24 = 28, + 16 = 44; set down 4 and carry 4."

$$\frac{\overline{8}4265}{48}$$
$$3726$$

We then shift the slip again, and proceed as before; the mental process being, "4 + 12 = 16; + 8 = 24; + 56 = 80; set down 0 and carry 8."

$$\frac{\overline{8}4265}{048}$$
$$3726$$

We then shift the slip again, and so on; the last step being reached when the sum stands thus, with 5 to carry:—

PRACTICAL HINTS ON TEACHING 275

$$\overline{8}4265$$
$$\underline{3726}$$
$$9580048$$

Hence the mental process of the last step is, "5 + 15 = 20; set it down." We then remove the slip, and the result appears thus:—

$$56248$$
$$\underline{3726}$$
$$209580048$$

A similar method will serve for multiplying decimals: all we have to remember is, to bring the marked digit of the slip vertically over whatever decimal place we wish to carry the working to. For example, if we wish to multiply together ·63624 and ·25873; and if, in order to have the answer correct to 3 places, we wish to carry the working to 4 places, we set the sum thus:—

$$0·63624$$
$$\overline{0·25873}$$

We then write 426360 on a separate slip of paper, and place it so that its marked digit comes vertically over the 4th decimal place in the answer, thus:—

$$42\ 63\overline{6}0$$
$$\underline{0·25873}$$

The mental process of the first step will be "0 + 48 = 48; + 15 = 63; + 12 = 75; set down 5 and carry 7."

$$42\ 63\overline{6}0$$
$$\underline{0·25873}$$
$$5$$

We then shift the slip to the left and proceed as before, the last step being reached when the sum stands thus, with 1 to carry:—

$$\frac{\begin{array}{r}42636\,\overline{0}\\ 0{\cdot}25873\end{array}}{\cdot\quad 635}$$

Hence the mental process of the last step is "1 + 0 = 1; set it down." We then remove the slip, and the result appears thus:—

$$\frac{\begin{array}{r}0{\cdot}63624\\ 0{\cdot}25873\end{array}}{\cdot 1635}$$

Hence the answer, correct to 3 places, will be ·164. This method seems to me not only to save space and time, but also to avoid the risk of mistakes involved in writing all the intermediate lines of figures required in the old method, as well as the constant risk of losing one's place while carrying the eye obliquely from one figure to another figure several rows above it.

    Your obedient servant,
    CHARLES L. DODGSON,
    Senior Student and Mathematical Lecturer
    of Christ Church, Oxford.

27. Divisibility by Seven

[1884: LCAT 456: Princeton]

The proof sheet in the Parrish Collection carries Falconer Madan's signature as well as the comments by him, "*Not* Mathematical Questions (nor Educational Times), I *think*. Not? English Mechanic." It is signed, "C. L. Dodgson, Ch. Ch., Oxford." "Divisibility by Seven" is not included in the *Lewis Carroll Handbook*. The citation in *Lewis Carroll at Texas* indicating it as a proof of a contribution to the *Educational Times* is incorrect, as is the 1885 date obtained from an earlier edition of the LCH. Dodgson submitted it as a response to Letter 1274 posed by Mr. Askew that was published in *Knowledge* on 30 May 1884. Dodgson's letter, number 1324, appeared in *Knowledge* on 4 July 1884.[1] He wrote in his diary on 10 June, "Sent a corrected proof of my letter on Divisibility by 17 to *Knowledge*."[2] The 17 is incorrect, a transcription error. However, in Dodgson's diary entry of 3 June he wrote about inventing rules for dividing numbers by 17 and 19.

❦

[ ]—Mr. Askew, in 1274, May 30, asks for a proof of a method for ascertaining the divisibility of a number by 7, which he states to have been discovered by Mr. Rickard, of Birmingham. Probably many have discovered it: my father did, for one, and taught it to me some thirty years ago. The test-number is equally useful for 7, 11, and 13. The method, as worked by my father, gives, in the case of a number divisible by all three factors, the other factor as well, without further labour: and in this respect it has an advantage over that of Mr. Rickard.

If a number, N, be marked off from the right-hand end in periods of three digits; and if *a, b, c,* &c., be the periods; and if M be the difference between the sums of the alternate periods; we have, writing *r* for 1000,

1. *Knowledge,* vol. VI, 15.
2. *The Diaries of Lewis Carroll,* edited by Roger L. Green (New York: Oxford University Press, 1954), vol. II, 426.

$$N = a + br + cr^2 + dr^3 + \&c.$$
$$M = a - b + c - d + \&c.$$
$$\therefore N - M = b(r + 1) + c(r^2 - 1) + d(r^3 + 1) + \&c.$$

and is divisible by $(r + 1)$; hence, if M be divisible by $(r + 1)$ or any factor of it, so also is N. And in this case $r + 1 = 1001 = 7 \times 11 \times 13$.

My father's rule was to set the right-hand period under the next, and subtract, setting the remainder under the next, and so on. In the last period, the subtraction is *downwards* if the lower number be the larger. In this instance, since we have 1 to carry into the last period, the 931 must be read as 932. The ultimate remainder, 924, is the test-number; and, since this is divisible by 7 and 11, so also is the whole number.

8	026	518	423
031	095	423	
924			

If the test-number chanced to be zero, the second line would be the quotient produced by dividing the given number by 1001; *i.e.*, it is the factor remaining after dividing out 7, 11, and 13. For let us call the second line "V;" writing three ciphers at the end, we get 1000V; and we know that, if this be deducted from the upper line, the remainder = V. Hence $N = 1001V = 7 \times 11 \times 13 \times V$. In the above example, if the left-hand period were 932 instead of 8, the test-number would be zero.

If the periods be *single* digits, *i.e.*, if $r = 10$, we get a test for divisibility by 11, and at the same time the quotient after dividing out 11. The rule is to set the last digit under the next, and subtract, setting the remainder under the next, and so on. In this instance the test-number = 0; hence the given number = 11×5852053.

```
6 4 3 7 2 5 8 3
0 5 8 5 2 0 5 3
```

With periods of two digits, we get a test for divisibility by 101; and so for four or more digits. C. L. DODGSON
Ch. Ch., Oxford.

P.S.—The sum of *all* the periods gives us, for periods of 1, 2, 3, &c., digits, a test for divisibility by 9, 99, 999 (= 27 × 37), &c., or for any factors of these numbers. This method may also be worked by a rule analogous to that given above; *e.g.,* to test for 999, mark off in periods of three, write 000 over the right-hand period, and subtract, writing the remainder over the next, and so on. Hence, also, if the test-number chanced to be zero, the upper line (omitting the 000) would be the quotient produced by dividing the given number by 999.

Probably similar rules may be made for most primes. I have myself made fairly simple rules for 17 and for 19; but such processes are rather curious than useful.

28. To Find the Day of the Week for Any Given Date

[1887: LCH 197]

This short article is included here because it is essential to a proper view of the scope of Dodgson's work on problems related to divisibility by seven. It is the only one of the three articles in *Nature* that he submitted as Lewis Carroll. Dodgson referred to it in his diary entry of 8 March 1887, "Discovered a Rule for finding the day of the week for any given day of the month. There is less to remember than in any other Rule I have met with."[1] This somewhat bland statement belies the difficulties that calendar problems posed in the nineteenth century. Reliable rules were few; most determinations were dependent on almanacs.[2]

※

Having hit upon the following method of mentally computing the day of the week for any given date, I send it you in the hope that it may interest some of your readers. I am not a rapid computer myself, and as I find my average time for doing any such question is about 20 seconds, I have little doubt that a rapid computer would not need 15.

Nature, vol. 35 (31 March 1887), 517.

1. *The Diaries of Lewis Carroll*, edited by Roger L. Green (New York, Oxford University Press, 1954), vol. II, 449. Dodgson's interest in methods to aid remembering are well known. His mnemonic system, "Memoria Technica. for Numbers," published in 1877 is fully described in Morton N. Cohen's article, "Lewis Carroll's *Memoria Technica*," *Library Chronicle of the University of Texas at Austin*, new series no. 11 (1979), 77–88.

2. John H. Conway's article, "Tomorrow is the Day after Doomsday," appearing in *Eureka*, October 1973, 28–31, details his rule—which is a simplification of Dodgson's rule—for working out the day of the week for any given date. Martin Gardner, in his copy of the article, corrects the erroneous date Conway cited for the item in *Nature*. In Dodgson's time, an authoritative text on the secular and ecclesiastical calendars was *The Book of Almanacs* by Augustus De Morgan, published in 1851 by Taylor, Walton and Maberly in London.

TO FIND THE DAY OF THE WEEK

Take the given date in 4 portions, viz. the number of centuries, the number of years over, the month, the day of the month.

Compute the following 4 items, adding each, when found, to the total of the previous items. When an item or total exceeds 7, divide by 7, and keep the remainder only.

The Century-Item.—For Old Style (which ended September 2, 1752) subtract from 18. For New Style (which began September 14) divide by 4, take overplus from 3, multiply remainder by 2.

The Year-Item.—Add together the number of dozens, the overplus, and the number of 4's in the overplus.

The Month-Item.—If it begins or ends with a vowel, subtract the number, denoting its place in the year, from 10. This, plus its number of days, gives the item for the following month. The item for January is "0"; for February or March (the 3rd month), "3"; for December (the 12th Month), "12."

The Day-Item is the day of the month.

The total, thus reached, must be corrected, by deducting "1" (first adding 7, if the total be "0"), if the date be January or February in a Leap Year: remembering that every year, divisible by 4, is a Leap Year, excepting only the century-years, in New Style, when the number of centuries is not so divisible (*e.g.* 1800).

The final result gives the day of the week, "0" meaning Sunday, "1" Monday, and so on.

EXAMPLES

1783, *September* 18

17, divided by 4, leaves "1" over; 1 from 3 gives "2"; twice 2 is "4."

83 is 6 dozen and 11, giving 17; plus 2 gives 19, *i.e.* (dividing by 7) "5." Total 9, *i.e.* "2."

The item for August is "8 from 10," *i.e.* "2"; so, for September, it is "2 plus 3," *i.e.* "5." Total 7, *i.e.* "0," which goes out.

18 gives "4." Answer, "*Thursday.*"

1676, *February* 23

16 from 18 gives "2."
76 is 6 dozen and 4, giving 10; plus 1 gives 11, *i.e.* "4." Total "6."
The item for February is "3." Total 9, *i.e.* "2."
23 gives "2." Total "4."
Correction for Leap Year gives "3." Answer, "*Wednesday.*"

Lewis Carroll

29. Question 9636

[1893: LCH 136: NY Public]

Of Dodgson's contributions to the *Educational Times,* Question 9636, originally appearing in the 1 July 1888 issue, is the only one that had such a large gap—five years—between its initial publication and the publication of a correct solution. Apparently, the reason has to do with the quality of the solution by Professor G. B. M. Zerr. Dodgson disagreed with it, claiming correctly that it is not a proof but merely an instance of the theorem. Remarkably, Dodgson singled out this same problem and its solution in the preface to the second edition of *Curiosa Mathematica, Part II: Pillow Problems* (1893), where he wrote,

> The reader will, I think, be interested to see a curiously illogical solution which has been proposed, by a correspondent of the *Educational Times,* for Problem 61, viz. . . . .
>
> Now if we denote by 'α', the property "which cannot be arranged in A. P. [arithmetic progression], and whose sum is a multiple of 3," and, by 'β', the property "the sum of whose squares is also the sum of another set of 3 squares, the 2 sets having no common term," we see that all, that this writer has succeeded in proving, is that *certain selected Numbers,* which have property 'α', have also property 'β': but this does not prove my Theorem, viz. that *any Numbers whatever,* which have property 'α', have also property 'β'.[1]

Dodgson provides a proof of his theorem in the solutions section of his book as well as some instances, e.g., one set is {1, 4, 4}; another is {5, 2, 2}; the sum of the squares of both sets is 33. The problem itself was thought out on 1/12/81. There is no mention of this problem in his preface to the fourth edition, dated March 1895, yet a correct solution by W. S. Foster, Professor Nash, and others appeared in *MQS*, vol. LXI, 1894, p. 86.[2]

Mathematical Questions and Solutions from the "Educational Times," vol. LIX, W. J. C. Miller, ed. (London: Francis Hodgson, 1893), 94–95; vol. LXI (1894), 86.

1. C. L. Dodgson, *Curiosa Mathematica, Part II: Pillow Problems,* 4th ed. (London: Macmillan, 1895; reprint, New York: Dover, 1958), x–xi.

2. This solution is not mentioned in the entry for Question 9636 in LCH. However, R. C. Archibald does include it in his "Bibliography of Lewis Carroll: Additions," *Notes and Queries* 179 (1940), 134.

284　　　　　　　　　　ITEM 29

(CHARLES L. DODGSON, M.A.)—If 3 numbers, not in arithmetical progression, be such that their sum is a multiple of 3, prove that the sum of their squares is also the sum of another set of 3 squares, the two sets having no common term.

Solution by Professor G. B. M. Zerr.

Let $3m, 21m, 30m$ be the three numbers; then we have

$$3m + 21m + 30m = 3 \times 18m.$$

Also

$$(3m)^2 + (21m)^2 + (30m)^2 = (6m)^2 + (15m)^2 + (33m)^2$$
$$= (5m)^2 + (13m)^2 + (34m)^2 = (10m)^2 + (17m)^2 + (31m)^2$$
$$= (14m)^2 + (23m)^2 + (25m)^2.$$

[Mr. Dodgson states that, in this solution, Prof. Zerr "takes a single special instance of 3 numbers, and seems to think that the theorem, since it is true in this single instance, is thereby proved to be true universally." He submits the following theorem, and asks whether Professor Zerr would consider the appended proof a sound logical one.

"(*Theorem.*) If 3 numbers be such that their sum is a multiple of 7, the sum of their squares is a multiple of 9.

"(*Proof.*) Let $m, 2m, 11m$ be the 3 numbers. Then $m + 2m + 11m = 7 \times 2m$. Also, $m^2 + (2m)^2 + (11m)^2 = 126m^2 = 9 \times 14m^2$."

We shall be glad to have a further solution of the Question.]

Solution by W. S. Foster, Professor Nash, *and others.*

If $a + b + c = 3m$, then $a^2 + b^2 + c^2 = (2m - a)^2 + (2m - b)^2 + (2m - c)^2$, and $2m - a$ is not equal to a or b or c unless those quantities are in Arithmetical Progression.

30. Question 12650

[1895: LCH 136: NY Public]

Two different solutions to this problem were given, one by D. Biddle, Professor Radhakinshuan, *et al.,* and the other by Professor Bourne. Technically, the problem can be classified as a number puzzle (and indeed it is), but its origin lies in Dodgson's concern with divisibility problems. He developed at least three such puzzles, including Question 12650 and "Number-Guessing" (item 31).

He wrote in his diary on 23 January 1895, "And I have improved my number-guessing puzzle."[1] Using some of the ideas in the solutions provided by the respondents, Dodgson went on to create the even better puzzle, "Number-Guessing," which appeared a year and a half later.

※

(C. L. DODGSON, M.A.)—To discover the rule by which the following puzzle is worked. It is best exhibited as a dialogue.

A. Think of a number less than 90.—B. I have done so.

A. Tack on to it any digit you like, from 0 to 9. Which shall it be?—B. I have tacked on a 7.

A. Now divide by 3. What is the remainder?—B. It is 2.

A. Tack on to the quotient any digit you like.—B. I have tacked on 4.

A. Divide by 3. What is the remainder?—B. It is 1.

A. And what is the third figure from the end?—B. It is 8.

A. (Instantly rejoins) Then the number you thought of was 76.

Solution by D. Biddle, Professor Radhakinshuan, *and others.*

Let $10x + y = $ the required number, whilst a, b, c, d, e are the given figures in the order of their declaration. Then we have

---

Mathematical Questions and Solutions from the "Educational Times," vol. LXIII, W. J. C. Miller, ed. (London: Francis Hodgson, 1895), 92–93.

1. The Diaries of Lewis Carroll, edited by Roger L. Green (New York: Oxford University Press, 1954), vol. II, 516.

$$\tfrac{1}{3}(100x + 10y + a - b)$$
$$= 33x + 3y + \tfrac{1}{3}(x + y + a - b) = Q_1,$$

and

$$\tfrac{1}{9}\{1000x + 100y + 10a - 10b + 3(c - d)\}$$
$$= 111x + 11y + a - b + \tfrac{1}{9}\{x + y + a - b + 3(c - d)\}$$
$$= Q_2.$$

We therefore know that $x + y + \{a - b + 3(c - d)\}$ is divisible by 9. Moreover, since 89 has the maximum $(x + y)$, we know that $(x + y)$ cannot exceed 17. Another known fact is that the third figure from the end of Q_2, represented by e, must either be x or $x + 1$. It is x when $x + y \not> 8$, except when the third figure is 0, which only occurs when the required number is 90. When there is no third figure from the end, the required number consists of y only, and does not exceed 8.

Consequently, we are able to give the following rule for instantly declaring the number:—Sum $a - b + 3(c - d)$ and add the resulting digits together. If positive, deduct this from 9 and also from 18. If negative, add it to 0 and also to 9. Then we know that $x + y =$ one or other of the two numbers arrived at, and e shows which, being itself x or $x + 1$, as shown above. The finding of y is then easy. But, unless the memory be exceptionally good, it is advisable to jot down a, b, c, d as they are declared, putting them in the form $a - b + 3(c - d)$, and finding the two possible values of $(x + y)$ in readiness to give the correct answer immediately on the declaration of e. All this can be done by A whilst B is engaged in his work.

[Professor Bourne gives the Solution as follows:—

Let the numbers tacked on be t_1 and t_2, and the respective remainders be r_1 and r_2; subtract r_1 from t_1 and to the result add, if necessary, 10 or 20, so as to make the result a multiple of 3; call this $3p$. Similarly, from t_2 and r_2 obtain $3q$.

Take the digit in the tens' place of $3q$ and subtract it from p, adding 10 or 20 if necessary, as before; call the result $3l$.

Let the figure third from the right in the final quotient be a; then the required original number is 3 $(3a\ +$ ten's digit of $3l$) $+$ ten's digit of $3p$.

The respective quantities a, l, q are the digits, read from the left, of the final quotient, and by working backwards from this, the reason of the rule is seen.

The work can be thus exhibited:—

If t_1 and t_2 are	9 and 8,
and r_1 and r_2 are	1 and 1,
then $3p$ and $3q$ are	18 and 27.
Also p is	6.
Hence, $3l$ is	24.

If the value of a is 7, multiply 7 by 3, and add in the 2 of the 24; multiply the result by 3, and add in the 1 of the 18; the final result is 70.]

31. Number-Guessing

[1896: LCH 280a: Berg]

At the bottom of the sheet a dealer or collector has added, "'Lewis Carroll.' autograph C. L. Dodgson" in pencil. The puzzle is dated "6/2/96." Dodgson's comment at the bottom of the "Number-Guessing" sheet, "This is an improvement on the puzzle containing the direction 'Multiply by 3. . . .'" refers to an earlier number puzzle, dated 3 February 1896, in manuscript form in the Weaver collection. Dodgson referred to this puzzle twice in his diaries; in the earlier entry he wrote that he was devising an original kind of number-guessing puzzle, one that gave a choice of numbers to operate with. In the later one, he had to admit that his puzzle was not insolvable because Edward Frank Sampson had "made out the principle, and guessed one himself."[1]

The "Number-guessing Puzzle" Dodgson constructed on 3 February 1896 had these directions:

1. Think of a number. *Example:* 29
2. Multiply by 3. Is the result even or odd? 87 (odd)
3. If odd, add 1 or 5 or 9. (add 5) 92
   If even, add 4 or 8 or 12.
4. Divide by 2 and add 1. 47
5. Multiply by 3. Is the result odd or even? 141 (odd)
   [increment 1][2]
6. If odd, add 3 or 5 or 7. (add 5) 146
   If even, subtract 2 or 4 or 6.

The Lewis Carroll Circular, no. 2, Trevor Winkfield, ed. (Leeds: privately printed, November 1974), 37.

1. C. L. Dodgson, unpublished Diaries, 3 February 1896, 15 February 1896.

2. LCAT 621. Dodgson's manuscript includes an analysis of the directions as well as a way of remembering the four pairs of answers that indicate the four possible increments:

(0) (1) (2) (3)
EO   OO   EE   OE , namely zErO, fOOt, fEEt, mOrE.

Number-guessing

6/2/96

A. "Think of a number."
B. [thinks of 23]
A. "Multiply by 3. Is the result odd or even?"
B. [obtains 69] "It is odd."
A. "Add 5, or 9, whichever you like."
B. [adds 9, & obtains 78]
A. "Divide by 2, & add 1."
B. [obtains 40]
A. "Multiply by 3. Is the result odd or even?"
B. [obtains 120] "It is even".
A. "Subtract 2, or 6, whichever you like."
B. [subtracts 6, & obtains 114]
A. "Divide by 2, & add 29, or 38, or 47, whichever you like."
B. [adds 38, & obtains 95]
A. "Add 19 to the original number, & tack on any figure you like".
B. [tacks on 5, & obtains 425]
A. "Add the previous result."
B. [obtains 520]
A. "Divide by 7, neglecting remainder."
B. [obtains 74]
A. "Again divide by 7. How often does it go?"
B. "Ten times".
A. "The number you thought of was 23."

[This is an improvement on the puzzle containing the direction "Multiply by 3. Is the result odd or even?" & afterwards "Divide by 2". Four times, in the course of it, B has the choice of certain numbers, & need not say which he uses! I don't think this phenomenon occurs in any other such puzzle.]

"Lewis Carroll." autograph C L Dodgson

7, 8.	Divide by 2; add a number from 50 to 80 for increment 0 40 to 70 for increment 1 30 to 60 for increment 2 20 to 50 for increment 3 and call the result the reserved number.	114
9.	Add 19 to the original number and tack on any figure. [Here 1 tacked on.]	481
10.	Add the reserved number.	595
11.	Divide by 7 twice and name the final quotient.	12
12.	Deduct 5 from the quotient, multiply by 4, add the increment to obtain the original number.	29

32. Question 13614

[1897: LCH 136, Colindale]

Dodgson was intrigued by Christopher Zeller's solution to the problem of finding the day of the week for any given date, which he read in W. W. Rouse Ball's second or third edition of *Mathematical Recreations and Problems*. Compared with the tabular method Dodgson had published in *Nature* ten years earlier (item 28 above), Zeller's algorithm solved the secular calendar problem completely.[1] Working with Zeller's equation in Question 13614, Dodgson worked out a formula for the number of days in any month except February, and he was requesting assistance from the mathematical community by submitting this question. Apparently no reply was ever published, probably because none were received. Archibald writes that the question was submitted "by the late Lewis Carroll" whereas "the earlier problems were proposed by 'C. L. Dodgson, M.A.'"[2] However, Dodgson used his own name for this contribution to the *Educational Times* and was alive when this question was published.[3]

※

(C. L. DODGSON, M.A.)—Zeller's formula for the day of the week corresponding to any given date, viz., the pth day of the qth month of the year N, Old Style, is

$$p + 2q + \{3(q + 1)/5\} + N + \{N/4\} - \{N/100\} + \{N/400\} + 2,$$

Educational Times, vol. L (1 September 1897), 391.

1. The first edition of Ball's book that appeared in February 1892 did not mention Zeller's work. It was included in the second edition published later the same year. A complete discussion of the arithmetization of the calendar will appear in the 1993 *Proceedings of the Canadian Society for the History and Philosophy of Mathematics*, edited by J. J. Tattersall.

2. R. C. Archibald, "Bibliography of Lewis Carroll: Additions," *Notes and Queries*, vol. 179, no. 8 (24 August 1940), 135.

3. In an unpublished letter to the editor of the *Educational Times*, dated 30 July 1887 (MS Columbia), Dodgson asked whether, in previous letters that he had written to him in connection with a problem in Chances, he had used his own or an assumed name.

where {N/4} means the integral number of 4's contained in N. Taking a hint from this, I have succeeded in evolving an *algebraical formula for the number of days in the qth month*. I take a certain algebraical function of q, and divide it by 5, and call the remainder r. Then a certain algebraical function of r will give the number of days in the qth month. This works correctly for March and all following months; and even for January, by calling it the 13th month of the preceding year; but I cannot manage *February*.

33. Brief Method of Dividing a Given Number by 9 or 11

[1897: LCH 284, LCAT 535: Princeton]

These galley sheets are unsigned, but Dodgson is certainly the author. From his correspondence number 98281, which appears at the top, it was printed between 28 September and 30 November of 1897. Dodgson states that he formulated the rule on September 28.[1]

An article having the same title was published in *Nature* on 14 October of the same year (item 34). The two differ somewhat. The article in *Nature* begins with a five-line introductory paragraph and ends with a three-line concluding paragraph, neither of which are in the galley. Furthermore, the galley includes additional material on division by numbers adjacent to a multiple of ten that does not appear in the *Nature* article. The latter was oriented toward teachers of arithmetic in schools; Dodgson's mathematical training pushed him to generalize the methods he described, but he realized that they would be inappropriate at an elementary level and thus omitted them from the published article.

Dodgson believed that his rule was new and he used it in his teaching at a girls' school in Brighton. He recorded these facts in his diary on 13 October 1897 and in a letter to George Davis dated 16 October of the same year.[2] With the letter he also sent the part of his rule that dealt only with finding the quotient and remainder when dividing a given number

---

1. The entry in Dodgson's diary for 28 September 1897 reads:

Die creta notandus! I have actually *superseded* the rules discovered yesterday! My new rules require to ascertain the 9-remainder and the 11-remainder, which the others did *not* require; but the new ones are much the quickest. I shall send them to *The Educational Times*, with date of discovery.

The Diaries of Lewis Carroll, edited by Roger L. Green (New York: Oxford University Press, 1954), vol. II, 539. This date is further corroborated by Dodgson's correspondence numbering system; see Warren Weaver, *Lewis Carroll Correspondence Numbers* (Scarsdale, N.Y.: privately printed, 1940), 14.

2. *The Letters of Lewis Carroll,* edited by Morton N. Cohen (New York: Oxford University Press, 1979), vol. II, 1138.

by 9, intending it for Davis's young daughter, Nellie, whom Dodgson had briefly met on a train ride.[3]

※

Years ago I had discovered the curious fact that, if you put a "0" over the unit-digit of a given number, which happens to be a multiple of 9, and subtract all along, always putting the remainder over the next digit, the final subtraction gives remainder "0," and the upper line, omitting its final "0," is the "9-quotient" of the given number (i.e. the quotient produced by dividing it by 9).

Having discovered this, I was at once led, by analogy, to the discovery that, if you put a "0" *under* the unit-digit of a given number, which happens to be a multiple of 11, and proceed in the same way, you get an analogous result.

In each case I obtained the quotient of a division-sum by the shorter and simpler process of *subtraction:* but, as this result was only obtainable in the (comparatively rare) case of the given number being an exact multiple of 9, or of 11, the discovery seemed to be more curious than useful.

Lately, it occurred to me to examine cases where the given number was *not* an exact multiple. I found that, in these cases, the final subtraction yielded a number which was sometimes the actual remainder produced by division, and which always gave materials from which that remainder could be found. But, as it did not yield the quotient (or only by a very "bizarre" process, which was decidedly longer and harder than actual division), the discovery still seemed to be of no practical use.

But, quite lately, it occurred to me to try what would happen if, after discovering the remainder, I were to put it, instead of a "0," over or under the unit-digit, and then subtract as before. And I was charmed to find that the old result followed: the final subtraction yielded remainder "0," and the new line, omitting its unit-digit, was the required quotient.

3. *Ibid.,* 1140.

BRIEF METHOD OF DIVIDING BY 9 OR 11

Now, there are shorter processes, for obtaining the 9-remainder of the 11-remainder of a given number, than my subtraction-rule (the process for finding the 11-remainder is another discovery of mine). Adopting these, I brought my rule to completion on September 28, 1897 (I record the exact date, as it is pleasant to be the discoverer of a new and, as I hope, a practically useful, truth).

(1) Rule for finding the quotient and remainder produced by dividing a given number by 9.

To find the 9-remainder, sum the digits: then sum the digits of the result: and so on, till you get a single digit. If this be less than 9, it is the required remainder: if it be 9, the required remainder is 0.

To find the 9-quotient, draw a line under the given number, and put its 9-remainder under its unit-digit: then subtract downwards, putting the remainder under the next digit, and so on. If the left-hand end-digit of the given number be less than 9, its subtraction ought to give remainder "0": if it be 9, it ought to give remainder "1," to be put in the lower line, and "1" to be carried, whose subtraction will give remainder "0." Now mark off the 9-remainder at the right-hand end of the lower line, and the rest of it will be the 9-quotient.

Examples. $\quad \dfrac{9/75309\ 6}{83677/3}, \quad \dfrac{9/\ 94613\ 8}{105126/4}, \quad \dfrac{9/58317\ 3}{64797/0}.$

(2) Rule for finding the quotient and remainder produced by dividing a given number by 11.

To find the 11-remainder, begin at the unit-end, and sum the 1st, 3rd, &c., digits, and also the 2nd, 4th, &c., digits; and find the 11-remainder of the difference of these sums. If the former sum be the greater, the required remainder is the number so found: if the former sum be the lesser, it is the difference between this number and 11: if the sums be equal, it is "0."

To find the 11-quotient, draw a line under the given number, and put its 11-remainder under its unit-digit: then subtract, putting the remainder under the next digit, and so on. The final subtraction ought to give remainder "0." Now mark off the 11-remainder at the right-hand end of the lower line, and the rest of it will be the 11-quotient.

Examples.

$$\frac{11/73210\ 8}{66555/3},\quad \frac{11/85347\ 1}{77588/3},\quad \frac{11/59426\ 3}{54023/10},\quad \frac{11/47568\ 4}{43244/0}$$

These new rules have yet another advantage over the rule of actual division, viz. that the final subtraction supplies a *test* of the correctness of the result: if it does not give remainder "0," the sum has been done wrong: if it does, then either it has been done right, or there have been *two* mistakes—a rare event.

Mathematicians will not need to be told that rules, analogous to the above, will necessarily hold good for the divisors 99, 101, 999, 1001, &c. The only modification needed would be to mark off the given number in periods of 2 or more digits, and to treat each period in the same way as the above rules have treated single digits. Here, for example, is the whole of the working needed for dividing a given number of 17 digits by 999 and by 1001:—

$$\frac{999/75410836428139\ \dot{2}14}{75486322750890/104},$$

$$\frac{1001/75410836428139\ \dot{2}14}{75335500927212/\ \ 2}$$

The same principle will apply to any number adjacent to a multiple of 10, provided we can ascertain, without division, the required *Remainder*.

For instance, 41 is a factor of 99999: so that we can find the 41-Remainder by first finding the 99999-Remainder, and then dividing it by 41. We may then proceed as in the 11-Rule, except that we must regard each digit in the quotient-part of the lower line as 4-times what it really is, when we use it as a subtrahend. We begin by marking off the given number in periods of 5: we then add these periods together; and, if their sum contains more than 5

BRIEF METHOD OF DIVIDING BY 9 OR 11

digits, we treat it in the same way. Hence it will be best to do the addition, first, *above* the given number, and only put its final result, which is the true Remainder, below.

Examples:—

$$\begin{array}{r} 1\dot{4}7705; \\ 41/3\dot{2}7501\dot{8}7652\dot{2}09641\dot{1}585 \\ \hline 7987850646880400282/23 \end{array} \quad ; \quad \begin{array}{r} 41/47706 \\ \hline 1163/23 \end{array}$$

$$\begin{array}{r} 3\dot{2}3329 \\ 41/58\dot{9}017632\dot{4}51639927381562837 \\ \hline 1436628371833268115564787\underline{4/3} \end{array} \quad ; \quad \begin{array}{r} 41/23332 \\ \hline 569/3 \end{array}$$

34. Brief Method of Dividing a Given Number by 9 or 11
[1897: LCH 197]

When he published "Brief Method," Dodgson firmly believed that he was the originator of the method. But from letters to the editor of *Nature*, he found his rule was not original.

> Another correspondent, Mr. Otto Sonne, says that my Rules, both for 9 and for 11, are to be found in a school-book, by a Mr. Adolph Steen, which was published at Copenhagen in 1847. So I fear I must reduce my claim, from that of being the first to discover them, to that of being the first to publish them in English.[1]

The fragment of "Curiosa Mathematica, Part III" in *The Lewis Carroll Picture Book,* 247–50, is practically identical to this item, with the exception of the last two examples.

※

I shall be grateful if you will allow me to communicate, through your columns, to mathematicians generally, but specially to those engaged in teaching arithmetic, two new rules, which effect such a saving of time and trouble that I think they ought to be regularly taught in schools.

Years ago I had discovered the curious fact that, if you put a "0" over the unit-digit of a given number, which happens to be a multiple of 9, and subtract all along, always putting the remainder over the next digit, the final subtraction gives remainder "0," and the upper line, omitting its final "0," is the "9-quotient" of the given number (*i.e.* the quotient produced by dividing it by 9).

Having discovered this, I was at once led, by analogy, to the

---

Nature, vol. LVI (14 October 1897), 565–66.

1. "Abridged Long Division," *Nature*, vol. 57 (20 January 1898), 269. L. E. Dickson, in his authoritative three-volume history of the theory of numbers (I, 342), cites Dodgson's 1897 publication in *Nature* as the source. There is no mention of Adolph Steen.

discovery that, if you put a "0" *under* the unit-digit of a given number, which happens to be a multiple of 11, and proceed in the same way, you get an analogous result.

In each case I obtained the quotient of a division-sum by the shorter and simpler process of *subtraction:* but, as this result was only obtainable in the (comparatively rare) case of the given number being an exact multiple of 9, or of 11, the discovery seemed to be more curious than useful.

Lately, it occurred to me to examine cases where the given number was *not* an exact multiple. I found that, in these cases, the final subtraction yielded a number which was sometimes the actual remainder produced by division, and which always gave materials from which that remainder could be found. But, as it did not yield the quotient (or only by a very "bizarre" process, which was decidedly longer and harder than actual division), the discovery still seemed to be of no practical use.

But, quite lately, it occurred to me to try what would happen if, after discovering the remainder, I were to put it, instead of a "0," over or under the unit-digit, and then subtract as before. And I was charmed to find that the old result followed: the final subtraction yielded remainder "0," and the new line, omitting its unit-digit, was the required quotient.

Now there are shorter processes, for obtaining the 9-remainder or the 11-remainder of a given number, than my subtraction-rule (the process for finding the 11-remainder is another discovery of mine). Adopting these, I brought my rule to completion on September 28, 1897 (I record the exact date, as it is pleasant to be the discoverer of a new and, as I hope, a practically useful, truth).

(1) Rule for finding the quotient and remainder produced by dividing a given number by 9.

To find the 9-remainder, sum the digits: then sum the digits of the result: and so on, till you get a single digit. If this be less than 9, it is the required remainder: if it be 9, the required remainder is 0.

To find the 9-quotient, draw a line under the given number, and put its 9-remainder under its unit-digit: then subtract downwards, putting the remainder under the next digit, and so on. If the left-hand end-digit of the given number be less than 9, its subtraction ought to give remainder "0": if it be 9, it ought to give remainder "1," to be put in the lower line, and "1" to be carried, whose subtraction will give remainder "0." Now mark off the 9-remainder at the right-hand end of the lower line, and the rest of it will be the 9-quotient.

Examples.

$$\frac{9/75309\ 6}{83677/3}, \frac{9/\ 94613\ 8}{105126/4}, \frac{9/58317\ 3}{64797/0}.$$

(2) Rule for finding the quotient and remainder produced by dividing a given number by 11.

To find the 11-remainder, begin at the unit-end, and sum the 1st, 3rd, &c., digits, and also the 2nd, 4th, &c., digits: and find the 11-remainder of the difference of these sums. If the former sum be the greater, the required remainder is the number so found: if the former sum be the lesser, it is the difference between this number and 11: if the sums be equal, it is "0."

To find the 11-quotient, draw a line under the given number, and put its 11-remainder under its unit-digit: then subtract, putting the remainder under the next digit, and so on. The final subtraction ought to give remainder "0." Now mark off the 11-remainder of the right-hand end of the lower line, and the rest of it will be the 11-quotient.

Examples.

$$\frac{11/73210\ 8}{66555/3}, \frac{11/85347\ 1}{77588/3}, \frac{11/59426\ 3}{54023/10}, \frac{11/47568\ 4}{43244/0}.$$

These new rules have yet another advantage over the rule of actual division, viz. that the final subtraction supplies a *test* of the correctness of the result: if it does not give remainder "0," the sum has been done wrong: if it does, then either it has been done right, or there have been *two* mistakes—a rare event.

Mathematicians will not need to be told that rules, analogous to the above, will necessarily hold good for the divisors 99, 101, 999,

1001, &c. The only modification needed would be to mark off the given number in periods of 2 or more digits, and to treat each period in the same way as the above rules have treated single digits. Here, for example, is the whole of the working needed for dividing a given number of 17 digits by 999 and by 1001:—

$$\frac{999/75\dot{4}10\dot{8}36\dot{4}28\dot{1}39\ \dot{2}14}{75486322750890/104}.$$

$$\frac{1001/75\dot{4}10\dot{8}36\dot{4}28\dot{1}39\ \dot{2}14}{75335500927212/\ \ 2}.$$

But such divisors are not in common use; and, for the purposes of school-teaching, it would not be worth while to go beyond the rules for division by 9 and by 11.

CHARLES L. DODGSON.

Ch. Ch., Oxford.

35. Rule for Finding Easter-Day for Any Date till A.D. 2499

[1892–97(?): Stern/Goodacre]

This item does not appear in the standard bibliographic references of Dodgson's work. It was intended to be in chapter VII of Book III, "Other Mental Recreations," of a projected volume on games and puzzles that was never published. The pages are extensively corrected in his own hand. Dodgson has taken Gauss's method for finding the date of Easter Sunday, a problem that had not been handled mathematically before Gauss tackled it, and simplified it so that the date can be calculated mentally. Dodgson cites both Gauss's original publication in the second volume of Zach's *Monatliche Correspondenz* of August 1800 as well as W. W. Rouse Ball's *Mathematical Recreations and Problems,* where the reference to Gauss is given.[1] More than likely, Dodgson took the principle of the rule in English translation from Ball's book. The first and second editions of this book were published in 1892; the third in 1896. Dodgson wrote in his diary on 7 May 1897, "Heard from Mr. W. W. R. Ball, to whom I had written abt the 'Easter' Rule in his 'Math. Recreations.' "[2] It is most likely that Dodgson worked on his version of the rule between 1892 and 1897.

I.
Introductory.

The *principle* of this Rule is due to Gauss, whose proof of it is given in Zach's *Monatliche Correspondenz,* August 1800, Vol. II,

---

1. Gauss wrote three articles on the subject of Easter Day and related matters. The most important one is the article Ball cited, although not by its title, "Berechnung des Osterfestes." It gained wider circulation when it was republished in volume 6, pp. 73–79 of Gauss's *Werke,* edited by C. Schering (Göttingen: Königlichen Gesellschaft der Wissenschaften, 1874).

2. C. L. Dodgson, unpublished Diaries, 7 May 1897.

pp. 221–230, whence it is quoted by Mr. W. W. Rouse Ball, in his *Mathematical Recreations,* published by Macmillan and Co. The only *original* feature, in my version of the Rule, consists in its greater simplicity. By my method the result can be calculated mentally, without much difficulty, in about half-a-minute: by the method of Gauss, it would certainly require a much longer time, and also much greater powers of mental calculation.

Before learning the Rule itself, the Reader should make himself expert in certain necessary Arithmetical Processes, as here specified.

2.
Some Necessary Arithmetical Processes.
(1)

To add 15 to a given No.
Go up in two steps—10 and 5.

[Thus, if the given No. be 187, say "197, 202."]

(2)

To find the Remainder that would be left if a given No. were divided by 4.
Divide the last 2 digits.

(3)

To find the Remainder that would be left if a given No. were divided by 7.

Name the successive *dividends.* This is all the mental *soliloquy* required. The *remainder* in each dividend (which of course serves as the tens-digit of the next dividend) is easily seen by mere inspection.

[Thus, if the given No. be 4325, say "43; 12; 55; 6."]

It will be well to *cast out sevens,* whenever that can conveniently be done.

[Hence, if a dividend be a multiple of 7, say "goes out," and ignore it.

Thus, if the given No. be 4225, say "42 goes out; 25; 4." If it be 4769, say "47; 56 goes out; 9; 2."]

(4)

To find the Remainder that would be left if a given No. were divided by 19.

If the given No. be not greater than 30, the Remainder is easily seen by mere inspection.

If it be greater than 30, take enough of its digits to make a number greater than "1." If this number be even, halve it, and add the next digit; if odd, take its lesser half, and add the next digit, with a "1" prefixed. Imagine the result to be substituted for the digits so used, and proceed as before.

[Thus, if the given No. be 88, say "4 and 8 is 12."
If it be 98, say "4 and 18 is 22; 3."
If it be 147, say "7 and 7 is 14."
If it be 157, say "7 and 17 is 24; 5."
If it be 687, say "3 and 8 is 11; 5 and 17 is 22; 3."]

Cast out nineteens, whenever you can.

[Thus, if the given No. be 1992, ignore the first 2 digits, and say "9 and 2 is 11."
If it be 5749, say "2 and 17 is 19, which goes out; 2 and 9 is 11."
If it be 998, say "4 and 19 gives 4; 2 and 8 is 10."
If it be 7994, say "3 and 19 gives 3; 1 and 19 gives 1; 14."]

If you have to add 18, or 17, &c, call it "19 *minus* 1," or "19 *minus* 2," &c, and ignore the "19."

[Thus, if the given No. be 789, say "3 and *minus* 1 is 2; 1 and 9 is 10."
If it be 967, say "4 and *minus* 3 is 1; 17."]

But this method must not be used if the number, to which the 18, &c, is to be added, is *less than the number which would be deducted*.

[Thus, if the given No. be 567, do *not* say "2 and *minus* 3," but say "2 and 16 is 18; 9 and 7 is 16."]

(5)

To multiply a given No., of 2 digits, whose sum is no greater than 9, by 11.

Put the sum of the digits between them.

(6)

To find the Defect of a given No. from the lowest multiple of 30 which contains it.

The given No. must be either (α) a multiple of 30, or (β) not more than 10 off the lowest multiple of 30 which contains it, or (γ) more than 10 off it.

In case (α), or (β), the Defect may be seen by mere inspection.

[Thus, if the given No. be 180, say "Defect is 0." If it be 208, say "Defect is 7."]

In case (γ), take the *Excess* of the given No. *above* the next *lower* Multiple of 30, and deduct from 30.

[Thus, if the given No. be 189, say "9 above; Defect is 21." If it be 192, say "12 above; Defect is 18."]

3.
Rule for Finding Easter-Day for Any Given Date till A.D. 2499.

The phrase "4-Rem," used with reference to a certain No., means "the Remainder that would be left if the No. were divided by 4; and similarly for the phrases "7-Rem" and "19-Rem."

Three Nos. are required, two of which are known, by memory, as soon as the Date is named; the third has to be calculated.[3]

The Rule may conveniently be divided into three parts, as follows:—

---

3. Dodgson added the handwritten notation, "Let us call them a, h, k," here.

(1) Name the given Date, and then recall, by memory, the values of a and h belonging to it. For Old Style the values are always 15 and 6. For New Style, they are given in the following Table:—

No. of hundreds in Date	15	16	17	18	19	20	21	22	23	24
Value of a	8	8	7	7	6	6	6	5	4	5
" h	2	2	3	4	5	5	6	0	1	1

Picture this in your mind's eye, and say over the Nos. of hundreds, till you reach the given one; then name the values of a and h in each column.

[Thus if the given Date be "A.D. 1582" (which, for our present purpose is O.S., as N.S. did not begin till October), "1582; Old Style; a and h are 15 and 6."

If it be "A.D. 1583, N.S.," say "1583; New Style; 15; a and h are 8 and 2."

If it be "A.D. 1583, O.S." say "1583; Old Style; a and h are 15 and 6."

If it be "A.D. 1948, N.S.," say "1948; New Style; 15, 16, 17, 18, 19; a and h are 6 and 5."]

(2) Again name the given Date, and find its 4-Rem and its 7-Rem: then take "4-Rem s twice 7-Rem": double; add h: the 7-Rem of this result is k.

[Thus, if the given Date be "A.D. 1582," say "1582; 4-Rem; 82; 2; 7-Rem; 15, 18, 42; 0; 2 and 0 is 2; 4; and 6 is 10; k is 3."

If it be "A.D. 1583, N.S.," say "1583; 4-Rem; 83; 3; 7-Rem; 15, 18, 43; 1; 3 and 2 is 5; 10; and 2 is 12; k is 5." If it be "A.D. 1583, O.S.," say "1583; 4-Rem; 83; 3; 7-Rem; 15, 18, 43; 1; 3 and 2 is 5; 10; and 6 is 16; k is 2."

If it be "A.D. 1948, N.S.," say "1948; 4-Rem; 48; 0; 7-Rem; 19, 54, 58; 2; 0 and 4 is 4; 8; and 5 is 13; k is 6."

If it be "A.D. 1948, O.S.," say "1948; 4-Rem; 48; 0 7-Rem; 19, 54, 58; 2; 0 and 4 is 4; 8; and 6 is 14; k is 0."]

(3) Name a and k: name given Date: find its 19-Rem: multiply by 11; add a: find Defect of this from lowest multiple of 30 which

contains it: find highest multiple of 7 contained in Defect, and add k. If this result falls short of Defect, then either deduct 2 and call it "April," or (if this cannot be done) add 29 and call it "March." If this result does *not* fall short of Defect, then either deduct 9 and call it "April," or (if this cannot be done) add 22 and call it "March."

> [Thus if the given Date be "A.D. 1582," and if a and k be known to be 15 and 3, say "a and k are 15 and 3; 1582; 7 and *minus* 1 is 6; 3 and 2 is 5; 55; and 15 is 65, 70; 10 above; Defect is 20; 14 and 3 is 17, which falls short; deduct 2; April 15."
>
> If it be "A.D. 1583, N.S.," and if a and k be known to be 8 and 5, say "a and k are 8 and 5; 1583; 7 and *minus* 1 is 6; 3 and 3 is 6; 66; and 8 is 74; 14 above; Defect is 16; 14 and 5 is 19, which does *not* fall short; deduct 9; April 10."
>
> If it be "A.D. 1583, O.S.," and if a and k be known to be 15 and 2, say "a and k are 15 and 2; 1583; 7 and *minus* 1 is 6; 3 and 3 is 6; 66; and 15 is 76, 81; Defect is 9, 7 and 2 is 9, which does *not* fall short; add 22; March 31."
>
> If it be "A.D. 1948, N.S.," and if a and k be known to be 6 and 6, say "a and k are 6 and 6; 1948; 9 and *minus* 5 is 4; 2 and 8 is 10; 110; and 6 is 116; Defect is 4; 0 and 6 is 6, which does *not* fall short; add 22; March 28."
>
> If it be "A.D. 1948, O.S.," and if a and k be known to be 15 and 0, say "a and k are 15 and 0; 1948; 9 and *minus* 5 is 4; 2 and 8 is 10; 110; and 15 is 120, 125; 5 above; Defect is 25; 21 and 0 is 21, which falls short; deduct 2; April 19."]

4.
Aids to Memory, &c.
(1)

To remember the given Date, while working the Rule.

If you are handy at making *Memoria Technica* words, you will find *that* a very useful method. Otherwise, you had better *write it down,* as it is certainly a trial, for the temper, to find, after carefully working out the Easter-Day, that the Date, for which you have calculated it, has vanished from your memory!

[The Dates, worked out as examples in the preceding Section, may conveniently be remembered by the following words:—
853 "kilt"; 1654 "box-leaf"; 1881 "choke boy."]

(2)

To remember the *ah*-Table.

The first 6 columns are the most useful. In them, note that the values for *a* are "2 eights, 2 sevens, 2 sixes," and that, in the first 3 columns, *a* and *h* add up to 10, and in the other 3 to 11.

In the last 4 columns, the values for *a* and *h* are given in the third and fourth lines of the following *Memoria Technica* stanza:—

> List my song to!
> 'Tis as wrong to
> Save a flea
> As rob a bee.

The values for the *sent* Century, viz. 7 and 4, had better be fixed firmly in the memory *as a separate item*.

To remember the value of k till it is wanted.

This can be done with one hand. For "0" keep the hand *open:* for "1," double in the 1st finger, and put the thumb on it; similarly for "2," "3," and "4;" for "5" double in the thumb, and put the fingers on it; for "6," clench the fist, with the thumb *outside*.

(3)

If it should happen, in working the second part of the Rule, that "4-Rem *s* twice 7-Rem" is a multiple of 7, this would make k equal to *h;* so go on *at once* to the *third* part.

[Thus, if the given Date be "A.D. 1731, N.S.," say "15, 16, 17, *a* and *h* are 7 and 3; 4-Rem; 31; 3; 7-Rem; 1731; 17, 33, 51; 2; 3 and 4 is 7; *a* and k are 7 and 3; 1731; 8 and *minus* 6 is 2; [4]22; and 7 is 29; Defect is 1; 0 and 3 is 3, which does *not* fall short; add 22; March 25."]

---

4. Dodgson added the handwritten correction "1 and 1 is 2;" here.

(4)

There is one single Date (and *only* one, so far as I know) for which this Rule fails. In the year A.D. 1954, New Style, Easter-Day will fall on the 18th of April: the Rule gives it as the 25th. I cannot in the least account for this very curious anomaly.

5.
Examples worked as Specimens.

(1)
A.D. 853.

"Old Style; a and h are 15 and 6; 4-Rem; 53; 5; 13; 1; 7-Rem; 853; 8, 15, 13; 6; 1 and 12 is 13; 26; and 6 is 32; k is 4.

a and k are 15 and 4; 853; 4 and 5 is 9; 4 and 13 is 17; 187; and 15 is 197, 202; Defect is 8; 7 and 4 is 11, which does *not* fall short; deduct 9; April 2."

(2)
A.D. 1654. [N.S.]

"15, 16; a and h are 8 and 2; 4-Rem; 54; 5, 14; 2; 7-Rem; 1654; 16, 25, 44; 2; 2 and 4 is 6; 12; and 2 is 14; k is 0.

a and k are 8 and 0; 1654; 8 and 5 is 13; 6 and *minus* 5 is 1; 11; and 8 is 19; Defect is 11; 7 and 0 is 7, which falls short; deduct 2; April 5."

(3)
A.D. 1654 [O.S.]

"Old Style; a and h are 15 and 6; 4-Rem; 54; 5, 14; 2; 7-Rem; 1654; 16, 25, 44; 2; 2 and 4 is 6; 12; and 6 is 18; k is 4.

a and k are 15 and 4; 1654; 8 and 5 is 13; 6 and *minus* 5 is 1; 11; and 15 is 26; Defect is 4; 0 and 4 is 4, which does *not* fall short; add 22; March 26."

(4)
A.D. *1881. [N.S.]*

"*a* and *h* are 7 and 4; 4-Rem; 81; 1; 7-Rem; 1881; 18, 48, 61; 5; 1 and 10 is 11; 22; and 4 is 26; *k* is 5.

a and *k* are 7 and 5; 1881; 9 and 8 is 17; 8 and 11 is 19; 0; and 7 is 7; Defect is 23; 21 and 5 is 26, which does *not* fall short; deduct 9; April 17."

(5)
A.D. *1881. [O.S.]*

"*a* and *h* are 15 and 6; 4-Rem; 81; 1; 7-Rem; 1881; 18, 48, 61; 5; 1 and 10 is 11; 22; and 6 is 28; *k* is 0.

a and *k* are 15 and 0; 1881; 9 and 8 is 17; 8 and 11 is 19; 0; and 15 is 15; Defect is 15; 14 and 0 is 14, which falls short; deduct 2; April 12."

6.
Examples for Practice.

For any Reader who does *not* possess "The Book of Almanacs" by Professor De Morgan, here are 100 miscellaneous Dates, which he can work as examples: the *answers* are given in the next Section.

1. 17.	26. 849.	51. 1655. N.S.	76. 1897. N.S.
2. 43.	27. 863.	52. 1718. "	77. " O.S.
3. 79.	28. 905.	53. 1749. "	78. 1900. N.S.
4. 108.	29. 947.	54. 1770. "	79. " O.S.
5. 131.	30. 992.	55. 1806. "	80. 1904. N.S.
6. 166.	31. 1007.	56. " O.S.	81. 1905. "
7. 215.	32. 1032.	57. 1815. N.S.	82. 1908. "
8. 246.	33. 1078.	58. 1825. "	83. 1911. "
9. 282.	34. 1100.	59. 1829. "	84. 1925. "
10. 327.	35. 1168.	60. " O.S.	85. 1930. "
11. 354.	36. 1181.	61. 1831. N.S.	86. 1936. "
12. 399.	37. 1215.	62. 1838 "	87. " O.S.
13. 403.	38. 1246.	63. 1843. "	88. 1949. N.S.

14. 438.	39. 1259.	64. 1847. "	89. 1957. "
15. 471.	40. 1324.	65. 1851. "	90. 1983. "
16. 519.	41. 1356.	66. " O.S.	91. 2009. "
17. 540.	42. 1372.	67. 1856. N.S.	92. 2021 "
18. 583.	43. 1428.	68. 1862. "	93. 2148. "
19. 629.	44. 1473.	69. 1865. "	94. 2177. "
20. 666.	45. 1491.	70. 1867. "	95. 2284. "
21. 695.	46. 1518.	71. " O.S.	96. 2295. "
22. 734.	47. 1533.	72. 1879. N.S.	97. 2351. "
23. 761.	48. 1589. N.S.	73. 1883. "	98. " O.S.
24. 788.	49. 1603. "	74. 1888. "	99. 2462. N.S.
25. 810.	50. " O.S.	75. 1893. "	100. 2499. "

7.
Answers.

1. Ap. 3.	26. Ap. 14.	51. Mar. 28.	76. Ap. 18.
2. " 14.	27. " 11.	52. Ap. 17.	77. " 13.
3. " 4.	28. Mar. 31.	53. " 6.	78. Ap. 15.
4. Mar. 25.	29. Ap. 11.	54. " 15.	79. Mar. 31.
5. Ap. 2.	30. Mar. 27.	55. " 6.	80. Ap. 3.
6. " 7.	31. Ap. 6.	56. " 1.	81. " 23.
7. " 2.	32. " 2.	57. Mar. 26.	82. " 19.
8. " 19.	33. " 8.	58. Ap. 3.	83. " 16.
9. " 16.	34. " 1.	59. " 19.	84. " 12.
10. Mar. 26.	35. " 2.	60. " 14.	85. " 20.
11. " 27.	36. " 5.	61. Ap. 3.	86. " 12.
12. Ap. 10.	37. " 19.	62. " 15.	87. Mar. 30.
13. Mar. 29.	38. " 8.	63. " 16.	88. Ap. 17.
14. " 27.	39. " 13.	64. " 4.	89. " 21.
15. " 28.	40. " 15.	65. " 20.	90. " 3.
16. " 31.	41. " 24.	66. " 8.	91. " 12
17. Ap. 8.	42. Mar. 28.	67. Mar 23.	92. 4.
18. " 18.	43. Ap. 4.	68. Ap. 20.	93. " 7.
19. " 16.	44. Mar. 29.	69. " 16.	94. " 20.
20. Mar. 27.	45. Ap. 3.	70. " 21.	95. " 6.
21. Ap. 11.	46. " 4.	71. Mar. 28.	96. " 7.
22. Mar. 28.	47. " 13.	72. Ap. 13.	97. " 15.
23. " 29.	48. " 2.	73. Mar. 25.	98. " 6.
24. " 30.	49. Mar. 30.	74. Ap. 1.	99. " 16.
25. " 31.	50. Ap. 24.	75. " 2.	100. " 7.

36. Abridged Long Division
[1898: LCH 197]

This article, written by Dodgson on 21 December 1897, is listed by the authors of the *Lewis Carroll Handbook* as his last work before his death on 14 January 1898. However, the last question he submitted to the *Educational Times* (Question 14122 on logic), which appeared in volume LII, p. 93, on 1 February 1899 might be a better candidate for that honor. "Abridged Long Division" is a generalization of "Brief Method of Dividing a Given Number by 9 or 11." The latter caused several letters to be written to the editor of *Nature* about various points in it. This article is much more of a formal mathematical treatise than "Brief Method" and was probably meant to be his final word on the subject. The fragment of "Curiosa Mathematica, Part III" in *The Lewis Carroll Picture Book*, pages 250–61, is identical to this article in all its important points. There is some additional material on pages 254–55 and 256–57.

※

A Brief Method of dividing a given Number by a Divisor of the Form ($h.10^n \pm k$), *where at least* one *of the two numbers,* h *and* k, *is greater than* 1.

My former paper on this subject, which appeared in *Nature* for October 14, 1897, dealt only with the case where $h = 1$ and $k = 1$. It elicited, from other correspondents of *Nature,* several interesting letters, which the editor kindly allowed me to see. One, from Mr. Alfred Sang, quotes Mons. L. Richard's *Stenarithmie,* as containing my Rule for dividing by 11. Mons. Richard's book, which I had not previously met with, does certainly contain the rule, but the author has failed to see that the test, which this Method furnishes, for the correctness of the working, is absolutely *definite*. He says "La dernière différence, ou cette différence augmontée de 1, égalera le chiffre de gauche du nombre proposé." So ambiguous a test as this would of course by useless. But the "dif-

Nature, vol. 57 (20 January 1898), 269–71.

ference" he is speaking of is really the *last but one:* the very *last* will always (as I stated in my former paper) be equal to zero. Another correspondent, Mr. Otto Sonne, says that my Rules, both for 9 and for 11, are to be found in a school-book, by a Mr. Adolph Steen, which was published at Copenhagen in 1847. So I fear I must reduce my claim, from that of being the first to discover them, to that of being the first to publish them in English.

The Method, now to be described, is applicable to three distinct cases:

(1) where $h > 1, k = 1$;
(2) where $h = 1, k > 1$;
(3) where $h > 1, k > 1$.

With certain limitations of the values of h, k, and n, this Method will be found to be a shorter and safer process than that of ordinary Long Division. These limitations are that neither h nor k should exceed 12, and that, when $k > 1$, n should not be less than 3: outside these limits, it involves difficulties which make the ordinary process preferable.

In this Method, two distinct processes are required—one, for dealing with cases where $h > 1$, the other, for cases where $k > 1$. The former of these processes was, I believe, first discovered by myself, the latter by my nephew, Mr. Bertram J. Collingwood, who communicated to me his Method of dealing with Divisors of the form $(10^n - k)$.

In what follows, I shall represent 10 by t.

Mr. Collingwood's Method, for Divisors of the form $(t^n - k)$, may be enunciated as follows:—

"To divide a given Number by $(t^n - k)$, mark off from it a period of n digits, at the units-end, and under it write k-times what would be left of it if its last period were erased. If this number contains more than n digits, treat it in the same way; and so on, till a number is reached which does not contain more than n digits. Then add up. If the last period of the result, *plus* k-times whatever was carried out of it, in the adding-up, be less than the Divisor, it is the required Remainder; and the rest of the result is the required Quotient. If it be not less, find what number of times it contains the Divisor, and add that number to the Quotient, and subtract that multiple of the Divisor from the Remainder."

For example, to divide 86781592485703152764092 by 9993 (*i.e.* by $t^4 - 7$), he would proceed thus:—

```
9993 ‖ 867 8159 2485 7031 5276 | 4092
           6074 7114 7399 9220   6932
              4 2522 9803 1799   4540
                  29 7660 8622   2593
                     208 3626    0354
                        1458     5382
                           1     0206
                                    7
```
Quot. 868 4238 2153 2104 0004 ‖ 4106 + 14 = 4120 Rem.

The new Method will be best explained by beginning with case (3): it will be easily seen what changes have to be made in it when dealing with cases (1) and (2).

The Rule for case (3), when the sign is " — ," may be enunciated thus:—

Mark off the Dividend, beginning at its units-end, in periods of n digits. If there be an overplus, at the left-hand end, less than h, do not mark it off, but reckon it and the next n digits as one period.

To set the sum, write the Divisor, followed by a double vertical; then the Dividend, divided into its periods by single verticals, with width allowed in each space for $(n + 2)$ digits. Below the Dividend draw a single line, and, further down, a double one, leaving a space between, in which to enter the Quotient, having its units-digit below that of the last period but one of the Dividend, and also the Remainder, having its units-digit below that of the last period of the Dividend. In this space, and in the space below the double line, draw verticals, corresponding to those in the Dividend; and make the last in the upper space double, to separate the Quotient from the Remainder.

For example, if we had to divide 5984407103826 by 6997 (*i.e.* $7 \cdot t^3 - 3$), the sum, as set for working, would stand thus:

ABRIDGED LONG DIVISION 315

$$
\begin{array}{c|c|c|c|c}
6997 \,\|\, 5984 & 407 & 103 & 826 \\
\text{Quot.} & & & \| & \text{Rem.}
\end{array}
$$

To work the sum, divide the 1st period by h: enter its quotient in the 1st Column below the double line, and place its Remainder above the 2nd period, where it is to be regarded as *prefixed* to that period. To the 2nd period, with its prefix, add k-times the number in the 1st Column, and enter the result at the top of the 2nd Column. If this number *is not* less than the Divisor, find what number of times it contains the Divisor, and enter that number in the 1st Column, and k-times it in the 2nd; and then draw a line below the 2nd Column, and add in this new item, deducting from the result t^n times the number just entered in the 1st Column; and then add up the 1st Column, entering the result in the Quotient. If the number at the top of the 2nd Column *is* less than the Divisor, the number in the 1st Column may be at once entered in the Quotient. The number entered in the Quotient, and the number at the foot of the 2nd Column, are the Quotient and Remainder that would result if the Dividend ended with its 2nd period. Now take the number at the foot of the 2nd Column as a new 1st period, and the 3rd period as a new 2nd period, and proceed as before.

The above example, worked according to this Rule, would stand thus:—

$$
\begin{array}{r|r|r|r|r}
 & & 6 & 5 & 3 \\
6997 \,\|\, 5984 & 407 & 103 & 826 \\
\text{Quot.} \; 855 & 281 & 849 & \| \; 6373 \;\text{Rem.} \\ \hline
854 & 8969 & 5946 & \\
1 & 3 & 849 & \\ \cline{2-2}
 & 1972 & & \\
 & 281 & &
\end{array}
$$

the Mental Process being as follows:—

316   ITEM 36

Divide the 5984 by 7, entering its Quotient, 854, in the 1st Column, and placing its Remainder, 6, above the 2nd period. Then add, to the 6407, 3-times the 854, entering the result in the 2nd Column, thus. "7 and 12, 19." Enter the 9, and carry the 1. "1 and 15, 16." Enter the 6, and carry the 1. "5 and 24, 29." Enter the 9, and carry the 2, which, added to the prefix 6, makes 8, which also you enter. Observing that this 8969 *is not* less than the Divisor, and that it contains the Divisor *once,* enter 1 in the 1st Column, and 3-times 1 in the 2nd, and then draw a line below, and add in this new item, remembering to deduct from the result 7-times t^3, *i.e.* 7000: the result is 1972. Then add up the 1st column, as far as the double line, and enter the result, 855, in the Quotient. Now take the 1972 as a new 1st period, and the 3rd period, 103, as a new 2nd period, and proceed as before.

The Rule for case (3), when the sign is "+," may be deduced from the above rule by simply changing the sign of k. This will, however, introduce a new phenomenon, which must be provided for by the following additional clause:—

When you add, to the 2nd period with its prefix, $(-k)$-times the number in the 1st Column, *i.e.* when you *subtract* k-times this number *from* the 2nd period with its prefix, it will sometimes happen that the subtrahend exceeds the minuend. In this case the subtraction will end with a *minus* digit, which may be indicated by an asterisk. Now find what number of Divisors must be added to the 2nd Column to cancel this *minus* digit, and enter that number, marked with an asterisk, in the 1st Column, and that multiple of the Divisor in the 2nd; and then draw a line below the 2nd Column and add in this new item.

As an example, let us take a new Dividend, but retain the previous Divisor, changing the sign of k, so that it will become 7003 (*i.e.* $7 \cdot t^3 + 3$). the sum, as set for working, would stand thus:—

```
7003 ‖ 6504 | 318 | 972 | 526
Quot.   |    |    ‖     Rem.
```

ABRIDGED LONG DIVISION

After working, it would stand thus:—

		1	4	5	
7003 ‖	6504 |	318 |	972 |	526	
Quot.	928 |	790 |	371 ‖	4413 Rem.	
	929	2*531	2602		
	1*	7 003	371 |		
		5 534			
		790 |			

the Mental Process being as follows:—

Divide the 6504 by 7, and enter the Quotient, 929, in the 1st Column, and the Remainder, 1, above the 2nd period. Then subtract, from the 1318, 3-times the 929, entering the result in the 2nd Column, thus. "27 from 8 I ca'n't, but 27 from 28, 1." Enter the 1, and carry the borrowed 2. "8 from 1 I ca'n't, but 8 from 11, 3." Enter the 3, and carry the borrowed 1. "28 from 3 I ca'n't, but 28 from 33, 5." Enter the 5, and carry the borrowed 3. "3 from 1, *minus* 2." Enter it, with an asterisk. Observing that, to cancel this *minus* 2, it will suffice to add *once* the Divisor, enter a (−1) in the 1st Column, and 7003 in the 2nd; and then draw a line below the 2nd Column, and add in this new item: the result is 5534. Then add up the 1st Column, and enter the result, 928, in the Quotient. Now take the 5534 as a new 1st period, and the 3rd period, 972, as a new 2nd period, and proceed as before.

The Rules for case (2) may be derived, from the above, by making $k = 1$: and those for case (3) by making $h = 1$. I will give worked examples of these; but it will not be necessary to give the Mental Processes.

By making $k = 1$, we get Divisors of the form $(h \cdot t^n \pm 1)$: let us take $(11t^4 - 1)$ and $(6t^5 + 1)$.

318 ITEM 36

		9	10	4	
109999 ‖ 107523	8168	9662	0985	
Quot. 9774	9813	0861	41846 Rem.	
	9774	107942	119474	
			1	
		9812		
		1	9475	
			861	

		3		3	
600001 ‖ 7239	51798	2 6004	13825		
Quot. 1206	58431	9 4595 ‖ 219230 Rem.			
	350592	4*7572			
		60 0001			
	58432				
	1*	56 7573			
		9 4595			

In this last example, there is no need to enter the Quotient, produced by dividing the 7239 by 7, in the 1st Column: we easily foresee that the number at the top of the 2nd Column *will be* less than the Divisor, so that there will be no new item in the 1st: hence we at once enter the 1206 in the Quotient.

By making $h = 1$, we get Divisors of the form $(t^n \pm k)$: let us take $(t^4 - 7)$ and $(t^5 + 12)$.

9993 ‖ 867	8159	2485	7031	5276	4092
Quot. 868	4238	2153	2104	0004	4120 Rem.
867	14228	32130	22088	19990	
1	7	21	14	14	
	4235	2151	2102	4	
	3	2	2		

100012 ‖	7185	6 2039	10327	53118
	7184	7 5822	00463	47562
	7185	3*5819	9*00355	
	1*	10 0012	9 00108	
		7 5831	463	
		9*		

The first of these two sums is the one I gave to illustrate Mr. Collingwood's Method of working with Divisors of the form $(t^n - k)$.

It may interest the Reader to see the 3 Methods of working the above example—ordinary Division, Mr. Collingwood's Method, and my version of it—compared as to the amount of labour which each entails in the working:—

	Ordinary Division.	Mr. C.'s Method.	My version of it.
Digits written:	202	82	44
Additions, or Subtractions:	204	97	25
Multiplications:	0	70	22

I am assuming that any one, working this example by ordinary Division, would begin by making a Table of Multiples of 9993, for reference: so that he would have *no* Multiplications to do. Still, the great number of digits he would have to write, and of Additions and Subtractions he would have to do, involving a far greater risk of error than either of the other Methods, would quite outweigh this advantage.

By whatever process a Question in Long Division has been worked, it is very desirable to be able to test, easily and quickly, the correctness of the Answer. The ordinary test is to multiply together the Divisor and Quotient, add the Remainder, and observe whether these together make up the given Number, as they ought to do.

Thus, if N be the given number, D the given Divisor, Q the Quotient, and R the Remainder, we ought to have

$$N = D . Q + R.$$

This test is specially easy to apply, when $D = (h . t^n \pm k)$; for then we ought to have

$$\begin{aligned} N &= (h . t^n \pm k) . Q + R; \\ &= (hQ . t^n + R) \pm kQ. \end{aligned}$$

Now $hQ . t^n$ may be found by multiplying Q by h, and tacking on n ciphers. Hence $(hQ . t^n + R)$ may be found by making R occupy the place of the n ciphers. If R contains less than n digits it must have ciphers prefixed; if more, the overplus must be carried on into the next period, and added to hQ.

Having found our "Test," viz. $(hQ . t^n + R)$, we can write it on a separate slip of paper, and place it below the working of the example, so as to come vertically below N, which is at the top. When the sign in D is "$-$," we must add kQ to N, and see if the result $= T$; when it is "$+$," we must add kQ to T, and see if the result $= N$.

Now it has been already pointed out that when, in the new Method, the 1st and 2nd Columns have been worked, the 1st period of the Quotient and the number at the foot of the 2nd Column are the Quotient and Remainder that would result if the Dividend ended with its second period. Hence the Test can be at once applied, before dealing with the 3rd Column. This constitutes a very important new feature in my version of Mr. Collingwood's Method. Every two adjacent columns contain a separate Division-sum, which can be tested *by itself.* Hence, in working my Method, as soon as I have entered the 1st period of the Quotient, I can test it, and, if I have made any mistake, I can correct it. But the hapless computor, who has spent, say, an hour, in working some gigantic sum in Long Division—whether by the ordinary process or by Mr. Collingwood's Method—and who has chanced to get a figure wrong at the very outset, which makes every subsequent figure wrong, has no warning of the fatal error till he has worked out the whole thing "to the bitter end," and has begun to test his Answer. Whereas, if working by *my* Method, he would have been warned of his mistake almost as soon as he made it, and would have been able to set it right before going any further.

As an aid to the Reader, I will give the Mental Process in full, for the 2nd and 3rd Columns of the first of the examples worked above.

The Divisor is 6997 (where $h = 7$, $k = 3$). Here you are supposed to have just entered the 281 in the Quotient. The Dividend, for these 2 Columns, is 1972 | 103 ; the Quotient is 281, and the Remainder 5946. The Test is $hQ.t^n + R$. (*i.e.* 7 × 281000 + 5946), the Mental Process being as follows. You write, on a separate slip of paper, the last 3 digits of R, viz. 946, and carry the 5 into the next period, adding it to the 7 × 281: thus. "5 and 7, 12." Enter the 2, and carry the 1. "1 and 56, 57." Enter the 7, and carry the 5. "5 and 14, 19." Enter it. Having got your Test, try whether ($N + kQ$) is equal to it. This you compute, comparing it with your Test, digit by digit, as you go on, thus. "3 and 3, 6." Observe it in the Test. "0 and 24, 24." Observe the 4, and carry the 2. "3 and 6, 9." Observe it. "1972 and 0, 1972." Observe it. The Test is satisfied.

$$
\begin{array}{r|r|r|}
 & 6 & 5 \\
| & 407 \;| & 103 \;| \\
| & 281 \;| & \;| \\
\hline
 & 8969 & 5946 \\
 & 3 & \\
\cline{2-2}
 & 1972 & \\
 & 281 & \\
\hline
\text{Test} \;| & 1972 \;| & 946 \;|
\end{array}
$$

For Divisors of the form ($t^n \pm k$) there is no need to write out the Test: the numbers, which compose it, already occur in the working, and may be used as they stand.

CHARLES L. DODGSON.

Ch. Ch., Oxford, December 21, 1897.

Cryptology

Key – Alphabet.

a b c d e f g h i j k l m n o p q r s t u v w x y z

a b c d e f g h i j k l m n o p q r s t u v w x y z a

Message – Alphabet.

Rules.

The correspondents agree on a 'key-word', which must be kept secret.

To translate a message into cipher, write the key-word over it, letter for letter, repeating it as often as may be necessary. Find the first letter of the key-word in the 'key-alphabet', and the first letter of the message in the 'message-alphabet': bring them into a column by sliding one alphabet under the other, and copy the letter over 'a': this is the first letter of the cipher.

Translate the cipher into English by same rule.

The Telegraph Cipher. A version of the cipher of the same name, discussed as item 40.

Introduction

Charles Dodgson invented four ciphers between 1858 and 1868. Public interest in ciphers had been sparked by the electric telegraph which, in 1843, had been opened from Paddington to Slough on the Great Western Railway. The telegraph had increased both the volume of communications and the need to protect them from unauthorized access. Intellectuals and politicians—dilettantes of many stripes—felt challenged to create an unbreakable cipher.

Ciphers have a long history. The oldest surviving cipher key, from a Coptic monastery in Egypt in the seventh century, is in the Metropolitan Museum of Art in New York. In a cipher, the basic unit is a letter or a group of letters of the same length. Many ciphers use a key or a keyword, a word that defines which cipher alphabets will be used to transform the plaintext, the message to be enciphered, into ciphertext.

Polyalphabetic ciphers were invented in the fifteenth century by Leon Battista Alberti, architect, lawyer, organist, writer, artist, athlete—a multi-talented individual and model Renaissance man. The most famous polyalphabetic cipher, the Vigenère cipher, is actually not the one invented by Blaise de Vigenère and described in his book, *Traicté des Chiffres* (1585), but a simpler and less secure form of it that emerged in the mid-nineteenth century after Vigenère's system had fallen into disuse. Nevertheless, even in its degraded form, it is the archetype of the most famous cipher system in history. It uses 26 cipher alphabets—standard alphabets presented horizontally, each beginning with the letter after the one above it. The plain alphabet, the standard 26 letters, is above the table, while another standard alphabet presented vertically on the left side is the key alphabet. To encipher a message, the keyword is placed above the plaintext as many times as needed. Using the plaintext letter in the alphabet at the top and the key letter from the alphabet on the left side, the ciphertext letter is at the intersection of the column and row. In the example below, the keyword VIGILANCE is repeated over the plaintext. In the tableau of letters, the intersection of the column headed by V and the row headed by M gives

---

The ideas developed in this introduction are the result of joint research by myself and Stanley H. Lipson, Professor of Mathematics and Computer Science at Kean College of New Jersey.—Ed.

the cipher letter H, etc. This encipherment can be represented by the equation, Key + Plain = Cipher.

Keyword: V I G I L A N C E V I G I L A N C E
Plaintext: M E E T M E O N T U E S D A Y E V E
Ciphertext: H M K B X E B P X P M Y L L Y R X I

Note that a different cipher letter results for the same letter of plaintext, e.g., V is encoded by H and by P, illustrating a polyalphabetic encipherment. In monalphabetic substitution, each letter is transformed into a unique cipher letter that remains the same throughout the encipherment.

Dodgson's Alphabet Cipher, from which the example above is taken, is a Vigenère cipher. It was reinvented by Dodgson in 1868. He repeated the top alphabet at the bottom and the left-side alphabet on the right side, and added grids to the table for ease of use.

Another polyalphabetic substitution cipher was reinvented in 1857 by retired Admiral Sir Francis Beaufort of the British Royal Navy and bears his name. Its origin goes back to the early eighteenth century and is ascribed to Giovanni Sestri (1710). Like the Vigenère, it uses an alphabetic square, but repeats the alphabet on all four sides, having an "A" in all four corners of the tableau. Here is an example of a Beaufort cipher that Dodgson reinvented and named the Telegraph Cipher.

Keyword: W A R W A R W A R W A
Plaintext: M E E T M E A T S I X
Ciphertext: K W N D O N W H Z O D

The table is entered at row M, the plaintext letter, and followed to the tabular entry of the key letter W; then the column heading above that entry gives the cipher letter K. This encipherment can be represented as Key − Plain = Cipher. Again, a different cipher letter results from the same letter of plaintext, e.g., E is encoded as both W and N.

Both the Alphabet Cipher and the Telegraph Cipher were published anonymously and undated. There is no mention of the Alphabet Cipher in his diary entry of 22 April 1868 when Dodgson wrote, "Sitting up at night I invented a new cipher, which I think of calling The Telegraph Cipher."[1] Dodgson's biographer and relative, Stuart Collingwood, and his bibliographer, Falconer Madan, dated them both 1868. From Dodgson's

---

1. *The Diaries of Lewis Carroll,* edited by Roger L. Green (New York: Oxford University Press, 1954), vol. II, 268–69.

statement, it would seem that the Telegraph Cipher followed the Alphabet Cipher chronologically. Dodgson communicated his invention to George Ward Hunt, First Lord of the Admiralty. No response is known.

Dodgson's enciphering scheme in the Telegraph Cipher is unusual. Unlike the Vigenère and Beaufort tables, it uses two sliding alphabets: a key letter alphabet A–Z and a plaintext alphabet A–Z, A. Dodgson's instructions for using the cipher are to find the key letter on the key alphabet and slide the plaintext alphabet under it so that the plaintext letter appears under the key letter. Then, using the (appropriate) "A" {index} on the plaintext alphabet, the cipher appears above the "A" on the key alphabet. This scheme produces a Beaufort enciphered text.

It is entirely probable that Dodgson was aware of Beaufort's 1857 cipher. What is curious, however, is that the equivalence of Beaufort's polyalphabetic tableau and Dodgson's cryptographic slide (as we will call the sliding alphabet) was not made until 1883 by Auguste Kerckhoffs, a Dutchman, in a book entitled *La Cryptographie Militaire,* considered one of the best books on cryptology. He named his cryptographic slide, the St.-Cyr system, after the French military academy where it was taught. Kerckhoffs's instructions showed the equivalence of the cryptographic slide and the Vigenère tableau. However, it was only in 1910 that Felix Delastelle showed that the slide could produce other ciphers like the Beaufort as well. He published his work in a book entitled *Traité Élémentaire de Cryptographie.*

Codes produced by polyalphabetic systems were considered best from the mid-nineteenth century until 1939. These ciphers were considered "unbreakable" (at least until 1917), when in reality they had been broken in 1863 by the Prussian Major Friedrich Kasiski. In his book, *Die Geheimschriften und die Dechiffrir-Kunst,* Kasiski demonstrated that when the keyword is repeated enough times, two identical pieces of plaintext will be enciphered into the same ciphertext. The number of letters between the two pieces of identical ciphertext gives the number of times the keyword has repeated. By analyzing these intervals, the length of the keyword becomes known. The number of letters in the keyword gives the number of alphabets used in the polyalphabetic encipherment. Now the ciphertext letters that were enciphered with each of the keyword letters can be grouped separately, i.e., all the Es transformed to the same ciphertext letter by one of the key letters, and so on. These groupings can now be solved using the simpler monalphabetic cipher techniques. However, Kasiski's work went unnoticed, so the world did not know that all periodic polyalphabetic ciphers (with repeating keywords) were not secure!

We know that Dodgson produced two ciphers ten years earlier, in February 1858—both appear in his unpublished diaries. Roger L. Green had this to say about their omission from his edited volume of the diaries: "[The cipher is too long and complicated to reproduce here. Moreover, Dodgson printed an improved and much shorter method, *The Alphabet Cipher,* in 1868 . . . and another variant, *The Telegraph Cipher,* in the same year.]"[2] Apparently Green was unaware that there were actually two different ciphers in the entries of February 23 and 26. Although these are also polyalphabetic ciphers, they are unrelated to the Alphabet and Telegraph ciphers.

Dodgson did not name the ciphers in his diaries. In describing the second of these, which we call the Matrix Cipher (item 30), he wrote on 26 February 1858,

> Invented another cypher, far better than the last: it has these advantages.
> (1) The system is easily carried in the head.
> (2) The key-word is the only thing necessarily kept secret.
> (3) Even one knowing the system cannot possibly read the cipher without knowing the key-word.
> (4) Even with the English to the cipher given, it is impossible to discover the key-word.[3]

Dodgson's instructions in the diary to construct the cipher start with writing the alphabet in the form of a matrix:

A	F	L	Q	W
B	G	M	R	X
C	H	N	S	Y
D	I	O	T	Z
E	K	P	V	*

The alphabet he uses is Latin; I and J and U and V are used interchangeably. The asterisk simply fills in the matrix. With a keyword of GROUND he enciphered the word SEND:

> Measuring from G to S we find it to be "2nd column 1st line," & write 21. In re-translating we begin at G, & go "2 columns to the right, & 1 line further down," & this gives us S again.
> Measuring from R to E gives 23. from O to N—04. from V to D—24. . . . we write 21.23.04.24.[4]

2. *Ibid.,* vol. I, 140.
3. C. L. Dodgson, unpublished diaries, 26 February 1858.
4. *Ibid.*

Dodgson next applied the theory of congruences, or "modular arithmetic":

> "In counting from one letter to another, call the column that the key-letter is in "0", & if you have to go beyond the last, begin again with the 1st—& so with the lines."[5]

To understand what he was actually constructing, define the group D to be $Z_5 \times Z_5$ under addition: consider the set of pairs (x y) where x and y are chosen from {0,1,2,3,4}. Addition of pairs is defined by:

(a b) + (c d) = ((a+c) mod 5 (b+d) mod 5), i.e.
(2 0) + (2 3) = (4 3); (3 4) + (4 1) = (2 0).

Mathematically, cipher systems enjoy group properties. Indeed, they are early applications of abstract groups. The fact that this Matrix Cipher system forms a group means, among other things, that addition and subtraction are inverse operations and have unique results. Furthermore, we observe that encipherment in the Vigenère system, Cipher = Plain + Key, is the inverse of decipherment in the Variant Beaufort system, Plain = Cipher + Key. This can be seen better by solving the second equation for Cipher, giving Cipher = Plain − Key. In the Beaufort system, encipherment, Cipher = Key − Plain, is the same as decipherment, Plain = Key − Cipher, i.e., they are self-inverse.

Dodgson's matrix provides a convenient mapping from the alphabet to D as follows: number the rows and columns from 0 to 4 and map each letter into the pair consisting of its row and column numbers. So G → (1 1) and S → (3 2). This mapping is an isomorphism under addition of pairs, which effectively means that whether working with letters or with pairs, the results will be consistent.

Dodgson's encipherment of S with the key letter G is equivalent to the following:

Key Letter G → (1 1)
Plaintext S → (3 2)
Ciphertext → (2 1)

Since (2 1) = (3 2) − (1 1), we have Cipher = Plain − Key.

Dodgson initially represented the cipher letters as number pairs without mapping them back to the alphabet. Decipherment gives:

5. *Ibid.*

330 INTRODUCTION

$$\text{Plain} = \text{Cipher} + \text{Key}$$
$$= (2\ 1) + (1\ 1) = (3\ 2) \to S.$$

These patterns (Cipher = Plain − Key for encipherment and Plain = Cipher + Key for decipherment) are precisely a Variant Beaufort system, with the arithmetic taking place in the group D. The tableau associated with it is not easily recognizable in spite of the fact that it contains a great deal of regularity. (See Figure 1.)

```
A B C D E    F G H I K    L M N O P    Q R S T U    W X Y Z *
B C D E A    G H I K F    M N O P L    R S T U Q    X Y Z * W
C D E A B    H I K F G    N O P L M    S T U Q R    Y Z * W X
D E A B C    I K F G H    O P L M N    T U Q R S    Z * W X Y
E A B C D    K F G H I    P L M N O    U Q R S T    * W X Y Z

F G H I K    L M N O P    Q R S T U    W X Y Z *    A B C D E
G H I K F    M N O P L    R S T U Q    X Y Z * W    B C D E A
H I K F G    N O P L M    S T U Q R    Y Z * W X    C D E A B
I K F G H    O P L M N    T U Q R S    Z * W X Y    D E A B C
K F G H I    P L M N O    U Q R S T    * W X Y Z    E A B C D

L M N O P    Q R S T U    W X Y Z *    A B C D E    F G H I K
M N O P L    R S T U Q    X Y Z * W    B C D E A    G H I K F
N O P L M    S T U Q R    Y Z * W X    C D E A B    H I K F G
O P L M N    T U Q R S    Z * W X Y    D E A B C    I K F G H
P L M N O    U Q R S T    * W X Y Z    E A B C D    K F G H I

Q R S T U    W X Y Z *    A B C D E    F G H I K    L M N O P
R S T U Q    X Y Z * W    B C D E A    G H I K F    M N O P L
S T U Q R    Y Z * W X    C D E A B    H I K F G    N O P L M
T U Q R S    Z * W X Y    D E A B C    I K F G H    O P L M N
U Q R S T    * W X Y Z    E A B C D    K F G H I    P L M N O

W X Y Z *    A B C D E    F G H I K    L M N O P    Q R S T U
X Y Z * W    B C D E A    G H I K F    M N O P L    R S T U Q
Y Z * W X    C D E A B    H I K F G    N O P L M    S T U Q R
Z * W X Y    D E A B C    I K F G H    O P L M N    T U Q R S
* W X Y Z    E A B C D    K F G H I    P L M N O    U Q R S T
```

FIGURE 1
Variant Beaufort System

Inspection reveals that like the familiar alphabet squares, the top row contains what appears to be the normal alphabet but is actually the five columns of the original matrix written sequentially. In the Vigenère (or Beaufort), subsequent rows are each rotated one letter to the left, returning to the original alphabet after twenty-six rows.

In the tableau that represents Dodgson's cipher, the rotations take place within a five-by-five block derived from one of the original columns. At the end of five rotations, instead of returning to the original position, the entire block is rotated to the left. The resulting pattern is characteristic of $Z_5 \times Z_5$.

In spite of Dodgson's enthusiastic optimism, this matrix cipher remains quite vulnerable to attack. The symmetry of $C = P - K$ and $K = P - C$ provides a relatively quick means of entry under these conditions:

1. If the matrix is known to have been used and at least a portion of the message is known in cleartext, enciphering the plaintext using the ciphertext will yield the key.

2. If the matrix is known to have been used, enciphering a repeated probable word with the ciphertext at successive start positions (sufficient to cover the length of the probable word) will reveal the key embedded in the text produced.

3. There is no reason to suspect that Dodgson thought of mixing the alphabet before building the matrix, but even if he had, and a portion of the plaintext is known, proceeding as in (1) above will yield a "pseudo" key that is sufficient to recover the rest of the plaintext.

4. If the matrix is built from a mixed alphabet and no plaintext is known, enciphering a probable word at each position and testing the potential "pseudo" key can still provide entry.

Since either the accidental ommission or repetition of a letter could hinder the decipherment, Dodgson suggests the occasional insertion of a pair of numbers in parentheses, the second of which would indicate the current position in the key. Thus, if the keyword were GROUND, (2.5) would mean that the current key letter is N: ". . . or I may insert (7.11). To make out the 11th letter of "ground", we must of course count twice through, or as a simpler way, divide by "6" and take the remainder."[6]

Dodgson realized that this provided a way to switch positions within the keyword while in the middle of the message: "If I wish to go on with any other letter than D, I invent a 2nd parenthesis—e.g. (2.5)(1.2) would mean "go on with R."[7]

6. Ibid.   7. Ibid.

Continuing in the effort to protect the cipher, Dodgson used this method to begin in the middle of the keyword and found a way to insert nulls at the beginning and end of the message. He followed the parenthesized pair with a letter whose encipherment indicated the nulls. In his example, he wrote:

(1.2) Q.

The "2" indicated starting with the second letter of GROUND. Enciphering Q by R yields 0.4, which is to be interpreted as meaning no nulls at the beginning of the message and four at the end. Finally he stated that by purposely misspelling words, either by leaving out letters or by inserting extra letters, the recovery of the key from the plaintext would be rendered impossible. His completed example, using the same keyword is:

(2.3)(V)10.14.20.00.00.01.33.40.42.40.01.20.23.02.

Almost as an afterthought, Dodgson recognized that the number pairs may be mapped back to the alphabet: "An improvement on this again is instead of 1.4 to write D . . . & so on."[8]

The Matrix Cipher followed closely on the heels of Beaufort's 1857 reinvention of the cipher that bears his name. Of great significance is that Dodgson's cipher appears to be the first example of a system that can be classified as Variant Beaufort. More important, it certainly is the first cipher system that is equivalent to one using a nonstandard arithmetic.

Dodgson was enthusiastic about The Key-Vowel Cipher, the name we have given to the first of the two ciphers that he invented in 1858. He thought it could not be broken, even by someone who knew the encipherment system, without knowing the keyword. He described it in his diary entry of 23 February.

Invented a system of cipher, which I think looks promising, as it may be carried entirely in the head.

(1) Agree on some word as the key: for every vowel in the key you write a letter representing some letter in the message, & for every consonant you write any letter at random.

(2) The rule of the vowels is as follows: Suppose M the letter to be
represented: imagine it written K L M N O then if "a" is the key-vowel, write K, & so on.

(with a e i o u above K L M N O)

---

8. *Ibid.*

CRYPTOLOGY

(3) Begin with some unmeaning letters, the number regulated by the 1st vowel of the key [a=1, e=2, i=3, o=4, u=5] & end with some, regulated by the last key-vowel.[9]

Commenting on this cipher, Dodgson wrote, "The great advantage which I think this has is, that no one, even if he knows the system, can read a piece of cipher done by it, without also knowing the key-word."[10] This is the example he gave:

Let the key be *imagine*.

As there are 4 vowels every 4 letters of the message will use it once, producing 7 letters of cypher.

Let the message be *come directly*.

We begin with 3 unmeaning letters, e.g. L A T—then for each letter of the key we proceed as follows—

for (i)   symbol for c   = C
    (m)   unmeaning —   H

$$\begin{bmatrix} a & e & i \\ m & n & o \end{bmatrix}$$

    (a)   symbol for o   = M
    (g)   unmeaning —   E
    (i)   symbol for m   = M
    (n)   unmeaning —   S
    (e)   symbol for e   = D

this gives us LATCHMEMSD.[11]

It is not apparent that the Key-Vowel Cipher is really a Vigenère in which only five of the possible twenty-six alphabets are used. Consider the following tableau:

   A B C D E F G H I J K L M N O P Q R S T U V W X Y Z
A Y Z A B C D E F G H I J K L M N O P Q R S T U V W X
E Z A B C D E F G H I J K L M N O P Q R S T U V W X Y
I A B C D E F G H I J K L M N O P Q R S T U V W X Y Z
O B C D E F G H I J K L M N O P Q R S T U V W X Y Z A
U C D E F G H I J K L M N O P Q R S T U V W X Y Z A B

Using the vowels in the keyword IMAGINE to index the rows and the letters of the plaintext COMEDIRECTLY to index the columns, we locate

---

9. C. L. Dodgson, unpublished diaries, 23 February 1858.
10. *Ibid.*   11. *Ibid.*

the letters of the ciphertext CMMDDGRDCRLX in the table. The random interior nulls are HESFEXODM; the exterior nulls are LAT on the left and DN on the right. So the entire message for transmission is LATCHMEMSDDFGERXDCORDLMXDN.

The encipherment below (without the nulls) can be represented as Key Vowel + Plain = Cipher.

   Key Vowel I A I E I A I E I A I E
   Plaintext C O M E D I R E C T L Y
   Ciphertext C M M D D G R D C R L X

Note that different cipher letters can result from the same letter of plaintext, e.g., M can be encoded by K, L, M, N, or O.

Decipherment, again without the nulls, is given by Cipher − Key Vowel = Plain. Even with the vulnerability to attack given by the periodicity of the keyword disclosed five years later by Kasiski, a cipher of this complexity was, practically speaking, unbreakable in 1858. However, it contains a serious flaw limiting its usefulness for military purposes. If multiple copies of a message were sent to different people using the same keyword, the ciphertexts would differ only in the randomly assigned nulls. Direct comparison of the texts would immediately produce the solution. This may have been one of Dodgson's reasons for abandoning this cipher and inventing the Matrix Cipher three days later. He made precisely the same claims for the Matrix Cipher as he had for the Key-Vowel Cipher, with one difference. For the Matrix Cipher Dodgson added the claim that even if the plaintext is known, it is not possible to recover the keyword. It appears that this became for him an all-important concern, sufficient to cause him to abandon the Key-Vowel Cipher rather than try to improve it.

Each of the two ciphers uses nulls at the beginning and end of the message to protect it. In both ciphers Dodgson avoided natural word divisions, which allow easier entry into polyalphabetic encipherment systems. It was precisely these that were permitting his contemporary, Charles Babbage, entry into polyalphabetics. Dodgson certainly knew Babbage and expressed an interest in his calculating machine, but it is almost certain that he was unacquainted with the latter's 1854 publications in cryptography. Dodgson and Charles Babbage were remarkable Victorian cryptologists. Babbage, the Lucasian Professor of Mathematics at Cambridge University, is perhaps best known for his Difference and Analytical Engines, which foreshadowed the modern computer. Although little of Babbage's work in cryptography was ever published, and Dodgson produced just four ciphers, their inventions show that their knowledge of cryptographic methods was "state of the art" for the time.[12]

Both of Dodgson's 1858 ciphers seem to represent a serious attempt at secure communication, but they would have been considered far too complicated for practical use by the Foreign Office. Ten years later, in the Alphabet and Telegraph Ciphers, he would reinstate both natural word divisions and punctuation. During the intervening years his focus appears to have changed from secure ciphers to those that were simpler, compact, and could be applied more automatically. He used these later ciphers for encoding messages to his child friends.

It seems clear, however, that in 1858 Charles Dodgson's Matrix Cipher was the first cipher system whose construction was based on mathematical principles. Moreover, the Matrix Cipher is even more unusual because it incorporates the encipherment of instructions about the decipherment, the "embellishments," in with the ciphertext itself. Quite remarkably, this is the first time to our knowledge that the notion of treating instructions as if they were data appears. Almost exactly 80 years later John von Neumann would reintroduce this as the "stored program" concept upon which modern computers so heavily depend.[13]

12. A complete discussion of Charles Babbage's work on ciphers can be found in Ole I. Franksen's book, *Mr. Babbage's Secret: The Tale of a Cypher—and APL* (Englewood Cliffs, N.J.: Prentice-Hall, 1985).

13. For a definitive history of codes and ciphers, the reader is referred to David Kahn, *The Codebreakers* (New York: Macmillan, 1967). A complete analysis of Dodgson's ciphers can be found in three papers by Stanley Lipson and Francine Abeles in *Cryptologia:* "The Matrix Cipher of C. L. Dodgson," vol. xiv, no. 1 (1990), 28–36; "Some Victorian Periodic Polyalphabetic Ciphers," vol. xiv, no. 2 (1990), 128–34; "The Key-Vowel Cipher Of Charles L. Dodgson," vol. xv, no. 1 (1991), 18–24.

37. The Key-Vowel Cipher
[1858: LCH 321: Marx]

This cipher, dated 23 February 1858, is not named as such in the *Circular* or in the *Lewis Carroll Handbook*. It is described in LCH only as "a long extract from the unpublished Diaries on a cipher." The entry in the LCH gives the "Notes" section of the *Circular* as its location, but it actually appears in the "Addenda to *The Lewis Carroll Circular #1*" in the second issue of the *Circular*. Apparently, Winkfield did not notice that there were two different ciphers in two distinct diary entries. He reproduced them together, with numerous insignificant errors corrected here, under the incorrect date, 15 February 1858.

※

Invented a system of cipher, which I think looks promising, as it may be carried entirely in the head—

(1) Agree on some word as the key: for every vowel in the key you write a letter representing some letter in the message, & for every consonant you write any letter at random.

(2) The rule of the vowels is as follows: suppose M the letter to be represented: imagine it written

$$a \quad e \quad i \quad o \quad u$$
$$K \quad L \quad M \quad N \quad O,$$

then if "a" is the key-vowel, write K, & so on.

(3) Begin with some unmeaning letters, the number regulated by the 1st vowel of the key [a = 1, e = 2, i = 3, o = 4, u = 5] & end with some, regulated by the last key-vowel.

The Lewis Carroll Circular, no. 2, Trevor Winkfield, ed. (Leeds: privately printed, November 1974), 65–67.

Example

Let the key be *imagine*—

As there are 4 vowels every 4 letters of the message will use it once, producing 7 letters of cypher.

Let the message be *come directly*.

We begin with 3 unmeaning letters, e.g. LAT—then for each letter of the key we proceed as follows—

for (i) symbol for c = C
 (m) unmeaning   H
 (a) symbol for o = M $\begin{bmatrix} a & e & i \\ m & n & o \end{bmatrix}$
 (g) unmeaning   E
 (i) symbol for m = M
 (n) unmeaning   S
 (e) symbol for e = D

This gives us LATCHMEMSD.

           i m a g i n e
To retranslate this suppose it written LAT/CHMEMSD.

Discard first 3 letters, & every other one with a consonant over it—

 i
As C has an i over it = *c*.

 a   a e i
M we write M & run the alphabet on till we get under i—this gives—*o*.

     i
the next M having an i over it = *m*

  e i
& D by the same rule gives *e*.

  The cypher in full will be

      i m a g i n e i m a g i n e i m a g i n e
     lat/c h m e m s d d f g e r x d c o r d l m x/d n
       c   o   m ed   i   r ec   t l   y

ending with 2 unmeaning letters.

The great advantage which I think this has is, that no one, even if he knows the system, can read a piece of cypher done by it, without also knowing the key-word. And this the correspondents may change as often as they like—or even use a whole sentence as key.

38. The Matrix Cipher

[1858: LCH 321: Marx]

This second cipher from the unpublished diaries is dated 26 February 1858. It too is not named in the *Circular* or in the *Lewis Carroll Handbook*. The error in the location of item 37 is true for this cipher, as well. Winkfield omitted half of the first page of the diary entry containing Dodgson's description of the cipher.[1]

Invented another cypher, far better than the last: it has these advantages. (1) The system is easily carried in the head.

(2) The key-word is the only thing necessarily kept secret.

(3) Even one knowing the system cannot possibly read the cipher without knowing the key-word.

(4) Even with the English to the cipher given, it is impossible to discover the key-word.

Write the alphabet thus:

A	F	L	Q	W
B	G	M	R	X
C	H	N	S	Y
D	I	O	T	Z
E	K	P	V	*

Suppose the key-word to be ground & the 1st word of the message send. Measuring from G to S we find it to be "2nd column 1st line," & write 21. In re-translating we begin at G, & go "2 columns to the right, & 1 line further down," & this gives us S again.

The Lewis Carroll Circular, no. 2, Trevor Winkfield, ed. (Leeds: privately printed, November 1974), 67–68.

1. This error was noted and the missing lines reproduced in S. H. Lipson and F. Abeles, "The Matrix Cipher of C. L. Dodgson," *Cryptologia,* vol. XIV, no. 1 (January 1990), 28.

Measuring from R to E gives 23, from O to N—04. from V to D—24, i.e. for "send" we write 21.23.04.24.

In counting from one letter to another, call the column that the key-letter is in "0," & if you have to go beyond the last, begin again with the 1st—& so on with the lines.

The accidental omission or repetition of a letter would confuse the message: to remedy this put a number every now & then in a parenthesis, to show what letter of the key has been reached.

If there are more words than one in the key, put 2 nos. in the parenthesis, to show what word & what letter of that word has been reached. If the key consists of only one word, still put the first number, at random—e.g. when I have reached "n" in the key given, I may insert (2.5.) this will look as if there were 2 words in the key—or I may insert (7.11.). To make out the 11th letter of "ground," we must of course count twice through, or as a simpler way, divide by "6" and take the remainder—If I wish to go on with any other letter than D, I invent a 2nd parenthesis—e.g. (2.5.) (1.2.) would mean "go on with R."

We may also begin & end each paragraph with a few superfluous symbols, which the correspondent will reject. To show how many there [are] at each end I add some letter of the alphabet to the parenthesis at the beginning of it—e.g. if I write (1.2.) Q, we interpret Q by means of the 2nd key-letter—viz: R, & find it to be "0th column 1st line"—this will mean "reject *no* symbol at the beginning of the paragraph but *one* at the end."

If besides this we mis-spell words, by putting in or leaving out letters, it will be impossible from the cipher & corresponding English to discover the key.

As an example: if "*ground*" is the key, and "*send him here*" the message, this may be written: (2.3) (V) 10.14.20.00.00.01.33.40. 42.40.01.20.23.02. to translate we begin with the 3rd key-letter, "O—" measuring from "O" to "V" gives "1.1." i.e. we reject the 1st & last symbols, & translate the others, beginning 0.14. − V.20. − & so on. Hence we get "*send h[k]im here*"—usually we should not write the "k" down, but omit it when we come to it.

An improvement on this again is instead of 1.4 to write D (being the 4th letter of the 1st column) & so on.

39. *The Alphabet Cipher*

[1868: LCH 63, LCAT 348: Bodleian]

Originally printed on card measuring 7⅛ by 4 15/16 inches, with the tableau on the front and the explanation on the back, the card is anonymous and undated. It is almost certainly by Dodgson and probably completed in April of 1868. The copy here is a reproduction and transcription of the original card.

EXPLANATION.

Each column of this table forms a dictionary of symbols representing the alphabet: thus, in the A column, the symbol is the same as the letter represented; in the B column, A is represented by B, B by C, and so on.

To use the table, some word or sentence should be agreed on by two correspondents. This may be called the "key-word," or "key-sentence," and should be carried in the memory only.

In sending a message, write the key-word over it, letter for letter, repeating it as often as may be necessary: the letters of the key-word will indicate which column is to be used in translating each letter of the message, the symbols for which should be written underneath: then copy out the symbols only, and destroy the first paper. It will now be impossible for any one, ignorant of the key-word, to decipher the message, even with the help of the table.

For example, let the key-word be *vigilance,* and the message "meet me on Tuesday evening at seven," the first paper will read as follows:—

The Penguin Complete Lewis Carroll (Harmondsworth, England: Penguin, 1982), 1156–57.

	A	B	C	D	E	F	G	H	I	J	K	L	M	N	O	P	Q	R	S	T	U	V	W	X	Y	Z	
A	a	b	c	d	e	f	g	h	i	j	k	l	m	n	o	p	q	r	s	t	u	v	w	x	y	z	A
B	b	c	d	e	f	g	h	i	j	k	l	m	n	o	p	q	r	s	t	u	v	w	x	y	z	a	B
C	c	d	e	f	g	h	i	j	k	l	m	n	o	p	q	r	s	t	u	v	w	x	y	z	a	b	C
D	d	e	f	g	h	i	j	k	l	m	n	o	p	q	r	s	t	u	v	w	x	y	z	a	b	c	D
E	e	f	g	h	i	j	k	l	m	n	o	p	q	r	s	t	u	v	w	x	y	z	a	b	c	d	E
F	f	g	h	i	j	k	l	m	n	o	p	q	r	s	t	u	v	w	x	y	z	a	b	c	d	e	F
G	g	h	i	j	k	l	m	n	o	p	q	r	s	t	u	v	w	x	y	z	a	b	c	d	e	f	G
H	h	i	j	k	l	m	n	o	p	q	r	s	t	u	v	w	x	y	z	a	b	c	d	e	f	g	H
I	i	j	k	l	m	n	o	p	q	r	s	t	u	v	w	x	y	z	a	b	c	d	e	f	g	h	I
J	j	k	l	m	n	o	p	q	r	s	t	u	v	w	x	y	z	a	b	c	d	e	f	g	h	i	J
K	k	l	m	n	o	p	q	r	s	t	u	v	w	x	y	z	a	b	c	d	e	f	g	h	i	j	K
L	l	m	n	o	p	q	r	s	t	u	v	w	x	y	z	a	b	c	d	e	f	g	h	i	j	k	L
M	m	n	o	p	q	r	s	t	u	v	w	x	y	z	a	b	c	d	e	f	g	h	i	j	k	l	M
N	n	o	p	q	r	s	t	u	v	w	x	y	z	a	b	c	d	e	f	g	h	i	j	k	l	m	N
O	o	p	q	r	s	t	u	v	w	x	y	z	a	b	c	d	e	f	g	h	i	j	k	l	m	n	O
P	p	q	r	s	t	u	v	w	x	y	z	a	b	c	d	e	f	g	h	i	j	k	l	m	n	o	P
Q	q	r	s	t	u	v	w	x	y	z	a	b	c	d	e	f	g	h	i	j	k	l	m	n	o	p	Q
R	r	s	t	u	v	w	x	y	z	a	b	c	d	e	f	g	h	i	j	k	l	m	n	o	p	q	R
S	s	t	u	v	w	x	y	z	a	b	c	d	e	f	g	h	i	j	k	l	m	n	o	p	q	r	S
T	t	u	v	w	x	y	z	a	b	c	d	e	f	g	h	i	j	k	l	m	n	o	p	q	r	s	T
U	u	v	w	x	y	z	a	b	c	d	e	f	g	h	i	j	k	l	m	n	o	p	q	r	s	t	U
V	v	w	x	y	z	a	b	c	d	e	f	g	h	i	j	k	l	m	n	o	p	q	r	s	t	u	V
W	w	x	y	z	a	b	c	d	e	f	g	h	i	j	k	l	m	n	o	p	q	r	s	t	u	v	W
X	x	y	z	a	b	c	d	e	f	g	h	i	j	k	l	m	n	o	p	q	r	s	t	u	v	w	X
Y	y	z	a	b	c	d	e	f	g	h	i	j	k	l	m	n	o	p	q	r	s	t	u	v	w	x	Y
Z	z	a	b	c	d	e	f	g	h	i	j	k	l	m	n	o	p	q	r	s	t	u	v	w	x	y	Z
	A	B	C	D	E	F	G	H	I	J	K	L	M	N	O	P	Q	R	S	T	U	V	W	X	Y	Z	

```
v i g i l a n c e v i g i l a n c e v i g i l a n c e v i
m e e t m e o n t u e s d a y e v e n i n g a t s e v e n
h m k b x e b p x p m y l l y r x i i q t o l t f g z z v
```

the second will contain only "h m k b x e b p x p m y l l y r x i i q t o l t f g z z v."

The receiver of the message can, by the same process, retranslate it into English.

N.B. If this table be lost, it can easily be written out from memory, by observing that the first symbol in each column is the same as the letter naming the column, and that they are continued downwards in alphabetical order. Of course, it would only be necessary to write out the particular columns required by the key-word: such a paper, however, should not be preserved, as it would afford means for discovering the key-word.

40. The Telegraph Cipher

[1868: LCH 62, LCAT 347: NYU]

Originally printed on a small card measuring $4\frac{1}{2}$ by $3\frac{1}{16}$ inches, the directions are on the front and the "sliding" alphabets are on the reverse. The card is undated and anonymous. Dodgson refers to this cipher as a "new" one in his diary entry of 22 April 1868.[1] This reference is the basis for inferring that the Alphabet Cipher is the "old" one that the Telegraph Cipher improves upon. The copy here is a transcription of the original card. The illustration on p. 324 shows a shorter version of this cipher, in Dodgson's hand, from the Pierpont Morgan Library.

Dodgson alludes to one of these two ciphers—probably the Alphabet Cipher—in a letter to Agnes [Dolly] Argles on April 17(?) and to the Telegraph Cipher in a second letter to her on April 22. In a letter to her older sister, Edith, on 29 April 1868, Dodgson enclosed the "new" cipher as well as a "Cipher-Poem" constructed with it using the keyword "fox."[2]

A version of this cipher appeared in *The Lewis Carroll Circular,* no. 1 (May 1973), Trevor Winkfield, ed.

DIRECTIONS FOR USE.

Cut this card in two along the line.

In order to send messages in this cipher, a key-word (or sentence) must be agreed on between the correspondents: this should be carried in the memory only.

To translate a message into cipher, write the key-word, letter for letter, over the message, repeating it as often as may be necessary: slide the message-alphabet along under the other, so as to bring the first letter of the message under the first letter of the

---

1. *The Diaries of Lewis Carroll,* edited by Roger L. Green (New York: Oxford University Press, 1954), vol. II, 268–69.

2. See Morton N. Cohen, *The Letters of Lewis Carroll* (New York: Oxford University Press, 1979), vol. I, 116–19.

THE TELEGRAPH CIPHER

key-word, and copy the letter that stands over 'a': then do the same with the second letter of the message and the second letter of the key-word, and so on.

Translate the cipher back into English by the same process.

For example, if the key-word be "war," and the message "meet me at six," we write it thus:—

$$\left\{\begin{matrix} W\ A\ R\ W & A\ R & W\ A & R\ W\ A \\ M\ E\ E\ T & M\ E & A\ T & S\ I\ X \\ K\ W\ N\ D & O\ N & W\ H & Z\ O\ D \end{matrix}\right\}$$

The cipher sent, "kwndonwhzod," may be re-translated by the same process.

KEY-ALPHABET
A B C D E F G H I J K L M N O P Q R S T U V W X Y Z

A B C D E F G H I J K L M N O P Q R S T U V W X Y Z A
MESSAGE-ALPHABET

Curriculum

Charles L. Dodgson's worksheet of grades in algebra for five of his pupils.

Introduction

Dodgson published his first book on mathematics, *A Syllabus of Plane Algebraical Geometry,* in 1860. As its title suggests, the book was meant to serve the needs of students. He wrote in his diary on 12 May 1855, "I began arranging a scheme for teaching systematically the first part of Algebraic Geometry: a thing which no one hitherto seems to have attempted—I find it exceedingly difficult to do it in anything like a satisfactory way."[1]

In the introduction to the book, Dodgson sets out his reasons for writing it: first, to provide formal definitions, postulates and axioms for algebraic, more properly called analytic, geometry; second, to systematically arrange propositions in their proper order; and finally, to improve the notation used in solid algebraic (analytic) geometry. He states his goal to be the reduction of the entire subject to a complete and uniform system that would stand for analytic geometry in the way Euclid does for pure geometry. That he did not achieve what he set out to accomplish is less surprising than the idea that he thought he could.

Dodgson viewed his role as Mathematical Lecturer quite seriously and spent a great deal of time on publications designed to improve both the courses and the texts used in them. *A Guide to the Mathematical Student* from 1864 (item 43) contains the syllabus of the course in pure mathematics, including all the topics covered in the lectures that were required for an honours degree in mathematics at Christ Church. Dodgson's work included substantial involvement with examinations. To relieve some of the pressure, Dodgson gave up working with classmen in 1871 after Sampson joined him as a mathematical lecturer. He continued to work with passmen until he resigned his lectureship, a post he had held for twenty-six years, in 1881.

The pamphlet received a short review in the *Athenaeum* on 4 March 1865; the writer observed that the "Guide" permitted students to take instances in different subjects in order to give each subject its proper importance and frequency—with variety.[2]

Dodgson's approach to teaching reflects his fundamentally logical way of working, namely of presenting a system—here the curriculum—in a complete and consistent manner.

1. *The Diaries of Lewis Carroll,* edited by Roger L. Green (New York: Oxford University Press, 1954), vol. I, 50.

2. *The Athenaeum,* no. 1949, 312. Published in London, the *Athenaeum* was a popular journal covering events in literature, science, and the fine arts. In 1870, music and drama were added.

41. Circular to Mathematical Friends

[1862: LCH 32: Princeton]

In this "open letter" to his mathematical colleagues, dated June 1862, Dodgson asks for their reactions to the accompanying sheets. These display all of the topics taught in Pure Mathematics as well as a list of two thousand examples, referenced by topic, which were designed to be worked by students. The *Circular* received attention beyond the confines of Oxford. G. Richardson of St. John's College, Cambridge, for one, thought that the scheme of problems was of doubtful value. Another respondent was Francis Harrison of Oriel College, Oxford, who in a letter dated 10 September 1862 suggested some minor changes—mainly in the order of teaching certain topics. For example, he proposed that the equations of the conic sections and the equations of the normals and tangents to them should be given in polar coordinates and then in rectangular coordinates, not the other way around.[1]

A version of the *Circular* has been published in *Jabberwocky,* 7, no. 1 (1977–78), 7–8.

Ch. Ch., Oxford, June 1862.

Sir,

May I beg the favour of your attention to the accompanying tables.

Their object is twofold:

First, to exhibit in a compendious form, the whole of the subject-matter of Pure Mathematics. This may be useful as an aid in laying out plans of reading and reviewing, and in showing the Student at a glance where he is on his course, how much is done, and how much remains to be done.

Secondly, to furnish a guide for working examples in the whole subject, so arranged as to secure that the most important subjects

---

1. Harrison's and Richardson's letters are included in Dodgson's mathematical papers in the Parrish Collection in the Princeton University Library.

shall have the largest share of attention. The cycle intended for this purpose will consist of two columns: one containing the numbers from 1 to 2000, the other, references to this list. The Student using this cycle will simply have to turn to the list for each reference, work as many examples as he thinks fit on the subject there indicated, and at the end of each day's work mark the point of the cycle which he has reached. Blanks will be left at intervals in the column of references, so that each individual teacher may adapt the cycle to his own views, by inserting additional references. In the accompanying pages, the small figures on the left of the line indicate how many of the 2000 examples it is proposed to assign to each subject.

Specimen of Cycle.

201	J. 4
202	H. 3
203	R. 8
204	B. 4
205	
206	D. 2

But before printing off these pages, or constructing the cycle to accompany them, I think it most desirable to obtain the opinions of other mathematical teachers on the following questions:

1. Are any subjects omitted from this list?
2. Are the subjects here arranged in the order in which it would usually be advisable that the Student should go through them?
3. Are any of them too minutely subdivided, or not subdivided sufficiently?
4. Is the proper relative number of examples assigned to each?

If you would do me the favour to look through the accompanying pages with a view to these four questions, making such additions or corrections as you think fit, and return them to me, (for which purpose I enclose a printed cover,) before the end of next October, you would greatly oblige

Your obedient Servant,
CHARLES L. DODGSON,
Student and Mathematical Lecturer of Ch. Ch., Oxford.

42. Proof Sheets Accompanying the "Circular to Mathematical Friends"

[Undated: LCH 32a: Bodleian]

These sheets form a printed draft of Part 1 of the *General List of [Mathematical] Subjects, and Cycle for Working Examples* that Dodgson published in 1863. This sixteen-page pamphlet, known to be in the Hartley collection, is unavailable. Dodgson remarked in his diary entry of 11 July 1862, "Sent off first batch of proof-sheets of *General List* etc. with printed letters [presumably the Circular] and stamped covers addressed to myself: 75 of the list."[1]

GENERAL LIST OF SUBJECTS.

28	A. Arithmetic.
36	B. Euclid I, II.
48	C. Algebra; to Quadratic Equations.
43	D. Euclid III, IV.
50	E. Algebra; from Quadratic Equations to Binomial Theorem.[1]
34	F. Euclid V, VI.
147	G. Linear Algebraical Geometry. Plane " to end of Trigonometry (1st time).
96	H. Geometrical Conic Sections.
50	I. Higher Plane Pure Geometry.
105	J. Algebra; from Binomial Theorem to Theory of Equations.

---

1. *The Diaries of Lewis Carroll,* edited by Roger L. Green (New York, Oxford University Press, 1954), vol. I, 184.

[1] i.e. From Quadratic Equations *exclusive* to Binomial Theorem *inclusive*. The same rule of interpretation applies to J. K. &c. [Dodgson used an asterisk instead of a numbered footnote.—Ed.]

118	K. Plane Algebraical Geometry; from end of Trigonometry to Circle.
101	L. ” ” Trigonometry (2nd time).
130	M. ” ” from Circle to Parabola.
141	N. Differential Calculus (1st time).
40	O. Calculus of Finite Differences (1st time).
65	P. Euclid XI, XII, and higher Pure Solid Geometry.
39	Q. Solid Algebraical Geometry; to end of Stereometry.
70	R. ” ” ; from end of Stereometry to Cylinder.
51	S. Higher Plane Algebraical Geometry.
150	T. Integral Calculus (1st time).
45	U. Solid Algebraical Geometry; Quadratic Loci.
71	V. Higher Algebra.
200	W. Differential Calculus (2nd time).
145	X. Integral Calculus (2nd time).
31	Y. Calculus of Finite Differences (2nd time).
37	Z. Calculus of Variations.

SUBJECTS SUBDIVIDED

A.

Arithmetic.

1	1. Addition, Subtraction, Multiplication, and Division; (Simple.)
2	2. Greatest Common Measure and Least Common Multiple.
2	3. Square root and Cube root.
3	4. Vulgar Fractions; addition, subtraction, multiplication, and division.
3	5. Decimal Fractions; addition, subtraction, multiplication, and division.
2	6. Circulating Decimals.
1	7. Reduction from one denomination to another.
1	8. Addition, Subtraction, &c. (Compound).
3	9. Reduction of Fractions (vulgar and decimal) of higher de-

	nomination to lower; and of lower denomination to fractions (vulgar and decimal) of higher.
1	10. Practice.
2	11. Mensuration, Superficial and Solid.
1	12. Duodecimals.
2	13. Rule of Three; Direct, Inverse, and Double. Proportional parts.
2	14. Interest; Simple and Compound. Discount. Equation of payments. Stocks.
2	15. Miscellaneous, viz.: Exchange. Profit and Loss. Partnership, &c.

B.

Euclid I, II.

4	1. Book I.
3	2. Book II.
8	3. Deductions from Book I. Problems.
10	4. " " Theorems.
5	5. " Book II. Problems.
6	6. " " Theorems.

C.

Algebra; to Quadratic Equations.

2	1. Addition, Subtraction, Multiplication, and Division.
2	2. Greatest Common Measure and Least Common Multiple.
4	3. Fractions.
3	4. Involution and Evolution.
3	5. Fractional Indices.
4	6. Equations, one unknown quantity; Simple.
4	7. " " Quadratic.
4	8. " two or more unknown quantities: Simple.
4	9. " " Quadratic.
	Problems leading to Equations,

3	10. One unknown quantity; Simple.
4	11. " Quadratic.
3	12. Two or more unknown quantities; Simple.
4	13. " Quadratic.
2	14. Theory of Equations (1st time).
2	15. Miscellaneous.

D.

Euclid III, IV.

6	1. Book III.
4	2. Book IV.
9	3. Deductions from Book III. Problems.
11	4. " " Theorems.
6	5. " Book IV. Problems.
7	6. " " Theorems.

E.

Algebra; from Quadratic Equations to Binomial Theorem.

2	1. Inequalities.
10	2. Ratio, Proportion, and Variation.
9	3. Series; Arithmetical, Geometrical, and Harmonical.
9	4. Permutations and Combinations.
5	5. Binomial Theorem.
6	6. Logarithms, use of.
5	7. Chances (1st time).
4	8. Miscellaneous.

F.

Euclid V, VI.

2	1. Book V.
7	2. Book VI.
10	3. Deductions from Book VI. Problems.
12	4. " " Theorems.

G.

Linear Algebraical Geometry.
Plane   "   "   to
end of Trigonometry (1st time).

	Linear Algebraical Geometry.
5	1. Representation and discussion of lengths absolute.
4	2. " " " with direction.
3	3. " of positions of Points by means of lengths; and discussion of such lengths.
8	4. Interpretation of Equations; and discussion of Points.

	Plane Algebraical Geometry.
8	5. Representation and discussion of magnitudes absolute.
4	6. " " " with direction.
10	7. Goniometry: i.e., representation of angles, with direction, by means of ratios; and discussion of such ratios.
10	8. Angles; relations between goniometrical ratios of an angle.
6	9. " co-indicants of particular angles.
18	10. " relations between goniometrical ratios of two or more angles.
7	11. " inverse function.
8	12. " elimination of goniometrical ratios.
5	13. Theory of Projection (Plane).
12	14. Trigonometry; properties of Triangles.
6	15. " " Quadrilateral Figures inscribed in Circles.
5	16. " " regular Polygons.
16	17. Heights and distances.
12	18. Miscellaneous, viz., Subsidiary angles, &c.

H.

Geometrical Conic Sections.

10	1. Ellipse.
9	2. Hyperbola.
8	3. Parabola.
6	4. Sections of Cone.
8	5. Problems on Ellipse.
10	6. Theorems "
7	7. Problems on Hyperbola.
9	8. Theorems "
6	9. Problems on Parabola.
8	10. Theorems "
4	11. Problems on Cone.
6	12. Theorems "
5	13. Miscellaneous, viz., mechanical methods of tracing curves, &c.

I.

Higher Plane Pure Geometry.

4	1. Harmonic Proportion.
5	2. Harmonic Pencils.
6	3. Transversals of Pencils.
4	4. Anharmonic Ratio.
5	5. Geometrical Involution.
4	6. Poles and Polars in relation to Circles.
4	7. Methods of Reciprocation.
5	8. Radical Axis and Centres of Similitude.
3	9. Principle of Continuity.
5	10. Projection.
5	11. Miscellaneous.

J.

Algebra; from Binomial Theorem to Theory of Equations.

6	1. Evolution of Binomial Surds.
12	2. Indeterminate Coefficients.
6	3. Continued Fractions.
10	4. Indeterminate Equations, (1st and 2nd degree).
7	5. Partial Fractions.
3	6. Scales of Notation.
7	7. Properties of Numbers.
7	8. Vanishing Fractions.
8	9. Converging and diverging Series.
9	10. Logarithms, construction of.
7	11. Interest, Discount, and Annuities.
6	12. Chances (2nd time), and Life-Annuities.
11	13. Theory of Equations (2nd time).
6	14. Miscellaneous.

K.

Plane Algebraical Geometry; from end of Trigonometry to Circle.

5	1. Determination of positions of Points, Lines, and Circles, by means of magnitudes; and discussion of such magnitudes.
1	2. Interpretation and classification of simple Equations.
4	3. Interpretation of Pairs of Equations. Representation and discussion of Points.
2	4. Investigation of Locus of single Simple Equations. Representation of Lines.
10	5. Lines; Problems.
12	6. ” Theorems.
8	7. Rectilinear Figures; Problems.
10	8. ” Theorems.
6	9. Pencils; Problems.

9	10. " Theorems.
10	11. Representation of Loci of Points fulfilling certain conditions.
2	12. Representation of Pairs of Lines. Criterion that Quadratic Equation should represent Pair of Lines.
4	13. Pairs of Lines; Problems.
5	14. " Theorems.
3	15. Representation of Circles. Criterion that Quadratic Equation should represent Circle.
10	16. Circles; Problems.
12	17. " Theorems.
5	18. Miscellaneous.

L.

Plane Algebraical Geometry;
Trigonometry (2nd time).

14	1. Circular measure. Area of Circle, &c.
35	2. Demoivre's Theorem; and theorems involving powers of goniometrical ratios.
14	3. Summation of series of goniometrical ratios.
21	4. Relations between angle and its goniometrical ratios. Gregorie's Series. Euler's and Machin's Series for π.
17	5. Miscellaneous; viz., resolution of sin θ and cos θ into factors, &c.

M.

Plane Algebraical Geometry; from Circle to Parabola.

6	1. Interpretation and classification of Quadratic Equations. Quadratic Locus;
6	2. General. Problems.
12	3. " Theorems.
8	4. " when $B^2 - 4AC \neq 0$, i.e. Central Locus; Problems.

18	5. " "	Theorems.
10	6. Central, when $B^2 - 4AC < 0$, i.e. Ellipse.	Problems.
20	7. " "	Theorems.
6	8. " when $B^2 - 4AC > 0$, i.e. Hyperbola.	Problems.
16	9. " "	Theorems.
6	10. General, when $B^2 - 4AC = 0$, i.e. Non-central Locus, or Parabola.	Problems.
16	11. " "	Theorems.
6	12. Miscellaneous.	

N.

Differential Calculus (1st time).

3 1. Elements of subject.
3 2. Differentiation of functions connected by addition, &c.
9 3. "   algebraical functions.
8 4. "   compound functions.
8 5. "   circular functions.
5 6. "   functions of many variables.
4 7. Successive differentiation. Leibnitz's Theorem.
4 8. Maclaurin's Theorem.
4 9. Theory of equicrescent variable. Taylor's Theorem.
6 10. Elimination of constants and functions by differentiation (1st time).
4 11. Relations between functions and derived functions; viz.

$$\frac{F(x_0 + h) - F(x_0)}{f(x_0 + h) - f(x_0)} = \frac{F'(x_0 + \theta h)}{f'(x_0 + \theta h)}, \&c.$$

6 12. Order of Infinitesimals.
7 13. Evaluation of quantities of the form $\frac{0}{0}$, &c.
8 14. Maxima and minima of explicit functions of *one* variable.
11 15. Geometrical application to end of   "
2 16. Symbols of direction extended.
8 17. Cissoid, Witch, &c.
10 18 Tangents &c. of plane curves.

5	19. Direction of curvature. Hessian.
5	20. Multiple points.
6	21. Tracing curves.
4	22. Curvature of plane curves.
5	23. Evolutes and involutes.
6	24. Miscellaneous.

O.

Calculus of Finite Differences (1st time).

3	1. Differentiation of functions.
4	2. Integration of functions by indeterminate coefficients.
3	3. " product of n terms in $A.\ P.$, and of reciprocal of the same.
4	4. Resolution of rational algebraical functions into these 2 forms.
6	5. Supplying deficient factors.
8	6. Integration of circular, exponential, and other functions.
12	7. Summation of Series by general methods.

P.

Euclid XI, XII,
and higher Pure Solid Geometry.

9	1. Book XI.
6	2. Book XII.
9	3. Deductions from Book XI. Problems.
12	4. " " Theorems.
8	5. " Book XII. Problems.
11	6. " " Theorems.
4	7. Higher Pure Solid Geometry. Problems.
6	8. " Theorems.

Q.

Solid Algebraical Geometry; to end of Stereometry.

2	1. Representation and discussion of volumes absolute.
2	2. " of magnitudes with direction.
3	3. Theory of Projection in Space.
10	4. Spherical Trigonometry; i.e., properties of solid angles.
2	5. Napier's Analogies.
1	6. Gauss' Theorems.
7	7. Solution of spherical Triangles; inscribed Circles; area of triangle and lune, &c.
1	8. Cagnolis' Theorem.
1	9. Llhuillier's Theorem.
6	10. Stereometry; i.e. properties of plane-sided Solids; inscribed Spheres; volume and diagonal of Parallelepipedon, &c.
4	11. Miscellaneous.

R.

Solid Algebraical Geometry; from end of Stereometry to Cylinder.

2	1. Determination of position, in Space, of Points, Lines, Planes, Spheres, and Cylinders, by means of certain magnitudes; and discussion of such magnitudes.
1	2. Interpretation and classification of Simple Equations.
2	3. " " Pairs of Equations.
4	4. " of sets of 3 Equations. Representation and discussion of Points.
2	5. Investigation of Locus of single Simple Equations. Representation of Planes.
6	6. Planes. Problems
6	7. " Theorems.
3	8. Plane-sided Solids. Problems.

4	9. " Theorems.
3	10. Representation of Superficial Loci of Points fulfilling certain conditions.
1	11. Representation of Pairs of Planes. Criterion that Quadratic Equation should represent Pair of Planes.
2	12. Pairs of Planes. Problems.
2	13. " Theorems.
1	14. Investigation of Locus of Pairs of Simple Equations. Representation of Lines.
5	15. Lines. Problems.
6	16. " Theorems.
1	17. Representation of Spheres. Criterion that Quadratic Equation should represent Sphere.
4	18. Spheres. Problems.
5	19. " Theorems.
1	20. Representation of Cylinders. Criterion that Quadratic Equation should represent Cylinder.
2	21. Cylinders. Problems.
3	22. " Theorems.
4	23. Miscellaneous.

S.

Higher Plane Algebraical Geometry.

2	1. Eccentric angles.
2	2. Similar Conic Sections.
7	3. Contact of Conics. Osculating circle. Centre of curvature, and Evolutes.
5	4. Anharmonic properties of Conics.
4	5. Method of reciprocal Polars.
4	6. Involution.
2	7. Pascal's Theorem.
3	8. Tangential coordinates.
2	9. Discussion of Locus of nth degree.
3	10. Interpretation and classification of Cubic Equations.
3	11. Discussion of Cubic Loci.

364 ITEM 42

2	12. Interpretation and classification of Biquadratic Equations.
3	13. Discussion of Biquadratic Loci.
4	14. Discussion of Transcendental Loci.
5	15. Miscellaneous.

T.

Integral Calculus (1st time).

4	1. Elements of subject.
18	2. Integration of rational algebraical functions.
18	3. " irrational "
12	4. " " " by rationalization.
11	5. " " " by reduction.
9	6. " exponential and logarithmic functions.
10	7. " circular functions.
20	8. Definite integration.
10	9. Rectification of plane curves.
10	10. Quadrature of plane surfaces.
8	11. " surfaces of revolution.
8	12. Cubature of solids of revolution.
12	13. Miscellaneous.

U.

Solid Algebraical Geometry;
Quadratic Loci.

1	1. Interpretation and classification of Single Quadratic Equation.
3	2. General Quadratic Locus Problems.
3	3. " Theorems.
3	4. Reduced Quadratic Locus, $(Px^2 + Qy^2 + Rz^2 + sx + Ty + vz + w = 0,)$ when neither P, Q, nor R $= 0$; i.e. Central Quadratic Locus. $(Px^2 + Qy^2 + Rz^2 + H = 0)$. Problems.
3	5. " Theorems.
2	6. Central Quadratic Locus, when H $= 0$, i.e. Cone. Problems.

2	7. " Theorems.
3	8. Central Quadratic Locus, when P, Q, R, and H have the same sign; i.e. Ellipsoid, and Prolate and Oblate Spheroid. Problems.
3	9. " Theorems.
1	10. Central Quadratic Locus, when one of them has a different sign from the root; i.e. Hyperboloid of one sheet. Problems.
1	11. " Theorems.
1	12. Central Quadratic Locus, when two of them have a different sign from the other two; i.e. Hyperboloid of two sheets. Problems.
1	13. " Theorems.
2	14. Central Quadratic Locus, when either P, Q, or R $= 0$; i.e. the Axicentral Locus, or Central Cylinder. Problems.
2	15. " Theorems.
2	16. Reduced Quadratic Locus, when one or more of the three, (P, Q, and R,) $= 0$; i.e. Non-central Locus. Problems.
2	17. " Theorems.
2	18. Non-central Locus, when *one* of the three, (P, Q, and R,) $= 0$; i.e. Paraboloid. Problems.
2	19. " Theorems.
1	20. Non-central Locus, when *two* of the three (P, Q, and R,) $= 0$; i.e. Parabolic Cylinder. Problems.
1	21. " Theorems.
2	22. Miscellaneous, (e.g. Cono-cuneus). Problems.
2	23. " Theorems.

V.

Higher Algebra.

4	1. Theory of equations (3rd time).
3	2. Transformation of equations.
2	3. Equal roots.
3	4. Limits of roots. Separation of roots.

2	5. Commensurable roots.
2	6. Depression of equations.
1	7. Reciprocal equations.
2	8. Binomial "
3	9. Cubic "
3	10. Biquadratic "
2	11. Sturm's Theorem.
2	12. Fourier's Theorem.
1	13. Lagrange's method of approximation.
1	14. Newton's "
1	15. Horner's method.
3	16. Symmetrical functions of roots.
1	17. Sums of powers of roots.
6	18. Determinants.
4	19. Expansion of functions in series.
5	20. Invariants. Covariants. Emanants. Evectants.
2	21. Contravariants.
3	22. Hyperdeterminant Calculus. Hermite's Law of Reciprocity.
2	23. Canonizants.
2	24. Binary Quantics, Quadrics, &c.
2	25. Ternary Quantics, Quadrics, &c.
3	26. Discriminants, &c.
2	27. Commutants.
4	28. Miscellaneous.

W.

Differential Calculus (2nd time).

3	1. Trigonometrical expressions. Roots of $+1$ and -1. Imaginary logarithms.
2	2. Limits of Maclaurin's and Taylor's Theorems.
5	3. Change of equicrescent variable.
4	4. Succsssive differentiation of functions of many independent variables.
2	5. Euler's Theorem of homogeneous functions.

3	6. Successive differentiation of implicit functions.
2	7. Bernoulli's Numbers.
2	8. Lagrange's Theorem.
2	9. Laplace's Theorem. Extension of Maclaurin's Theorem.
10	10. Elimination of constants and functions (2nd time).
5	11. Transformation of differential expressions into their equivalents in terms of other variables.
4	12. Expansion of functions of one variable. Accurate proofs of Maclaurin's and Taylor's Theorems.
4	13. Expansion of functions of two or more variables. Maxima and minima
6	14. Of implicit functions of 2 independent variables.
7	15. Of explicit ” ” ”
5	16. Of functions of 3 or more ” ”
5	17. ” ” not independent ”
5	18. Properties of Curves of the nth degree.
3	19. Contact of curves (plane).
6	20. Envelopes ”
2	21. Theory of reciprocation.
3	22. Caustics.
8	23. Application to curved surfaces.
10	24. Tangent-planes, &c. of ”
4	25. Application to curves in space.
8	26. Tangents, &c. of ”
3	27. Geodesic lines, &c.
4	28. Formation of surfaces by generators and directors.
3	29. Curved surfaces generated by right lines. Ruled surfaces.
3	30. ” ” Conical ”
3	31. ” ” Cylindrical ”
3	32. ” ” Developable ”
3	33. ” ” Skew ”
3	34. ” ” Conoidal ”
3	35. ” by circles. Surfaces of revolution.
3	36. ” ” Tubular ”
2	37. Curves in space. Curvature-angle of contingence.
2	38. ” Torsion.

2	39. "	The polar surface.
2	40. "	The osculating sphere.
2	41. "	Complex flexure.
2	42. "	The osculating surface.
2	43. "	The rectifying surface and line.
2	44. Curved surfaces.	Curvature. Euler's Theorem.
2	45. "	Umbilics.
2	46. "	Lines of curvature.
2	47. "	Dupin's Theorem.
2	48. "	Osculating surfaces.
8	49. Calculus of operations, Elements of.	
6	50. Laws of commutation, distribution, and iteration.	
3	51. Law of total differentiation.	
8	52. Miscellaneous.	

X.

Integral Calculus (2nd time).

4	1. Successive integration.
5	2. Rectification of non-plane curves.
5	3. Determination of the equation to a curve by means of a relation between the length and the coordinates to any point on it.
5	4. Involutes of plane curves.
5	5. Quadrature of curved surfaces.
5	6. Cubature of solids bounded by any curved surface.
4	7. Properties of multiple integrals.
4	8. Transformation of multiple integrals.
3	9. Curvilinear co-ordinates. Gauss' System.
3	10. " Lamé's System and Jacobi's modification.
4	11. Variation of definite integrals due to variation of parameters involved in element-function.
4	12. Variation of definite integrals due to variation of parameters involved in element-function and in the limits. Differential equations.

	13.	General principles.
		First order.
3	14.	Exact total differential equations.
3	15.	Homogeneous equations of 2 variables.
3	16.	The first linear differential equation.
4	17.	Partial differential equations of 1st degree.
3	18.	Integrating factors of differential equations.
3	19.	Singular solutions of "
2	20.	Differential equations of higher degrees.
4	21.	Particular processes.
		Higher orders;
5	22.	First degree; general properties.
5	23.	" linear differential equations.
4	24.	" " with constant coefficients.
		1st method.
4	25.	" " " 2nd "
4	26.	" " " 3rd "
5	27.	" " with variable coefficients.
4	28.	Higher degrees; total differential equations.
4	29.	" partial "
5	30.	Geometrical Problems involving diff. equations.
		1st order.
5	31.	" " 2nd "
	32.	Simultaneous differential equations. General principles.
3	33.	" Linear. 1st order.
3	34.	" " Higher
		orders.
		Integration of differential equations by series.
3	35.	Application of Taylor's Theorem.
3	36.	" Maclaurin's "
4	37.	Method of undetermined coefficients.
2	38.	Solution of Riccati's Equation.
6	39.	Miscellaneous.

Y.

Calculus of Finite Differences (2nd time).

2	1. Solution of equations of differences.	1st order.
2	2. " " "	2nd order.
2	3. " " "	nth order.
3	4. " mixed differences.	
4	5. Summation of Series; by particular assumptions.	
3	6. " by differentiation.	
2	7. " of recurring Series.	
4	8. Interpolation of Series.	
3	9. Generating functions.	
6	10. Miscellaneous.	

Z.

Calculus of Variations.

    1. General principles.

2   2. Variation of $\int_0^1 F(x, dx, d^2x, \ldots y, dy, d^2y, \ldots)$.

2   3. Variation of $\int_0^1 F(x, y, y', y'', \ldots)$.

2   4.   " $\int_0^1 F(x, dx, d^2x, \ldots y, dy, d^2y, \ldots z, dz, d^2z, \ldots)$.

2   5.   " $\int_0^1 F(x, y, y', y'', \ldots z, z', z'', \ldots)$.

2   6. Variation of a variation.
2   7.   "   of a product of differentials.
2   8.   "   of a definite double integral due to the variations of the limits.

    Maxima and minima.

    9. Critical values of definite integrals, whose element-functions involve variables and their differentials; general principles.

3	10.	"	relative max. and min.
2	11.	"	absolute "
2	12.	Geodesic lines;	equations to.
2	13.	"	properties of.
	14.	Critical values of definite integrals, whose element-functions involve derived functions; general principles.	
3	15.	"	particular cases.
	16.	Discriminating conditions; general principles.	
	17.	"	requisite data.
	18.	"	proof that $\delta \int u\,dx$ is an exact differential: its integral, &c.
2	19.	"	particular cases.
	20.	Critical values of a double definite integral; necessary criteria.	
3	21.	"	application of.
6	22.	Miscellaneous.	

43. *A Guide to the Mathematical Student in Reading, Reviewing, and Working Examples*
PART I: PURE MATHEMATICS.

[1864: LCH 38, LCAT 338: Bodleian]

The first eighteen pages of this pamphlet present all of the subject matter of pure mathematics, divided into twenty-six (not all different) topics. This part is substantially the same as the proofsheets, item 42 above. Each of the major topics is subdivided into anywhere from four to fifty-one subtopics. The number of times the topic appears in the Cycle, which appears on the following nine pages, is indicated by the number in the left margin. The Cycle gives the order for working examples on the subtopics in the General List. The purpose of the design, having students do several examples at each place the topic appeared in the Cycle, would ensure complete coverage of the material since more important subtopics appeared a greater number of times. For example, under topic X.9, "Integral Calculus (2nd time)," the subtopic "Curvilinear co-ordinates, Gauss' System, Lamé's System and Jacobi's modification" has the marginal number 2. In the Cycle, X.9 is listed as numbers 221 and 1085. The Cycle has places for 1702 items.[1] The level of the subject matter indicates that the *Guide* was meant for honours candidates in the Final Mathematical School.

The preface to the pamphlet, dated December 1864, reiterates the contents of the *Circular* (item 41), further evidence that this *Guide* is the result of the reworking of the proofsheets sent with the Circular. It seems likely that the "missing" pamphlet from the Hartley collection, number 33 in the *Lewis Carroll Handbook*, is really an early version of the *Guide* or an edited version of the proofsheets. Dodgson's diary entry of 10 February 1863 states, "Today (Feb: 10) I have sent to the Press the MS. for the Cycle: it has taken me many hours to work."[2] This reference to the early version appears to be the only reference to that work.

---

1. In "An Assessment of *General List of Subjects* 1863, and *A Guide to the Mathematical Student* 1864," *Jabberwocky* 7, no. 1, 1977–78, 9–18, Edward Wakeling points out that Dodgson actually listed 1621 subtopics in the Cycle. The gaps were left so that teachers could insert their own items.

2. *The Diaries of Lewis Carroll*, edited by Roger L. Green (New York: Oxford University Press, 1954), vol. I, 192.

A GUIDE TO THE MATHEMATICAL STUDENT

PREFACE

The object of the following pages is twofold:—

First, to exhibit, in a compendious form, the whole subject-matter of Pure Mathematics, arranged in the order in which it would usually be advisable that the student should go through it. This Syllabus may be useful as an aid in laying out plans of reading and reviewing, and in shewing the student at a glance where he is on his course, how much is done, and how much remains to be done.

Secondly, to furnish a guide for working examples in the whole subject, so arranged as to secure that the most important subjects shall have the largest share of attention. The Cycle intended for this purpose consists of two columns: one containing the numbers from 1 to 1702, the other, references to the Syllabus. It is intended that the student using it should turn to the Syllabus for each reference, and work two or three examples in the subject there indicated, (of course passing over all references to subjects he has not read,) and at the end of each day's work mark what point in the Cycle he has reached.

In the Syllabus, the small figures to the left of the line indicate how often each subject is referred to in the cycle: so that if the teacher should consider that the examples assigned to any subject are either too many or too few, he can remedy the defect by erasing references in the Cycle, or by inserting additional ones.

The present attempt is, no doubt, deficient and faulty in many respects: and any suggestions from Mathematical teachers for remedying its defects will be gratefully received by the compiler.

Christ Church, Oxford,
December, 1864

GENERAL LIST OF SUBJECTS.

30	A. Arithmetic.
20	B. Euclid I, II.
75	C. Algebra; to Quadratic Equations.

23	D. Euclid III, IV.
45	E. Algebra; from Quadratic Equations to Binomial Theorem[1]
16	F. Euclid V, VI.
114	G. Linear Algebraical Geometry. Plane " to end of Trigonometry (1st time).
45	H. Geometrical Conic Sections.
100	I. Algebra; from Binomial Theorem to Theory of Equations.
45	J. Higher Plane Pure Geometry.
110	K. Plane Algebraical Geometry; from end of Trigonometry to Quadratic Loci (constructed from Geometrical properties).
24	L. Plane Algebraical Geometry; Trigonometry (2nd time).
120	M. Plane Algebraical Geometry; Quadratic Loci (constructed from Equations).
135	N. Differential Calculus (1st time).
19	O. Calculus of Finite Differences (1st time).
20	P. Euclid XI, XII, and higher Solid Pure Geometry.
22	Q. Solid Algebraical Geometry; to end of Stereometry.
65	R. Solid Algebraical Geometry; from end of Stereometry to Quadratic Superficial Loci (constructed from Geometrical properties).
37	S. Higher Plane Algebraical Geometry.
135	T. Integral Calculus (1st time).
45	U. Solid Algebraical Geometry: Quadratic Superficial Loci (constructed from Equations).
77	V. Higher Algebra.
145	W. Differential Calculus (2nd time).
102	X. Integral Calculus (2nd time).
25	Y. Calculus of Finite Differences (2nd time).
35	Z. Calculus of Variations.

[1] i.e. From Quadratic Equations *exclusive* to Binomial Theorem *inclusive*. The same rule of interpretation applies to J, K, &c. [Dodgson used an * instead of a numbered footnote.—Ed.]

SUBJECTS SUBDIVIDED

A.

Arithmetic.

1	1. Addition, Subtraction, Multiplication, and Division; (Simple.)
2	2. Greatest Common Measure and Least Common Multiple.
2	3. Square root and Cube root.
3	4. Vulgar Fractions; addition, subtraction, multiplication, and division.
3	5. Decimal Fractions; addition, subtraction, multiplication, and division.
2	6. Circulating Decimals.
1	7. Reduction from one denomination to another.
1	8. Addition, Subtraction, &c. (Compound).
3	9. Reduction of Fractions (vulgar and decimal) of higher denomination to lower; and of lower denomination to fractions (vulgar and decimal) of higher.
1	10. Practice.
2	11. Mensuration, Superficial and Solid.
1	12. Duodecimals.
2	13. Rule of Three; Direct, Inverse, and Double. Proportional parts.
2	14. Interest, Simple and Compound. Discount. Equation of payments. Stocks.
4	15. Miscellaneous, viz.: Exchange. Profit and Loss. Partnership, &c.

B.

Euclid I, II.

	1. Book I.
	2. Book II.
6	3. Deductions from Book I. Problems.

7	4.	"	"	Theorems.
3	5.	"	Book II.	Problems.
4	6.	"	"	Theorems.

C.

Algebra; to Quadratic Equations.

2	1. Addition, Subtraction, Multiplication, and Division.
2	2. Greatest Common Measure and Least Common Multiple.
5	3. Fractions.
3	4. Involution and Evolution.
4	5. Fractional Indices.
9	6. Equations, one unknown quantity; Simple.
10	7. " " Quadratic.
6	8. " two or more unknown quantities; Simple.
6	9. " " Quadratic.
	Problems leading to Equations,
5	10. One unknown quantity; Simple.
6	11. " Quadratic.
5	12. Two or more unknown quantities; Simple.
6	13. " Quadratic.
2	14. Theory of Equations (1st time).
4	15. Miscellaneous.

D.

Euclid III, IV.

	1. Book III.
	2. Book IV.
6	3. Deductions from Book III. Problems.
8	4. " " Theorems.
4	5. " Book IV. Problems.
5	6. " " Theorems.

A GUIDE TO THE MATHEMATICAL STUDENT 377

E.

Algebra; from Quadratic Equations to Binomial Theorem.

2 | 1. Inequalities.
6 | 2. Ratio, Proportion, and Variation.
9 | 3. Series: Arithmetical, Geometrical, and Harmonical.
9 | 4. Permutations and Combinations.
5 | 5. Binomial Theorem.
6 | 6. Logarithms, use of.
4 | 7. Chances (1st time).
4 | 8. Miscellaneous.

F.

Euclid V, VI.

   1. Book V.
   2. Book VI.
8 | 3. Deductions from Book VI. Problems.
8 | 4.   "   "   Theorems.

G.

Linear Algebraical Geometry,
Plane   "   " to
end of Trigonometry (1st time).

Linear Algebraical Geometry.
5 | 1. Representation and discussion of lengths absolute.
    2.   "   "   " with direction.
    3.   "   of positions of Points by means of lengths; and discussion of such lengths.
3 | 4. Interpretation of Equations; and discussion of Points.

Plane Algebraical Geometry.
5 | 5. Representation and discussion of magnitudes absolute.
    6.   "   "   " with direction.

	7. Goniometry: i.e., representation of angles, with direction, by means of ratios; and discussion of such ratios.
12	8. Angles; relations between goniometrical ratios of an angle.
6	9. " goniometrical ratios of particular angles.
18	10. " relations between goniometrical ratios of two or more angles.
7	11. " inverse function.
5	12. " elimination of goniometrical ratios.
	13. Theory of Projection (Plane).
18	14. Trigonometry; properties of Triangles.
6	15. " " Quadrilateral Figures inscribed in Circles.
5	16. " " regular Polygons.
16	17. Heights and distances.
8	18. Miscellaneous, viz., Subsidiary angles, &c.

H.

Geometrical Conic Sections.

	1. Ellipse.
	2. Hyperbola.
	3. Parabola.
4	4. Problems on Parabola.
5	5. Theorems "
5	6. Problems on Ellipse.
8	7. Theorems "
5	8. Problems on Hyperbola.
8	9. Theorems "
5	10. Miscellaneous, viz., mechanical methods of tracing curves, &c.

I.

Algebra; from Binomial Theorem to Theory of Equations.

6	1. Evolution of Binomial Surds.
12	2. Indeterminate Coefficients.
6	3. Continued Fractions.

A GUIDE TO THE MATHEMATICAL STUDENT 379

10	4. Indeterminate Equations, (1st and 2nd degree).
7	5. Partial Fractions.
3	6. Scales of Notation.
7	7. Properties of Numbers.
7	8. Vanishing Fractions.
6	9. Converging and diverging Series.
4	10. Logarithms, construction of.
7	11. Interest, Discount, and Annuities.
6	12. Chances (2nd time), and Life-Annuities.
11	13. Theory of Equations (2nd time).
6	14. Miscellaneous.

J.

Higher Plane Pure Geometry.

4	1. Anharmonic and Harmonic Proportion.
5	2. Anharmonic ratio of a Pencil. Harmonic Pencils.
5	3. Geometrical Involution.
4	4. Poles and Polars in relation to Circles.
4	5. Methods of Reciprocation.
5	6. Radical Axis and Centres of Similitude.
5	7. Principle of Continuity.
5	8. Projection.
8	9. Miscellaneous.

K.

Plane Algebraical Geometry;
from end of Trigonometry to Quadratic Loci
(constructed from Geometrical properties).

1. Determination of positions of Points, Lines, and Circles, by means of magnitudes; and discussion of such magnitudes.
2. Interpretation and classification of simple Equations.

4 3. Interpretation of Pairs of Equations. Representation and discussion of Points.

4. Investigation of Locus of single Simple Equations. Representation of Lines.

10	5. Lines; Problems.
3	6. " Theorems.
7	7. Rectilinear Figures; Problems.
2	8. " Theorems.
3	9. Pencils; Problems.
9	10. " Theorems.
7	11. Representation of Loci of Points fulfilling certain conditions.
	12. Representation of Pairs of Lines. Criterion that Quadratic Equation should represent Pair of Lines.
3	13. Pairs of Lines; Problems.
2	14. " Theorems.
	15. Representation of Circles. Criterion that Quadratic Equation should represent Circle.
12	16. Circles; Problems.
6	17. " Theorems.
	18. Representation of Parabola. Criterion that Quadratic Equation should represent Parabola.
4	19. Parabola; easy Problems.
4	20. " Theorems.
	21. Representation of Ellipse. Criterion that Quadratic Equation should represent Ellipse.
6	22. Ellipse; easy Problems.
8	23. " Theorems.
	24. Representation of Hyperbola. Criterion that Quadratic Equation should represent Hyperbola.
6	25. Hyperbola; easy Problems.
8	26. " Theorems.
6	27. Miscellaneous.

L.

Plane Algebraical Geometry; Trigonometry (2nd time).

4	1. Circular measure. Area of Circle, &c.
6	2. Demoivre's Theorem; and theorems involving powers of goniometrical ratios.

A GUIDE TO THE MATHEMATICAL STUDENT 381

4	3. Summation of series of goniometrical ratios.
4	4. Relations between angle and its goniometrical ratios. Gregorie's Series. Euler's and Machin's Series for π.
6	5. Miscellaneous; viz., resolution of sin θ and cos θ into factors, &c.

M.

Plane Algebraical Geometry; Quadratic Loci (constructed from Equations).

6	1. Interpretation and classification of Quadratic Equations. Quadratic Locus;
8	2. General. Problems.
6	3. " Theorems.
12	4. " when $B^2 - 4 AC \neq 0$, i.e. Central Locus; Problems.
8	5. " " Theorems.
16	6. Central, when $B^2 - 4 AC < 0$, i.e. Ellipse. Problems.
10	7. " " Theorems.
12	8. " when $B^2 - 4 AC > 0$, i.e. Hyperbola. Problems.
8	9. " " Theorems.
16	10. General, when $B^2 - 4 AC = 0$, i.e. Non-central Locus, or Parabola. Problems.
10	11. " " Theorems.
8	12. Miscellaneous.

N.

Differential Calculus (1st time).

	1. Elements of subject.
3	2. Differentiation from first principles.
3	3. Differentiation of functions connected by addition, &c.
9	4. " algebraical functions.
8	5. " compound functions.
8	6. " circular functions.
5	7. " functions of many variables.

4	8. Successive differentiation. Leibnitz's Theorem.
4	9. Maclaurin's Theorem.
4	10. Theory of equicrescent variable. Taylor's Theorem.
6	11. Elimination of constants and functions by differentiation (1st time).
	12. Relations between functions and derived functions; viz. $$\frac{F(x_0 + h) - F(x_0)}{f(x_0 + h) - f(x_0)} = \frac{F'(x_0 + \theta h)}{f'(x_0 + \theta h)}, \&c.$$
6	13. Order of Infinitesimals.
7	14. Evaluation of quantities of the form $\frac{0}{0}$, &c.
8	15. Maxima and minima of explicit functions of *one* variable.
11	16. Geometrical application to end of "
	17. Symbols of direction extended.
8	18. Cissoid, Witch, &c.
10	19. Tangents &c. of plane curves.
5	20. Direction of curvature. Hessian.
5	21. Multiple points.
6	22. Tracing curves.
4	23. Curvature of plane curves.
5	24. Evolutes and involutes.
6	25. Miscellaneous.

O.

Calculus of Finite Differences (1st time).

2	1. Differentiation of functions.
2	2. Integration of functions by indeterminate coefficients.
	3. " product of *n* terms in *A.P.*, and of reciprocal of the same.
2	4. Resolution of rational algebraical functions into these 2 forms.
2	5. Supplying deficient factors.
5	6. Integration of circular, exponential, and other functions.
6	7. Summation of Series by general methods.

P.

*Euclid XI, XII,
and higher Solid Pure Geometry.*

    1. Book XI.
    2. Book XII.
2 | 3. Deductions from Book XI.   Problems.
3 | 4.   "   "   Theorems.
1 | 5.   "   Book XII. Problems.
2 | 6.   "   "   Theorems.
    7. Sections of Cone.
2 | 8. Problems on "
3 | 9. Theorems on "
   10. Higher Solid Pure Geometry.
3 | 11. Problems on   "
4 | 12. Theorems on   "

Q.

*Solid Algebraical Geometry;
to end of Stereometry.*

2 | 1. Representation and discussion of volumes absolute.
    2.   "   of magnitudes with direction.
    3. Theory of Projection in Space.
6 | 4. Spherical Trigonometry; i.e., properties of solid angles.
    5. Napier's Analogies.
    6. Gauss' Theorems.
5 | 7. Solution of spherical Triangles; inscribed Circles; area of triangle and lune, &c.
    8. Cagnolis' Theorem.   Llhuillier's Theorem.
4 | 9. Stereometry; i.e. properties of plane-sided Solids; inscribed Spheres; volume and diagonal of Parallel-epipedon, &c.
5 | 10. Miscellaneous.

R.

Solid Algebraical Geometry;
from end of Stereometry to Quadratic Superficial Loci
(constructed from Geometrical properties).

	1. Determination of position, in Space, of Points, Lines, Planes, Spheres, and Cylinders, by means of certain magnitudes; and discussion of such magnitudes.
	2. Interpretation and classification of Simple Equations.
	3. ” ” Pairs of Equations.
	4. ” of sets of 3 Equations.
4	5. Representation and discussion of Points.
	6. Investigation of Locus of single Simple Equations. Representation of Planes.
6	7. Planes. Problems.
6	8. ” Theorems.
3	9. Plane-sided Solids. Problems.
4	10. ” Theorems.
3	11. Representation of Superficial Loci of Points fulfilling certain conditions.
	12. Representation of Pairs of Planes. Criterion that Quadratic Equation should represent Pair of Planes.
2	13. Pairs of Planes. Problems.
2	14. ” Theorems.
	15. Investigation of Locus of Pairs of Simple Equations. Representation of Lines.
6	16. Lines. Problems.
4	17. ” Theorems.
	18. Representation of Spheres. Criterion that Quadratic Equation should represent Sphere.
4	19. Spheres. Problems.
5	20. ” Theorems.
	21. Representation of Cylinders. Criterion that Quadratic Equation should represent Cylinder.
2	22. Cylinders. Easy Problems.
3	23. ” Theorems.

A GUIDE TO THE MATHEMATICAL STUDENT 385

	24. Representation of Cones. Criterion that Quadratic Equation should represent Cones.
2	25. Cones. Easy Problems.
3	26. ” Theorems.
6	27. Miscellaneous.

S.

Higher Plane Algebraical Geometry.

2	1. Eccentric angles.
2	2. Similar Conic Sections.
4	3. Contact of Conics. Osculating circle. Centre of curvature, and Evolutes.
5	4. Anharmonic properties of Conics.
4	5. Method of reciprocal Polars.
4	6. Involution.
	7. Pascal's Theorem.
	8. Tangential coordinates.
	9. Discussion of Locus of nth degree.
	10. Interpretation and classification of Cubic Equations.
3	11. Discussion of Cubic Loci.
	12. Interpretation and classification of Biquadratic Equations.
3	13. Discussion of Biquadratic Loci.
4	14. Discussion of Transcendental Loci.
6	15. Miscellaneous.

T.

Integral Calculus (1st time).

	1. Elements of subject.
4	2. Integration from first principles.
8	3. Definite integration.
12	4. Integration of rational algebraical functions.
14	5. ” irrational ”
8	6. ” ” ” by rationalization.
7	7. ” ” ” by reduction.

9	8.	" exponential and logarithmic functions.
10	9.	" circular functions.
15	10.	Definite integrals and their properties.
10	11.	Rectification of plane curves.
10	12.	Quadrature of plane surfaces.
8	13.	" surfaces of revolution.
8	14.	Cubature of solids of revolution.
12	15.	Miscellaneous.

U.

Solid Algebraical Geometry; Quadratic Superficial Loci (constructed from Equations).

    1. Interpretation and classification of Single Quadratic Equations.

4	2. General Quadratic Superficial Locus	Problems.
3	3. "	Theorems.
3	4. Reduced Quadratic Locus, ($px^2 + qy^2 + rz^2 + sx + ty + vz + w = 0$,) when neither P, Q, nor R $= 0$; i.e. Central Quadratic Locus. ($px^2 + qy^2 + rz^2 + h = 0$). Problems.	
3	5. "	Theorems.
2	6. Central Quadratic Locus, when H $= 0$, i.e. Cone.	Problems.
2	7. "	Theorems.
3	8. Central Quadratic Locus, when P, Q, R, and H have the same sign; i.e. Ellipsoid, and Prolate and Oblate Spheroid.	Problems.
3	9. "	Theorems.
1	10. Central Quadratic Locus, when one of them has a different sign from the other three; i.e. Hyperboloid of one sheet.	Problems.
1	11. "	Theorems.
1	12. Central Quadratic Locus, when two of them have a different sign from the other two; i.e. Hyperboloid of two sheets.	Problems.
1	13. "	Theorems.

2	14. Central Quadratic Locus, when either P, Q, or R = 0; i.e. the Axicentral Locus, or Central Cylinder. Problems.
2	15. " Theorems.
2	16. Reduced Quadratic Locus, when one or more of the three, (P, Q, and R,) = 0; i.e. Non-central Locus. Problems.
2	17. " Theorems.
2	18. Non-central Locus, when *one* of the three, (P, Q, and R,) = 0; i.e. Paraboloid. Problems.
2	19. " Theorems.
1	20. Non-central Locus, when *two* of the three (P, Q, and R,) = 0; i.e. Parabolic Cylinder. Problems.
1	21. " Theorems.
2	22. Miscellaneous, (e.g. Cono-cuneus). Problems.
2	23. " Theorems.

V.

Higher Algebra.

4	1. Theory of equations (3rd time).
3	2. Transformation of equations.
2	3. Equal roots.
3	4. Limits of roots. Separation of roots.
2	5. Commensurable roots.
2	6. Depression of equations.
1	7. Reciprocal equations.
2	8. Binomial "
3	9. Cubic "
3	10. Biquadratic "
2	11. Sturm's Theorem. Fourier's Theorem.
2	12. Lagrange's and Newton's methods of approximation.
1	13. Horner's method.
3	14. Symmetrical functions of roots.
1	15. Sums of powers of roots.
6	16. Determinants.
5	17. Elimination.

4	18. Expansion of functions in series.
5	19. Invariants. Covariants. Emanants. Evectants.
2	20. Contravariants.
3	21. Hyperdeterminant Calculus. Hermite's Law of Reciprocity.
2	22. Canonizants.
2	23. Binary Quantics, Quadrics, &c.
2	24. Ternary Quantics, Quadrics, &c.
3	25. Discriminants, &c.
2	26. Commutants.
5	27. Miscellaneous.

W.

Differential Calculus (2nd time).

3	1. Trigonometrical expressions. Roots of $+1$ and -1. Imaginary logarithms.
2	2. Limits of Maclaurin's and Taylor's Theorems.
5	3. Change of equicrescent variable.
4	4. Successive differentiation of functions of many independent variables.
2	5. Euler's Theorem of homogeneous functions.
3	6. Successive differentiation of implicit functions.
2	7. Bernoulli's Numbers.
2	8. Lagrange's Theorem.
2	9. Laplace's Theorem.
	10. Extension of Maclaurin's Theorem.
10	11. Elimination of constants and functions (2nd time).
5	12. Transformation of differential expressions into their equivalents in terms of other variables.
	13. Expansion of functions of one variable. Accurate proofs of Maclaurin's and Taylor's Theorems.
4	14. Expansion of functions of two or more variables. Maxima and minima.
6	15. Of implicit functions of 2 independent variables.
7	16. Of explicit " " "

5	17.	Of functions of 3 or more " "
5	18.	" " not independent "
5	19.	Properties of Curves of the nth degree.
3	20.	Contact of curves (plane).
6	21.	Envelopes "
2	22.	Theory of reciprocation.
3	23.	Caustics.
10	24.	Curved surfaces, tangent planes, &c.
4	25.	Singular points of curved surfaces.
8	26.	Curves in space, tangents, &c.
3	27.	Geodesic lines, &c.
2	28.	Curved surfaces generated by right lines. Ruled surfaces.
2	29.	" " Conical "
2	30.	" " Cylindrical "
2	31.	" " Developable "
2	32.	" " Skew "
2	33.	" " Conoidal "
2	34.	" by circles. Surfaces of revolution.
2	35.	" " Tubular "
1	36.	Curves in space. Curvature-angle of contingence.
1	37.	" Torsion.
1	38.	" The polar surface.
1	39.	" The osculating sphere.
1	40.	" Complex flexure.
1	41.	" The osculating surface.
1	42.	" The rectifying surface and line.
1	43.	Curved surfaces. Curvature. Euler's Theorem.
1	44.	" Umbilics.
1	45.	" Lines of curvature.
1	46.	" Dupin's Theorem.
1	47.	" Osculating surfaces.
	48.	Calculus of operations, Elements of.
	49.	Laws of commutation, distribution, and iteration.
	50.	Law of total differentiation.
6	51.	Miscellaneous.

X.

Integral Calculus (2nd time).

3	1.	Successive integration.
3	2.	Rectification of non-plane curves.
2	3.	Determination of the equation to a curve by means of a relation between the length and the coordinates to any point on it.
3	4.	Involutes of plane curves.
3	5.	Quadrature of curved surfaces.
3	6.	Cubature of solids bounded by any curved surface.
2	7.	Properties of multiple integrals.
1	8.	Transformation of multiple integrals.
2	9.	Curvilinear co-ordinates. Gauss' System. Lamé's System and Jacobi's modification.
2	10.	Variation of definite integrals due to variation of parameters involved in element-function.
2	11.	Variation of definite integrals due to variation of parameters involved in element-function and in the limits.

Differential equations.

    12.   General principles.

          First order.

4	13.	Exact total differential equations.
4	14.	Homogeneous equations of 2 variables.
4	15.	The first linear differential equation.
4	16.	Partial differential equations of 1st degree.
2	17.	Integrating factors of differential equations.
2	18.	Singular solutions of "
4	19.	Differential equations of higher degrees.
3	20.	Particular processes.

          Higher orders;

    21.   First degree; general properties.

3	22.	" linear differential equations.
3	23.	" " with constant coefficients.
3	24.	" " with variable coefficients.
3	25.	Higher degrees; total differential equations.
3	26.	" partial "

4	27. Geometrical Problems involving diff. equations. 1st order.
4	28. " " 2nd "
	29. Simultaneous differential equations. General principles.
3	30. " Linear. 1st order.
2	31. " " Higher orders.
	Integration of differential equations by series.
2	32. Application of Taylor's and Maclaurin's Theorems.
2	33. Method of undetermined coefficients.
2	34. Solution of Riccati's Equation.
4	35. Application of Integral Calculus to Theory of Probabilities.
5	36. Elliptic Integrals.
6	37. Miscellaneous.

Y.

Calculus of Finite Differences (2nd time).

2	1. Solution of equations of differences. 1st order.
2	2. " " 2nd order.
1	3. " " nth order.
2	4. " mixed differences.
3	5. Summation of Series; by particular assumptions.
2	6. " by differentiation.
2	7. " of recurring Series.
3	8. Interpolation of Series.
2	9. Generating functions.
6	10. Miscellaneous.

Z.

Calculus of Variations.

	1. General principles.
2	2. Variation of $\int_0^1 \mathrm{F}(x,\ dx,\ d^2x,\ \ldots\ y,\ dy,\ d^2y,\ \ldots)$.

2	3. Variation of $\int_0^1 F(x, y, y', y'', \ldots)$.
2	4. " $\int_0^1 F(x, dx, d^2x, \ldots y, dy, d^2y, \ldots z, dz, d^2z, \ldots)$.
2	5. " $\int_0^1 F(x, y, y', y'', \ldots z, z', z'', \ldots)$.
1	6. Variation of a variation.
1	7. " of a product of differentials.
1	8. " of a definite double integral due to the variations of the limits.

Maxima and minima.

9. Critical values of definite integrals, whose element-functions involve variables and their differentials; general principles.

3	10. " relative max. and min.
2	11. " absolute "
3	12. Geodesic lines; equations to.
2	13. " properties of.

14. Critical values of definite integrals, whose element-functions involve derived functions; general principles.

3	15. " particular cases.

16. Discriminating conditions; general principles.

	17. " requisite data.
	18. " proof that $\delta\ \mathrm{H}\ u\ d\ x$ is an exact differential: its integral, &c.
2	19. " particular cases.

20. Critical values of a double definite integral; necessary criteria.

3	21. " application of.
6	22. Miscellaneous.

CYCLE
For Working Examples

1	M 6	34	N 25	67	H 10	100	
2	W 16	35	K 27	68	M 5	101	I 10
3	L 3	36	C 8	69	T 8	102	1
4	V 1	37	W 11	70	C 11	103	G 8
5	T 6	38	R 20	71	W 24	104	X 36
6	G 10	39	X 27	72	R 5	105	H 9
7	17	40		73	P 4	106	L 1
8	H 7	41	Y 1	74	X 19	107	W 8
9	X 13	42	M 4	75	G 14	108	I 12
10	K 16	43	T 11	76	11	109	M 8
11	10	44	Z 10	77	I 5	110	T 14
12	N 16	45	V 23	78	J 6	111	J 4
13	10	46	N 6	79	T 10	112	K 23
14	C 12	47	19	80		113	K 17
15	M 11	48	G 15	81	M 2	114	V 25
16	W 1	49	J 1	82	A 6	115	X 6
17	O 7	50	W 29	83	W 14	116	R 10
18	J 8	51	U 8	84	K 19	117	B 3
19	R 7	52	Z 2	85	7	118	W 12
20		53	T 4	86	N 5	119	S 4
21	T 5	54	M 7	87	24	120	
22	I 7	55	O 6	88	V 12	121	Y 10
23	2	56	X 1	89	D 3	122	C 3
24	U 2	57	Q 4	90	X 37	123	T 13
25	X 32	58	W 19	91	U 3	124	M 10
26	D 4	59	E 3	92	C 6	125	N 15
27	W 4	60		93	3	126	O 5
28	A 1	61	E 2	94	T 3	127	A 9
29	M 10	62	K 26	95	M 6	128	U 6
30	V 16	63	5	96	F 3	129	X 14
31	T 9	64	N 4	97	W 51	130	I 2
32	L 5	65	11	98	R 16	131	3
33	B 4	66	V 17	99	Z 10	132	W 27

133	Q 9	170	K 11	207	Y 5	244	L 4
134	F 4	171	25	208	M 6	245	A 4
135	T 15	172	C 7	209	W 17	246	N 8
136	G 10	173	W 28	210	I 14	247	U 5
137	18	174	L 2	211	U 14	248	X 3
138	V 4	175	F 3	212	H 7	249	C 5
139	Z 11	176	X 4	213	Z 15	250	4
140		177	R 13	214	S 11	251	P 12
141	M 1	178	Z 22	215	T 11	252	T 8
142	E 7	179	V 14	216	C 13	253	M 8
143	W 23	180		217	W 25	254	I 2
144	K 16	181	M 4	218	M 7	255	7
145	28	182	T 9	219	V 19	256	W 21
146	X 25	183	H 4	220		257	K 26
147	C 9	184	S 1	221	X 9	258	16
148	R 8	185	W 2	222	K 5	259	H 8
149	T 3	186	I 4	223	27	260	
150	H 6	187	8	224	N 9	261	N 7
151	S 15	188	U 4	225	22	262	19
152	N 13	189	X 30	226	E 4	263	R 23
153	V 27	190	J 9	227	6	264	X 15
154	W 43	191	T 10	228	Q 7	265	V 16
155	M 9	192	A 15	229	T 7	266	S 3
156	J 3	193	M 11	230	J 7	267	W 18
157	X 18	194	C 10	231	R 22	268	J 5
158	U 9	195	W 11	232	W 16	269	M 3
159	I 13	196	K 10	233	I 11	270	T 15
160		197	22	234	9	271	C 6
161	T 12	198	B 6	235	X 28	272	G 4
162	G 17	199	N 16	236	M 10	273	8
163	W 26	200		237	D 4	274	14
164	S 6	201	R 27	238	T 6	275	Z 5
165	N 18	202	V 20	239	V 7	276	F 4
166	V 21	203	X 16	240		277	X 23
167	D 6	204	G 14	241	G 10	278	U 19
168	M 12	205	1	242	16	279	Y 2
169	T 5	206	T 4	243	W 24	280	

281	W 6	318	R 7	355	M 9	392	X 36
282	M 5	319	W 32	356	W 7	393	R 16
283	B 4	320		357	C 7	394	T 13
284	T 4	321	O 1	358	K 7	395	C 12
285	Q 10	322	T 5	359	17	396	F 3
286	V 6	323	K 23	360		397	M 8
287	N 14	324	6	361	T 9	398	Z 21
288	6	325	D 5	362	J 8	399	J 9
289	R 11	326	N 21	363	X 13	400	
290	Z 12	327	5	364	A 14	401	L 3
291	O 7	328	M 4	365	Q 1	402	W 19
292	K 3	329	C 8	366	V 17	403	V 2
293	13	330	X 24	367	W 11	404	X 27
294	W 30	331	L 5	368	M 11	405	K 16
295	I 5	332	A 13	369	T 3	406	11
296	13	333	W 3	370	R 20	407	H 10
297	T 14	334	G 10	371	Y 8	408	T 15
298	M 2	335	11	372	N 15	409	M 12
299	H 9	336	T 12	373	G 18	410	C 9
300		337	I 4	374	14	411	W 24
301	X 35	338	6	375	X 26	412	R 19
302	S 2	339	P 6	376	C 11	413	A 7
303	C 15	340		377	S 4	414	N 18
304	W 15	341	S 15	378	W 26	415	G 17
305	U 15	342	N 16	379	E 8	416	8
306	E 3	343	M 10	380		417	X 17
307	5	344	V 9	381	T 11	418	I 8
308	T 10	345	W 41	382	M 7	419	11
309	G 17	346	R 17	383	B 3	420	
310	5	347	T 4	384	N 20	421	D 3
311	9	348	Q 4	385	K 10	422	T 10
312	N 4	349	X 5	386	5	423	M 6
313	23	350	I 1	387	V 1	424	V 27
314	V 18	351	I 12	388	W 51	425	W 36
315	M 6	352	U 18	389	U 2	426	U 10
316	X 2	353	N 25	390	I 2	427	P 11
317	J 2	354	H 5	391	3	428	X 31

429	S 5	466	13	503	W 14	540	
430	T 5	467	M 4	504	U 22	541	C 8
431	C 3	468	J 6	505	C 13	542	2
432	6	469	T 7	506	1	543	U 12
433	K 19	470	R 14	507	T 2	544	T 5
434	25	471	O 6	508	M 2	545	I 5
435	W 12	472	W 9	509	B 4	546	2
436	J 6	473	K 26	510	N 4	547	G 10
437	M 1	474	V 10	511	16	548	1
438	N 11	475	X 20	512	G 16	549	N 22
439	R 8	476	L 2	513	17	550	V 16
440		477	T 4	514	F 4	551	Z 22
441	E 2	478	Y 6	515	W 37	552	M 10
442	4	479	M 3	516	I 4	553	H 4
443	T 8	480		517	7	554	W 17
444	B 6	481	W 16	518	Q 4	555	R 10
445	W 4	482	D 4	519	T 9	556	T 15
446	G 10	483	S 6	520		557	Y 9
447	12	484	N 24	521	P 3	558	K 5
448	15	485	C 7	522	S 13	559	16
449	V 8	486	X 7	523	X 22	560	
450	X 19	487	G 14	524	E 3	561	X 25
451	Q 7	488	H 9	525	M 6	562	C 5
452	H 7	489	W 20	526	N 6	563	11
453	N 2	490	Q 10	527	14	564	N 5
454	19	491	I 14	528	D 6	565	M 7
455	M 10	492	T 12	529	W 21	566	W 15
456	I 13	493	M 5	530	A 5	567	I 10
457	T 6	494	A 2	531	R 5	568	9
458	U 7	495	Z 13	532	V 24	569	U 8
459	W 40	496	R 27	533	T 10	570	T 11
460		497	X 14	534	K 23	571	V 19
461	Y 10	498	V 3	535	27	572	F 3
462	S 14	499	J 1	536	J 3	573	L 1
463	X 11	500		537	X 1	574	X 16
464	C 14	501	K 22	538	M 8	575	O 7
465	N 10	502	9	539	W 11	576	W 24

577	M 11	614	N 23	651	S 3	688	V 17
578	J 4	615	J 7	652	X 24	689	M 5
579	T 3	616	T 4	653	J 9	690	T 15
580		617	M 12	654	G 17	691	J 2
581	G 8	618	C 7	655	10	692	R 25
582	14	619	Z 19	656	5	693	W 27
583	E 6	620		657	T 5	694	Y 10
584	1	621	G 18	658	I 1	695	U 4
585	P 9	622	9	659	8	696	N 11
586	N 3	623	W 5	660		697	21
587	15	624	K 14	661	W 51	698	P 12
588	R 9	625	20	662	M 10	699	X 36
589	A 15	626	I 13	663	N 25	700	
590	W 26	627	X 33	664	R 16	701	A 9
591	M 9	628	U 16	665	V 4	702	M 2
592	T 14	629	T 8	666	T 12	703	C 12
593	V 11	630	Y 4	667	U 9	704	3
594	B 3	631	M 6	668	A 3	705	W 11
595	X 37	632	N 18	669	W 3	706	K 25
596	K 10	633	V 22	670	K 26	707	H 7
597	7	634	W 34	671	16	708	T 6
598	N 19	635	L 5	672	X 35	709	L 4
599	7	636	F 4	673	C 15	710	V 21
600		637	X 15	674	B 5	711	N 6
601	C 6	638	R 26	675	M 8	712	20
602	U 3	639	O 2	676	T 9	713	R 20
603	H 8	640		677	Q 9	714	X 10
604	W 1	641	T 10	678	I 4	715	I 2
605	M 4	642	M 1	679	W 22	716	11
606	T 13	643	E 4	680		717	M 7
607	I 12	644	5	681	G 11	718	W 16
608	S 15	645	W 25	682	14	719	T 11
609	X 28	646	K 17	683	D 4	720	
610	D 3	647	11	684	N 16	721	E 3
611	W 18	648	N 8	685	4	722	2
612	V 25	649	Z 4	686	S 4	723	G 16
613	R 7	650	C 9	687	X 6	724	8

725	N 13	762	Q 7	799	7	836	G 8
726	S 11	763	H 5	800		837	17
727	Z 10	764	N 19	801	W 8	838	V 16
728	C 13	765	M 10	802	Y 5	839	W 35
729	W 23	766	C 10	803	G 18	840	
730	M 6	767	W 33	804	10	841	M 12
731	V 14	768	K 3	805	X 34	842	A 4
732	T 4	769	13	806	I 5	843	X 4
733	K 23	770	J 5	807	6	844	Q 4
734	5	771	X 32	808	N 16	845	T 12
735	X 23	772	R 17	809	M 3	846	Y 7
736	O 6	773	T 5	810	J 6	847	J 9
737	U 17	774	V 27	811	O 7	848	S 15
738	W 19	775	G 14	812	T 13	849	W 31
739	J 8	776	12	813	S 14	850	L 3
740		777	W 12	814	V 12	851	V 26
741	G 10	778	D 6	815	W 21	852	N 18
742	17	779	M 11	816	U 5	853	M 6
743	15	780		817	C 6	854	B 6
744	N 14	781	N 24	818	4	855	Z 7
745	5	782	I 7	819	X 27	856	R 19
746	I 3	783	13	820		857	W 15
747	14	784	U 2	821	N 22	858	C 8
748	T 7	785	T 3	822	Z 3	859	T 9
749	M 4	786	B 4	823	E 4	860	
750	F 3	787	X 26	824	6	861	K 11
751	X 13	788	L 2	825	M 8	862	26
752	R 8	789	P 4	826	T 8	863	P 8
753	V 18	790	C 7	827	D 5	864	G 14
754	W 24	791	W 26	828	R 23	865	9
755	K 22	792	M 9	829	W 6	866	X 37
756	10	793	H 9	830	I 2	867	I 4
757	A 11	794	N 15	831	X 5	868	9
758	S 5	795	R 27	832	K 16	869	M 5
759	T 10	796	V 1	833	25	870	F 4
760		797	T 14	834	H 10	871	W 4
761	E 7	798	K 27	835	T 15	872	A 10

873	U 20	910	M 7	947	C 9	984	M 10
874	V 5	911	N 10	948	S 1	985	C 5
875	T 10	912	Z 15	949	Z 2	986	7
876	Q 10	913	O 4	950	M 6	987	W 39
877	H 6	914	D 4	951	W 18	988	G 10
878	W 11	915	R 7	952	V 2	989	18
879	M 10	916	W 14	953	K 17	990	Y 1
880		917	Y 10	954	10	991	X 16
881	Y 8	918	T 11	955	T 7	992	Q 9
882	X 19	919	U 6	956	X 2	993	T 10
883	R 13	920		957	G 14	994	I 11
884	N 9	921	H 7	958	8	995	3
885	4	922	X 14	959	5	996	S 3
886	E 3	923	M 11	960		997	Z 12
887	8	924	C 10	961	E 2	998	J 7
888	K 16	925	B 3	962	4	999	M 9
889	6	926	W 24	963	W 16	1000	
890	Z 22	927	K 23	964	U 23	1001	W 26
891	T 5	928	A 6	965	H 8	1002	O 6
892	V 19	929	T 6	966	P 6	1003	N 5
893	W 17	930	G 4	967	T 3	1004	21
894	G 11	931	17	968	M 8	1005	R 5
895	1	932	N 25	969	I 13	1006	T 5
896	10	933	V 23	970	N 6	1007	K 27
897	J 3	934	F 3	971	19	1008	7
898	X 8	935	X 22	972	R 16	1009	V 17
899	M 4	936	M 2	973	X 20	1010	X 3
900		937	I 8	974	S 4	1011	M 1
901	C 11	938	1	975	W 7	1012	A 15
902	T 4	939	W 51	976	V 9	1013	W 3
903	S 6	940		977	T 15	1014	U 13
904	I 12	941	L 5	978	K 9	1015	N 14
905	W 29	942	J 1	979	22	1016	C 6
906	K 5	943	T 2	980		1017	H 9
907	19	944	R 11	981	D 3	1018	G 17
908	X 30	945	N 23	982	N 11	1019	14
909	V 20	946	7	983	16	1020	

1021	T 12	1058	11	1095	A 5	1132	X 1
1022	L 1	1059	W 20	1096	N 4	1133	S 14
1023	W 28	1060		1097	20	1134	C 10
1024	M 12	1061	S 15	1098	U 19	1135	W 15
1025	B 4	1062	X 15	1099	T 10	1136	M 3
1026	I 2	1063	V 10	1100		1137	T 14
1027	7	1064	R 8	1101	V 27	1138	G 16
1028	X 28	1065	H 4	1102	H 5	1139	10
1029	S 13	1066	T 13	1103	W 21	1140	
1030	T 8	1067	M 6	1104	S 5	1141	N 16
1031	E 5	1068	W 25	1105	C 15	1142	O 7
1032	3	1069	C 13	1106	X 35	1143	I 14
1033	R 10	1070	3	1107	M 8	1144	Z 5
1034	W 2	1071	N 18	1108	B 5	1145	U 15
1035	K 5	1072	K 26	1109	N 22	1146	H 7
1036	11	1073	I 4	1110	R 20	1147	W 30
1037	N 15	1074	Z 11	1111	T 5	1148	K 25
1038	Z 21	1075	L 2	1112	I 5	1149	8
1039	V 6	1076	T 11	1113	10	1150	V 8
1040		1077	V 15	1114	G 17	1151	T 3
1041	M 4	1078	M 5	1115	9	1152	M 2
1042	T 4	1079	W 19	1116	W 12	1153	E 4
1043	J 4	1080		1117	J 9	1154	N 24
1044	X 36	1081	E 6	1118	Q 1	1155	R 9
1045	G 15	1082	1	1119	X 18	1156	X 37
1046	12	1083	J 2	1120		1157	I 13
1047	10	1084	R 27	1121	Y 2	1158	L 4
1048	C 12	1085	X 9	1122	M 11	1159	W 1
1049	W 11	1086	T 6	1123	T 15	1160	
1050	U 14	1087	G 14	1124	F 4	1161	J 8
1051	D 6	1088	A 8	1125	W 47	1162	Q 4
1052	N 13	1089	D 4	1126	K 23	1163	T 7
1053	M 7	1090	W 24	1127	P 12	1164	C 7
1054	P 11	1091	M 10	1128	V 16	1165	1
1055	Y 3	1092	Z 22	1129	N 3	1166	N 6
1056	T 9	1093	K 16	1130	19	1167	8
1057	G 8	1094	10	1131	R 22	1168	M 6

1169	D 3	1206	W 11	1243	R 17	1280	
1170	W 45	1207	P 9	1244	X 27	1281	N 19
1171	G 18	1208	X 13	1245	I 4	1282	O 5
1172	8	1209	A 14	1246	11	1283	M 6
1173	V 18	1210	R 7	1247	K 16	1284	X 4
1174	T 4	1211	I 8	1248	14	1285	I 7
1175	U 8	1212	1	1249	N 18	1286	13
1176	A 13	1213	T 10	1250	J 5	1287	G 14
1177	X 6	1214	K 17	1251	M 8	1288	8
1178	M 4	1215	5	1252	T 9	1289	T 2
1179	N 25	1216	F 3	1253	W 24	1290	S 4
1180		1217	V 1	1254	Q 10	1291	W 18
1181	Y 10	1218	W 26	1255	D 5	1292	V 21
1182	B 3	1219	U 3	1256	X 17	1293	U 9
1183	R 26	1220		1257	G 10	1294	N 13
1184	W 32	1221	N 14	1258	5	1295	7
1185	K 20	1222	9	1259	15	1296	J 3
1186	3	1223	M 3	1260		1297	M 11
1187	I 2	1224	X 31	1261	Y 5	1298	E 2
1188	9	1225	E 7	1262	T 15	1299	6
1189	T 12	1226	3	1263	U 2	1300	
1190	S 11	1227	G 17	1264	A 9	1301	W 4
1191	H 10	1228	1	1265	W 27	1302	L 3
1192	X 23	1229	T 11	1266	M 7	1303	P 4
1193	M 10	1230	O 1	1267	N 11	1304	F 4
1194	W 16	1231	W 51	1268	16	1305	T 6
1195	V 25	1232	L 5	1269	V 4	1306	R 19
1196	Q 7	1233	H 9	1270	X 25	1307	I 2
1197	N 2	1234	N 15	1271	H 6	1308	X 19
1198	5	1235	4	1272	T 13	1309	K 10
1199	J 6	1236	M 12	1273	R 16	1310	11
1200		1237	C 11	1274	C 8	1311	N 21
1201	G 14	1238	T 5	1275	Z 19	1312	10
1202	T 8	1239	S 6	1276	W 23	1313	C 2
1203	C 6	1240		1277	K 7	1314	9
1204	Z 10	1241	V 19	1278	26	1315	M 10
1205	M 9	1242	W 17	1279	B 4	1316	W 42

ITEM 43

1317	Y 6	1354	U 7	1391	15	1428	K 16
1318	G 17	1355	T 5	1392	R 27	1429	5
1319	10	1356	Y 8	1393	J 1	1430	X 16
1320		1357	H 7	1394	T 11	1431	C 6
1321	H 8	1358	M 2	1395	K 27	1432	T 7
1322	Z 13	1359	W 38	1396	A 4	1433	M 9
1323	S 15	1360		1397	W 6	1434	W 26
1324	T 4	1361	R 8	1398	X 34	1435	R 20
1325	V 14	1362	I 6	1399	C 13	1436	V 16
1326	W 9	1363	12	1400		1437	J 2
1327	U 4	1364	X 14	1401	M 8	1438	X 37
1328	O 7	1365	L 2	1402	N 19	1439	G 10
1329	X 11	1366	J 7	1403	I 14	1440	
1330	M 4	1367	E 4	1404	4	1441	A 15
1331	T 10	1368	5	1405	T 15	1442	T 10
1332	I 5	1369	T 3	1406	G 9	1443	M 7
1333	3	1370	P 3	1407	17	1444	W 16
1334	R 23	1371	K 23	1408	H 5	1445	P 12
1335	W 14	1372	Z 4	1409	W 15	1446	I 8
1336	D 4	1373	N 5	1410	P 5	1447	13
1337	K 22	1374	16	1411	Q 4	1448	H 9
1338	N 6	1375	V 11	1412	V 22	1449	N 25
1339	22	1376	M 1	1413	X 36	1450	S 3
1340		1377	W 21	1414	R 5	1451	K 9
1341	C 14	1378	G 14	1415	E 3	1452	25
1342	X 7	1379	18	1416	8	1453	C 12
1343	M 5	1380		1417	Y 10	1454	X 28
1344	J 9	1381	X 26	1418	T 9	1455	U 21
1345	T 12	1382	C 7	1419	M 10	1456	D 3
1346	G 11	1383	S 2	1420		1457	W 19
1347	12	1384	B 6	1421	N 18	1458	R 10
1348	V 3	1385	T 8	1422	14	1459	T 4
1349	W 3	1386	D 6	1423	U 5	1460	
1350	Q 9	1387	W 11	1424	W 24	1461	V 27
1351	A 2	1388	M 6	1425	O 6	1462	M 12
1352	N 3	1389	V 17	1426	F 3	1463	N 23
1353	20	1390	N 4	1427	Z 22	1464	J 8

1465	Z 15	1502	1	1539	2	1576	N 4
1466	G 8	1503	N 9	1540		1577	19
1467	16	1504	6	1541	Z 21	1578	C 8
1468	C 3	1505	R 7	1542	J 4	1579	X 20
1469	X 10	1506	V 9	1543	L 5	1580	
1470	I 2	1507	W 34	1544	H 7	1581	G 14
1471	9	1508	M 6	1545	W 22	1582	8
1472	L 1	1509	T 12	1546	K 26	1583	O 2
1473	W 12	1510	K 17	1547	7	1584	Z 3
1474	B 3	1511	10	1548	X 24	1585	V 2
1475	T 14	1512	Z 12	1549	R 16	1586	W 17
1476	K 13	1513	H 10	1550	T 15	1587	M 2
1477	19	1514	X 5	1551	B 4	1588	H 9
1478	E 7	1515	C 10	1552	M 8	1589	T 8
1479	N 24	1516	S 13	1553	N 13	1590	S 5
1480		1517	W 51	1554	5	1591	E 3
1481	M 4	1518	G 15	1555	V 13	1592	6
1482	H 4	1519	10	1556	G 17	1593	W 35
1483	W 5	1520		1557	11	1594	P 9
1484	R 11	1521	J 9	1558	W 11	1595	R 9
1485	V 24	1522	P 11	1559	T 10	1596	X 30
1486	X 15	1523	Q 7	1560		1597	C 15
1487	T 5	1524	T 6	1561	F 4	1598	M 1
1488	C 5	1525	M 5	1562	A 12	1599	T 5
1489	U 22	1526	V 7	1563	C 11	1600	K 5
1490	N 11	1527	N 16	1564	X 2	1601	23
1491	2	1528	R 14	1565	M 3	1602	N 22
1492	Y 4	1529	C 7	1566	J 3	1603	I 2
1493	W 25	1530	X 35	1567	U 11	1604	13
1494	M 11	1531	U 18	1568	W 33	1605	W 24
1495	J 6	1532	W 46	1569	R 25	1606	G 18
1496	T 13	1533	I 7	1570	H 6	1607	10
1497	G 14	1534	10	1571	I 5	1608	Y 9
1498	4	1535	D 4	1572	4	1609	J 5
1499	A 3	1536	T 7	1573	T 11	1610	L 4
1500		1537	M 10	1574	K 16	1611	Z 8
1501	I 11	1538	E 4	1575	11	1612	V 26

1613	U 16	1636	R 27	1659	V 18	1682	V 10
1614	T 4	1637	B 5	1660		1683	M 9
1615	M 6	1638	Z 22	1661	K 3	1684	H 8
1616	A 11	1639	Q 10	1662	8	1685	K 20
1617	W 44	1640	X 33	1663	C 13	1686	T 15
1618	R 8	1641	C 9	1664	P 8	1687	U 23
1619	N 15	1642	M 10	1665	W 31	1688	W 3
1620		1643	W 18	1666	N 21	1689	M 8
1621	C 6	1644	U 17	1667	8	1690	J 2
1622	4	1645	D 5	1668	G 14	1691	N 14
1623	H 5	1646	N 7	1669	5	1692	20
1624	I 3	1647	18	1670	J 9	1693	R 26
1625	12	1648	L 2	1671	M 11	1694	C 7
1626	T 9	1649	E 5	1672	A 5	1695	W 15
1627	M 4	1650	4	1673	Y 7	1696	G 9
1628	W 20	1651	S 14	1674	T 10	1697	12
1629	K 22	1652	T 14	1675	R 19	1698	T 12
1630	6	1653	R 17	1676	I 9	1699	E 8
1631	V 19	1654	I 14	1677	2	1700	M 12
1632	G 17	1655	8	1678	Z 6	1701	I 4
1633	1	1656	W 26	1679	W 21	1702	11
1634	T 3	1657	M 7	1680	X 22		
1635	J 7	1658	T 13	1681	O 4		

Bibliography

Abeles, Francine. "Multiplication in Changing Bases: A Note on Lewis Carroll." *Historia Mathematica* 3 (1976): 183–84.

Abeles, Francine. "Ranking by Inversion: A Note on C. L. Dodgson." *Historia Mathematica* 6 (1979): 310–17.

Abeles, Francine. "C. L. Dodgson and Apportionment for Proportional Representation." *Bulletin of Indian Society for History of Mathematics* 3 (1981): 71–82.

Abeles, Francine. "The Mathematical-Political Papers of C. L. Dodgson." In *Lewis Carroll: A Celebration,* edited by Edward Guiliano, 195–210. New York: Clarkson N. Potter, 1982.

Abeles, Francine. "Determinants and Linear Systems: Charles L. Dodgson's View." *British Journal for History of Science* 19 (1986): 331–35.

Abeles, Francine. "An Early Database: Lewis Carroll's Letter Register." *Jabberwocky* 19 (1990): 9–15.

Abeles, Francine. "Lewis Carroll's Method of Trees: Its Origins in *Studies in Logic.*" *Modern Logic* 1 (1990): 25–35.

Abeles, Francine. "Charles L. Dodgson's Geometric Approach to Arctangent Relations for π." *Historia Mathematica* 20 (1993): 151–59.

Abeles, Francine, and Stanley Lipson. "Some Victorian Periodic Polyalphabetic Ciphers." *Cryptologia* XIV (1990): 128–34.

Acland, T. D. *The Discouragement of Elementary Mathematics in General Education at Oxford.* Oxford and London: James Parker and Co., 1867.

Archibald, R. C. "Bibliography of Lewis Carroll: Additions." *Notes and Queries* 179 (1940): 134–35.

Augarde, Tony. *The Oxford Guide to Word Games.* New York: Oxford University Press, 1984.

Ball, Walter William Rouse. *Mathematical Recreations and Problems.* 1st and 2d eds. London: Macmillan, 1892.

Baltzer, Heinrich Richard. *Théorie et applications des déterminants, avec l'indication des sources originales* (Translation by J. Hoüel of *Theorie und Anwendung der Determinanten.* Leipzig: S. Hirzel, 1857). Paris: Mallet-Batchelier, 1861.

Bartley, William W. III, ed. *Lewis Carroll's Symbolic Logic.* New York: Clarkson N. Potter, 1977.

Bill, E. G. W. *University Reform in Nineteenth-Century Oxford.* Oxford: Clarendon Press, 1973.

Bill, E. G. W., and J. F. A. Mason. *Christ Church and Reform, 1830–1867.* Oxford: Clarendon Press, 1970.

Black, Duncan. *The Theory of Committees and Elections.* London: Cambridge University Press, 1958.

Compiled with the editorial assistance of Evelyn J. Abeles.

Black, Duncan. "The Central Argument in Lewis Carroll's *The Principles of Parliamentary Representation*." *Papers on Non-Market Decision Making* 3 (1967): 1–17.

Black, Duncan. "Lewis Carroll and the Theory of Games." *American Economic Review* 59 (1969): 206–15.

Black, Duncan. "Evaluating Carroll's Theory of Parliamentary Representation," *Jabberwocky* 4 (1970), 19–21.

Black, Duncan. "Lewis Carroll and the Cambridge Mathematical School of P.R.; Arthur Cohen and Edith Denman." *Public Choice* 8 (1970): 1–28.

Boole, George. "Exposition of a General Theory of Linear Transformations, Part I." *Cambridge Mathematical Journal* III (1841): 1–119.

Boyer, Carl B. *The Concepts of the Calculus: A Critical and Historical Discussion of the Derivative and the Integral.* New York: Hafner, 1949. Reprint. New York: Dover, 1959.

Braithwaite, R. B. "Lewis Carroll as Logician." *Mathematical Gazette* XVI (1932): 174–78.

Brock, W. H. 1975. "Geometry and the Universities: Euclid and His Modern Rivals, 1860–1900." *History of Education* 4, (1975): 21–35.

Cajori, Florian. *A History of the Conceptions of Limits and Fluxions in Great Britain from Newton to Woodhouse.* Chicago and New York: Open Court, 1919.

Cajori, Florian. *A History of Mathematical Notations,* vol. II. Chicago: Open Court, 1928.

Catalogue of an Exhibition at Columbia University to Commemorate the One-Hundredth Anniversary of the Birth of Lewis Carroll. New York: Columbia University Press, 1932.

Cayley, Arthur. *The Collected Mathematical Papers of Arthur Cayley.* 11 vols. Cambridge: University Press, 1889–97.

Chapman, S. "University Training of Mathematicians." *Mathematical Gazette* 30 (1946): 61–70.

Clark, Anne. *Lewis Carroll: A Biography.* New York: Schocken Books, 1979.

Clifford, William K. *Mathematical Papers,* edited by Robert Tucker. London: Macmillan, 1882.

Cohen, Morton N., ed. *The Letters of Lewis Carroll.* 2 vols. New York: Oxford University Press, 1979.

Cohen, Morton N. "Lewis Carroll's *Memoria Technica*." *Library Chronicle of the University of Texas at Austin* new series no. 11 (1979), 77–88.

Cohen, Morton N. "Lewis Carroll: 'dishcoveries'—and more." In *Nineteenth-Century Lives,* edited by L. S. Lockridge *et al.,* 112–24. New York: Cambridge University Press, 1989.

Cohen, Morton N., ed. *Lewis Carroll: Interviews and Recollections.* Iowa City: University of Iowa Press, 1989.

Cohen, Morton N., and Anita Gandolfo, eds. *Lewis Carroll and the House of Macmillan.* New York: Cambridge University Press, 1987.

A College Tutor. *The Fifth Book of Euclid Treated Algebraically.* Oxford and London: John Henry and James Parker, 1858.

Collingwood, E. F. "A Century of the London Mathematical Society." *Journal of the London Mathematical Society* 41 (1966): 577–94.
Collingwood, Stuart D. *The Life and Letters of Lewis Carroll*. London: T. Fisher Unwin, 1898.
Collingwood, Stuart D., ed. *The Lewis Carroll Picture Book*. London: T. Fisher Unwin, 1899. Reprint. New York: Dover, 1961.
Conway, J. H. "Tomorrow is the Day after Doomsday." *Eureka* 36 (1973): 28–31.
Crofton, M. W. "Probability." *Encyclopedia Brittanica*, 9th ed. 1885.
De Morgan, Augustus. *Elements of Trigonometry and Trigonometric Analysis, Part I: The Connexion of Numbers and Magnitude: An Attempt to Explain the Fifth Book of Euclid*. London: Taylor and Walton, 1836.
De Morgan, Augustus. *The Differential and Integral Calculus*. London: Baldwin and Cradock, 1842.
De Morgan, Augustus. *The Book of Almanacs*. London: Taylor, Walton, and Maberly, 1851.
De Morgan, Augustus. "On the Early History of Infinitesimals in England." *Philosophical Magazine and Journal of Science,* Fourth Series 4 (1852): 321–30.
De Morgan, Augustus. *A Budget of Paradoxes*. Edited by D. E. Smith. 2d ed. 2 vols. Chicago and London: Open Court, 1915.
Dickson, Leonard E. *History of The Theory of Numbers*. 3 vols. Washington: Carnegie Foundation, 1919–1923. Reprint. New York: Chelsea, 1971.
Dummett, Michael. *Voting Procedures*. Oxford: Clarendon Press, 1984.
Eperson, D. B. "Lewis Carroll—Mathematician." *Mathematical Gazette* 17 (1933): 92–100.
Fishburn, Peter. *The Theory of Social Choice*. New Jersey: Princeton University Press, 1973.
Franksen, Ole I. *Mr. Babbage's Secret, The Tale of a Cypher—and APL*. Englewood Cliffs, N.J.: Prentice-Hall, 1985.
G. P. B. "Ciphers and Cipher-Writing." *Macmillan's Magazine* 23 (1871): 329–38.
Gardner, Martin. "Mathematical Games." *Scientific American* 202 (1960): 172–76.
Gardner, Martin. *Logic Machines and Diagrams*. 2d ed. Chicago: University of Chicago Press, 1982.
Garnett, William. "Alice Through the Convex Looking Glass." *Mathematical Gazette* IX (1918): 237–41, 249–52, 293–98.
Gattégno, Jean. *Lewis Carroll: Fragments of a Looking Glass*. New York: Thomas Y. Crowell, 1974.
Gauss, Carl F. "Berechnung des Osterfestes." *Zach's Monatliche Correspondenz* II (1800): 221–30.
Gauss, Carl F. *Werke*. Edited by C. Schering. Göttingen: Königlichen Gesellschaft der Wissenschaften, 1874.
Glaisher, J. W. L., ed. *The Collected Mathematical Papers of Henry John Stephen Smith*. 1894. 2 vols. Reprint. New York: Chelsea, 1965.
Grattan-Guinness, Ivor, ed. *P. E. B. Jourdain: Selected Essays on the History of Set Theory and Logics (1906–1918)*. Bologna: CLUEB, 1989.

Green, Roger L. "Lewis Carroll." In *Three Bodley Head Monographs,* 5–87. London: Bodley Head, 1968.

Green, Roger L., ed. *The Diaries of Lewis Carroll.* 2 volumes. New York: Oxford University Press, 1954.

Greenacre, Phyllis. *Swift and Carroll: A Psychoanalytic Study of Two Lives.* New York: International Universities Press, 1955.

Gridgeman, Norman. "Charles Lutwidge Dodgson." *Dictionary of Scientific Biography,* vol. IV, ed. Charles C. Gillispie. New York: Charles Scribner's Sons, 1971.

Guiliano, Edward. "Lewis Carroll: A Sesquicentennial Guide to Research." *Dickens Studies Annual: Essays on Victorian Fiction* 10 (1982): 263–310.

Guiliano, Edward, ed. *Lewis Carroll: A Celebration.* New York: Clarkson N. Potter, 1982.

Halsted, G. B. "Light from Non-Euclidean Spaces on the Teaching of Elementary Geometry." Appendix III to *The Science of Absolute Space* by John Bolyai. Roberto Bonola, *Non-Euclidean Geometry: A Critical and Historical Study of its Developments,* 59–71. New York: Dover, 1955.

Heath, Peter. "Lewis Carroll." *Encyclopedia of Philosophy,* vol. 2. Edited by Paul Edwards. New York: Macmillian, 1967.

Heath, Thomas L. *The Thirteen Books of Euclid's Elements.* 2d ed. 3 vols. Cambridge: Cambridge University Press, 1926. Reprint New York: Dover, 1956.

Howson, G. *A History of Mathematics Education in England.* Cambridge: Cambridge University Press, 1982.

Hudson, Derek. *Lewis Carroll.* London: Constable, 1954.

Johnston, R. H. "On the Method of Contractants." *American Mathematical Monthly* 67 (1960): 865.

Kahn, David. *The Codebreakers.* New York: Macmillan, 1967.

Keller, Evelyn Fox. "Lewis Carroll: A Study of Mathematical Inhibition." *Journal of American Psychoanalytic Society* 28 (1980): 133–60.

Kline, Morris. *Mathematical Thought from Ancient to Modern Times.* New York: Oxford University Press, 1972.

Knuth, Donald E. *The Art of Computer Programming,* vol. III: *Sorting and Searching.* Reading, Mass.: Addison-Wesley, 1973.

Lennon, Florence Becker. *Lewis Carroll.* London: Cassell, 1947.

Lipson, Stanley, and Francine Abeles. "The Matrix Cipher of C. L. Dodgson." *Cryptologia* XIV (1990): 28–36.

Lipson, Stanley, and Francine Abeles. "The Key-Vowel Cipher of Charles L. Dodgson." *Cryptologia* XV (1991): 18–24.

Lotkin, Mark. "Note on the Method of Contractants." *American Mathematical Monthly* 66 (1959): 476–79.

Lovett, Charles. *Alice in North Carroll-ina: Exhibition Notes.* Winston-Salem: privately printed, 1989.

Lyell, Charles. "State of the Universities." *Quarterly Review* 36 (1827): 216–68.

Lyons, Henry. *The Royal Society 1660–1940.* Cambridge: University Press, 1944.

Macfarlane, Alexander. *Lectures on the British Mathematicians.* New York: John Wiley, 1916.

Macmillan, R. H. "A New Method for the Numerical Evaluation of Determinants." *Royal Aeronautical Society Journal* 59 (1955): 772.
Mallett, C. E. *A History of the University of Oxford,* vol. III. London: Methuen, 1927.
Montgomery, R. J. *Examinations: An Account of Their Evolution as Administrative Devices in England.* London: Longmans, Green, 1965.
"Mr. Dodgson on Parallels." Review of *Curiosa Mathematica. Part I: A New Theory of Parallels,* in *Nature* 39 (1888): 124.
Muir, Thomas. *The Theory of Determinants in the Historical Order of Development,* vol. I, II, III. London: Macmillan, 1906, 1911, 1920.
Nagel, Ernest. "Impossible Numbers: A Chapter in the History of Modern Logic." In *Studies in the History of Ideas.* New York: Columbia University Press, 1935. 429–74.
"Pillow Problems." Review of *Curiosa Mathematica Part II: Pillow Problems,* in *Nature* 48 (1893): 564.
Potts, Robert. *Euclid's Elements of Geometry.* London: Longman, Roberts, and Green, 1840.
Pycior, Helena. "At the Intersection of Mathematics and Humor: Lewis Carroll's *Alices* and Symbolic Algebra." *Victorian Studies* 28 (1984): 149–70.
"Rev. C. L. Dodgson." *Nature* 57 (1898): 279–80.
"Review of Euclid and His Modern Rivals." *Nature* 20 (1879): 240–1.
Richards, Joan L. "Projective Geometry and Mathematical Progress in Mid-Victorian Britain." *Studies in the History and Philosophy of Science* 17 (1968): 297–325.
Richards, Joan L. "The Art and the Science of British Algebra: A Study in the Perception of Mathematical Truth." *Historia Mathematica* 7 (1980): 343–65.
Richards, Joan L. *Mathematical Visions: The Pursuit of Geometry in Victorian England.* New York: Academic Press, 1988.
Richmond, D. E. "The Theory of the Cheshire Cat." *American Mathematical Monthly* 41 (1934): 361–68.
Riemann, Bernhard. "On the Hypotheses Which Lie at the Bases of Geometry." Trans. W. K. Clifford. *Nature* 8 (1873): 14–17, 36–37.
Robbins, David P., and Howard Rumsey, Jr. "Determinants and Alternating Sign Matrices." *Advances in Mathematics* 62 (1986): 169–84.
Roos, David A. "The Aims and Intentions of *Nature.*" In *Victorian Science and Victorian Values: Literary Perspectives,* edited by James Paradis and Thomas Postelwait, 159–80. New York: New York Academy of Sciences, 1981.
Seneta, Eugene. "Pascal and Probability." In *Interactive Statistics,* edited by D. McNeil. Amsterdam: North Holland, 1979, 225–33.
Seneta, Eugene. "Lewis Carroll as a Probabilist and Mathematician." *Mathematical Scientist* 9 (1984): 79–84.
Seneta, Eugene. "Lewis Carroll's 'Pillow Problems': On the 1993 Centenary." *Statistical Science* 8 (1993): 180–86.
Stern, Jeffrey, ed. *Lewis Carroll's Library.* Charlottesville: Lewis Carroll Society of North America/University Press of Virginia, 1981.
Sommerville, Duncan M. Y. *Bibliography of Non-Euclidean Geometry Including the Theory of Parallels, the Foundations of Geometry and Space of n-Dimensions.* London:

Harrison and Sons, 1911.

Taylor, Alexander L. *The White Knight: A Study of C. L. Dodgson (Lewis Carroll).* London: Oliver and Boyd, 1952.

Three Problems by Oedipus. London: Harrison, 1895.

Todhunter, Isaac, ed. *The Elements of Euclid.* 1862. Reprint. London: J. M. Dent and Sons, 1933.

Tucker, Robert. "Mr. Dodgson on Parallels." *Nature* 39 (1889): 175.

Turner, Frank M. "The Victorian Conflict between Science and Religion: A Professional Dimension." *Isis* 69 (1978): 56–76.

Wakeling, Edward. "An Assessment of *General List of Subjects* 1863, and *A Guide to the Mathematical Student* 1864." *Jabberwocky* 7 (1977–78): 9–18.

Ward, W. R. *Victorian Oxford.* London: Frank Cass, 1965.

Weaver, Warren. "Lewis Carroll and a Geometrical Paradox." *American Mathematical Monthly* 45 (1938): 234–36.

Weaver, Warren. *Lewis Carroll Correspondence Numbers.* Scarsdale, New York: Privately printed, 1940.

Weaver, Warren. "Alice's Adventures in Wonderland, its Origin and its Author." *Princeton University Library Chronicle* XIII (1951): 1–17.

Weaver, Warren. "The Mathematical Manuscripts of Lewis Carroll." *Proceedings of the American Philosophical Society* 98 (1954): 377–81.

Weaver, Warren. "Lewis Carroll: Mathematician." *Scientific American* 194 (1956): 118.

Weaver, Warren. "Description of the Mathematical Manuscripts of Charles Lutwidge Dodgson (Lewis Carroll) in the Morris L. Parrish Collection." Princeton Univ. Library. Typescript, 1980.

Willerding, Margaret F. "Mathematics Through a Looking Glass." *Scripta Mathematica* 25 (1960): 209–19.

Williams, Sydney H., et al. *The Lewis Carroll Handbook.* Folkestone/Hamden: Dawson, Archon Books, 1979.

Winkfield, Trevor, ed. *The Lewis Carroll Circular* no. 1, 2. Leeds: privately printed, 1973–74.

Young, G. M., ed. *Early Victorian England 1830–1865,* vol. 1. London: Oxford University Press, 1963.

Zeller, C. "Kalender-Formeln." *Acta Mathematica* IX (1887): 131–36.

Charles L. Dodgson's Mathematics and Logic Books

Dodgson, Charles L. *A Syllabus of Plane Algebraical Geometry, Systematically Arranged with Formal Definitions, Postulates, and Axioms.* Oxford: James Wright, 1860.

Dodgson, Charles L. *An Elementary Treatise on Determinants with Their Application to Simultaneous Linear Equations and Algebraical Geometry.* London: Macmillan, 1867.

Dodgson, Charles L. *Euclid, Book V. Proved Algebraically So Far As It Relates to Commensurable Magnitudes to Which is Prefixed a Summary of All the Necessary Algebraical Operations, Arranged in Order of Difficulty.* Oxford: James Parker, 1874.

Dodgson, Charles L. *Euclid, Books I, II.* Oxford: Printed for Private Circulation, 1875. London: Macmillan, 1882.

BIBLIOGRAPHY

Carroll, Lewis. *A Tangled Tale*. London: Macmillan, 1885. Reprint. New York: Dover, 1958.

Dodgson, Charles L. *Euclid and His Modern Rivals*. 2d ed. 1885. London: Macmillan. Reprint. New York: Dover, 1973.[1] (First edition 1879).

Carroll, Lewis. *The Game of Logic*. London: Macmillan, 1887. Reprint. New York: Dover, 1958.

Dodgson, Charles L. *Curiosa Mathematica. Part I: A New Theory of Parallels*. London: Macmillan, 1888.

Dodgson, Charles L. *Curiosa Mathematica. Part II: Pillow Problems*. 4th ed. London: Macmillan, 1895. Reprint. New York: Dover, 1958. (First edition 1893).

Carroll, Lewis. *Symbolic Logic, Part I*. 4th ed. London: Macmillan, 1897. Reprint. New York: Dover, 1958. (First edition 1896).

1. Includes all of the pamphlet, *Supplement to "Euclid and His Modern Rivals"* (London: Macmillan, 1885), except Appendix VII: Reviews of 'Euclid and His Modern Rivals' with the author's remarks thereon.

Index

Italicized page references signify illustrations. Titles not otherwise qualified are those of Dodgson.

"Abridged Long Division" (item 36), 227, 232, 298, 312–21
Acland, T. D., 11n10
Additivity, 206
Adjugate matrices, 165, 175
Aesthetic values, Dodgson's, 18, 233
Agnosticism, 5–6, 7
Alberti, Leon Battista, 325
Algebra, 155–98
　The Fifth Book of Euclid Treated Algebraically on, 30, 52
　in geometry texts, 28, 31
　linear, 22, 157, 158, 159, 166
　syllabus for, 354, 355, 358, 366–67, 376, 377, 378–79, 387–88
　symbolical, 23–24
Algebra (item 14), 140, 157, 195–96
Algebra (Chrystal), 151–52
Algebra of invariants, 14
Algebraical Formulæ (item 11), 157, 181–83, 184
Algebraical Formulæ and Rules for the Use of Candidates for Responsions (item 13), 157, 190–94, 235
Algebraical geometry, 121
　Curiosa Mathematica, Part II on, 32
　Dodgson's symbols for, 122, 123–24, 162
　syllabus for, 356, 358–60, 362–64, 364–65, 377–78, 379–81, 383–85
　A Syllabus of Plane Algebraical Geometry on, 349
Algorithms, 230, 233–34
Alice books, 165. *See also* Dodgson, Charles L., literary work
Almanacs, 280
Alphabet Cipher (item 39), 326–27, 328, 335, 341–43, 344
Analytic geometry. *See* Algebraical geometry
Analytical Engine, 334
Anglican clergy, 3–5, 9
Archibald, R. C., 213, 283n2, 291

Archimedean axiom, 19
Argles, Agnes, 344
Argles, Edith, 344
Aristotle, 12
Arithmetic, 353–54, 375
Arithmetic (item 25), 227, 271–72
Arithmetic computation and theory, 225–321
Arithmetic. I (item 23), 227, 266–68
Arithmetic. II (item 24), 227, 269–70
Arithmetical Formulæ and Rules for the Use of Candidates for Responsions (item 21), 227, 235–39
Askew, H. (journal correspondent), 229, 277
Association for the Improvement of Geometrical Teaching, 12
Astronomical computations, 170
Athenaeum (journal), 2, 349
Axioms, Euclidian, importance of for Dodgson
　tacit, 31–32, 42

Babbage, Charles, 334–35
Ball, R. S., 15, 16
Ball, Walter William Rouse, 230, 291, 302, 303
Baltzer, Heinrich Richard, 15, 16, 157, 162, 164n12
Basmakova, I. G., 158
Baynes, Robert Edward, 3
Beaufort, Francis, Sir, 326
Beaufort cipher, 327. *See also* Variant Beaufort cipher
Bell, Eric Temple, 16
Beltrami, E., 150
"Berechnung des Osterfestes" (Gauss), 302n1
Berkeley, George, Bishop, 218
Bernoulli, Jean, 117
Biddle, D., 203, 209, 213, 216–17, 285–86
Black, Duncan, 167n16
Book of Almanacs, The (De Morgan), 280n2, 310
Boole, George, 14, 201
Bourne, Professor, 285, 286–87

413

"Brief Method of Dividing a Given Number by 9 or 11" (item 33), 227, 230–31, 232, 233, 293–301, 312
Brill, J., 206, 221
British Association for the Advancement of Science, 2, 3–4, 13, 15, 16, 27
British Foreign Office, 335
British universities, Euclid read at, 10
Brock, W. H., 12n11

Cajori, Florian, 20–21, 118
Calculus
 differential, 218, 360–61, 366–68, 381–82, 388–89
 infinitesimal, 20–21
 integral, 364, 368–69, 385–86, 390–91
Calculus of finite differences, 361, 370, 382, 391
Calculus of variations, 370–71, 391–92
Calendar problems. *See* Days of the week; Easter-Day calculation
Cambridge University, 8, 30
Cantor, Georg, 19n24, 206
"Carroll, Lewis," as pseudonym, 167, 291
Cauchy, Augustin, 22n29
Cayley, Arthur, 3, 14, 15, 16, 18, 150
Chio, F., 157
Choice and Chance (Whitworth), 201
Christ Church (Oxford), 9, 10, 11, 349
Christianity and mathematics, Dodgson's view of, 5–8
Chronological problems. *See* Days of the week, Easter-Day calculation
Chrystal, Professor, 151–52
"Cipher-Poem," 344
Ciphers (secret writing), 24, 325–45. *See also* Alphabet Cipher, Beaufort Cipher, Key-vowel Cipher, Matrix Cipher, Telegraph Cipher, Variant Beaufort Cipher, Vigenère Cipher
Ciphers (zeroes). *See* Zero
Circle-squaring. *See* Squaring the circle
Circular to Mathematical Friends (item 41), 350–51
Clarendon Report (1864), 11
Classics, study of the, 11–12
Classmen, 11, 349
Clergy, 3–5, 9
Clifford, William K., 14, 15, 16
Coalition games, 167. *See also* Election procedures
Codes. *See* Ciphers
Cohen, Morton N., 6n4, 7n5, 8n6–7, 9n8, 121, 213n1, 273n1, 280n1, 293n2, 344n2
Collingwood, Bertram J., 232–33, 313–14, 319, 320
Collingwood, Mary, 121

Collingwood, Stuart D., xii, 18n23, 42, 164n11, 181, 190, 227, 233n4, 235, 326
Commensurable magnitudes, 52–87, 88n1, 206, 208, 220, 221–23
Complemental minors, 165
Computers, foreshadowed by Victorian cryptographic methods, 334, 335
Condensation of Determinants (item 10), 157, 162–66, 170–80
Congruence theory. *See* Modular arithmetic
Conic sections, 357, 378
Connected minors, 166
Conway, John H., 280n2
Corollaries, 31–2
 Corollary, definition of, 37
Crofton, M. W., 207n8
Cryptographie Militaire, La (Kerckhoffs), 327
Cryptology, 323–45
Curiosa Mathematica, Part I, 10, 27, 34
 on algebra, 157
 on Euclid's tacit assumptions, 42
 frontispiece of, *33*
 non-Euclidean geometries and, 13
 on parallels, 18, 22
Curiosa Mathematica, Part II, 32
 on arithmetic, 227
 on probability, 201–2
 on puzzles, 169
 on Question 9636, 283
 on trigonometry, 117, 118
"Curiosa Mathematica, Part III," 227, 233, 273, 298, 312
Curriculum, 347–404
Cuthbertson, Francis, 42
Cyclical majorities, 166, 168, 202. *See also* Dodgson's function
Cyclical transposition, 174
Cyclostyled sheets, Dodgson's use of, 140, 197

Darwin, Charles, 2. *See also* Victorian era
Davis, George, 293–94
Davis, Nellie, 294
Days of the week, 229–30, 280–82, 291–92. *See also* Easter-Day calculation
"De binis quibuslibet functionibus homogeneis secundi ordinis . . ." (Jacobi), 170
De Morgan, Augustus, *23*, 29, 118, 201, 280n2, 310
Deacon's Orders, Dodgson taking, 9
Dedekind, R., 19n24
Delastelle, Felix, 327
Denumerable sets, 206–7
Determinants, *160–61*, 162–66, 170–80
Dice games. *See* Random experiments
Dickson, L. E., 232n3, 298n1

Differential calculus. *See* Calculus
"The Direction-Theory as Applied to Pairs of Lines," 22
Discussion of the Various Methods of Procedure in Conducting Elections, 169 n21
Disquisitiones Arithmeticae (Gauss), 232
Divisibility, 230–33, 285
  Archdeacon Dodgson on, *226*, 229, 277–78
"Divisibility by Seven" (item 27), 227, 229, 231, 232, 277–79
Dodgson, Charles, Archdeacon (father of Charles L. Dodgson), *228*
  on divisibility, *226*, 229, 277–78
  Pusey and, 9
  at Ripon Cathedral, 52
  on series summation problem, *20–21*
Dodgson, Charles L.
  Axioms, importance of, 6, 17–18
  Backward reasoning, 32, 165–6, 202, 231
  creativity in work of, xi, 10, 22, 24, 157, 159, 234
  lectureship at Christ Church, Oxford, 8, 10, 11, 349
  correspondence, 2, 3, 5–9, 118–19, 201, 291, 350
  literary work, xi–xii, 158, 164
  logic, importance of, xi, 1, 13, 18, 24, 150, 157, 312
  and his mathematical colleagues, 3, 16, 349
  mathematical views, 1, 3, 12–13, 15, 17, 19, 22, 24, 119, 122, 162–63, 166, 229, 233, 280, 349
  notation, system of, 117–18, 131, 159, 162
  paradoxical (fallacious) reasoning, 1, 24, 32 n13, 33–34, 205, 208
  philosophical views, xi–xii, 13, 17
  prose style, 1, 24, 158–59
  pseudonym, use of, 167, 291
  religious views of, 5–9
  sense of humor, 24, 207
  studentship, 9
  and students, preparation of, 10, 27, 30–31, 35, 42, 52, 88, 140, 190, 235, 349–50, 372
  working style of, 3, 24
Dodgson's function, 166
Dodgson's rule, 280 n2, 302 n2
Dummett, Michael, 166
Dwyer, P. S., 166

Easter-Day calculation, 302–11. *See also* Days of the week
Educational Times articles, 233, 273–76
Educational Times Questions
  7695 (item 17), 201, 203–5, 207–8, 209–13, 214, 219, 221
  8200: 205–6, 213–18
  8861: 206, 221, 222
  9588 (item 20), 201, 206–7, 221–23
  9636 (item 29), 227, 283–84
  9995 (item 16), 157, 169, 198
  11530 (item 9), 117, 119–20, 144, 150–54
  12650 (item 30), 227, 285–87
  13614 (item 32), 227, 230, 291–92
Election procedures, 166, 167, 168, 169 n21, 202–3
Electric pen, Dodgson's use of the, 140, 195
Electric telegraph, 325
Elementary Geometry (J. M. Wilson), 34 n15
Elementary Treatise on Determinants with Their Application to Simultaneous Linear Equations and Algebraical Geometry, 157, 158–59, 163–64
Elements of Geometry (Halsted), 16
Ellis, A. J., 217
English Mechanic (journal), 227
Enumerative combinatorics, 166
Enunciations of Euclid, I, II, The (item 2), 27, 30, 31, 42–51, 88
Enunciations of Euclid I–VI, The (item 4), 27, 31–32, 42, 88–113
Euclid and His Modern Rivals, 12, 13, 27, 32–34
  appendix to, 16
  on Cuthbertson, 42
  "The Direction-Theory as Applied to Pairs of Lines" and, 22
  Heath on, 34 n15
  Supplement to, 17, 27
Euclid, Book V, 27, 88 n1
Euclid, Books I, II, 26, 27
Euclidean geometry, 6–7, 10, 13, 23, 25–113, 128. *See also* Geometry
Euclid's Elements, 10, 12, 14, 27–28
  Books I–VI: 88–113
  Books I–II: 11, 30–31, 35–41, 355, 375, 376
  Books III–IV: 356, 376
  Book V, 19: 28–29, 30, 52–87
    Smith on, 27
    syllabus for, 356, 377
  Book VI: 30, 356, 377
  Book VII: 28
  Book X: 19
  Book XI: 362, 383
  Book XII: 362, 383
Euler's series, 118, 137
Examinations, university, 12
  at Cambridge, 30
  at Oxford, 10–11, 52, 88
  templates for, 266–68, 269–70, 271–72
Examples in Arithmetic (item 22), 227, 240–65, 266

Fairness in probability problems
  in proportional representation, 22, 167, 168
  in tournaments, 22, 167–69
Fifth Book of Euclid Treated Algebraically, The
  (item 3), 27, 29–30, 42, 52–87, 88, 158
Finite incommensurables. *See*
  Incommensurable magnitudes
Fishburn, Peter, 166
Fluxions. *See* Calculus
Formal Logic (De Morgan), 201
Formulæ (item 15), 157, *197*
Formulæ (Group C) (item 6), 140–43
Formulæ in Algebra (item 12), 157, 181, 184–89
Formulæ of Plane Trigonometry, The (item 5),
  117, 118, 121–39
"Formulæ of Pure Mathematics" (projected),
  117
Foster, W. S., 283, 284
Fowler, Thomas, 3, 29
Franksen, Ole I., 335n12
Free will, Dodgson on, 6

Games. *See* Tournaments; Number-guessing
  puzzles; Puzzles
Gardner, Martin, 32–33n14, 164n9, 280n2,
  302n2
Gauss, Carl Friedrich, 24, 120, 151–54, 232,
  302, 303
Geheimschriften und die Dechiffrir-Kunst, Die
  (Kasiski), 327
*General List of Subjects, and Cycle for Working
  Examples,* 121, 353
Geometry
  algebraical. *See* Algebraical geometry
  Euclidean, 6–7, 10, 13, 23, 25–113, 128. *See
    also* Euclid's Elements
  Greek, 28, 32
  non-Euclidean, 6, 13, 14, 15–18, 19
  plane. *See* Plane geometry
  projective, 14
  solid. *See* Solid geometry
Glaisher, J. W. L., 13n14, 159n5
God, 2, 7
Goniometry, 118, 122–26, 127, 132–39
"Great-go" examination, 10–11
Greats school (Oxford), 10
Greek culture, 12
Greek geometry, 28, 32
Green, Roger L., 29n7, 30n8, 42n1, 52n2,
  88n1, 162n8, 170n1, 181n1, 277n2,
  280n1, 285n1, 293n1, 326n1, 327–28,
  328n2, 4, 344n1, 349n1, 352n1, 372n2
Greenstreet, W. J., 203, 209
Gregory, James, 118
Gregory's series, 118, 136

Groups, abstract, 329
*Guide to the Mathematical Student in Reading,
  Reviewing, and Working Examples, A*
  (item 43), 140, 195, 197, 349, 372–404
Guiliano, Edward, 167n16

Hadamard, J., 150
Halsted, George Bruce, 15, 16
Harrison, Francis, 350
Heath, Thomas, 28n3, 29n4, 5, 30n9, 32,
  34n15
Heiberg, J. L., 27n2
Helmholtz, Hermann von, 15, 16
Henrici, Olaus, 14, 15, 16
Hermite, C., 150, 157
Herz-Fischler, Roger, 31
Hippolytos (3d cent.), 231–32
Hirst, Thomas Archer, 14
History of Probability (Todhunter), 201
Holy Orders, Dodgson's attitude toward, 9
Hoüel, J., 164n12
Hughlings, J. P., 201
Hunt, George Ward, 327

Incommensurable magnitudes
  omitted in *The Fifth Book of Euclid Treated
    Algebraically,* 29–30, 52–53, 56n10,
    57, 58n13
  Question 9588 on, 206, 221–23
  Response to "Infinitesimal or Zero?" on, 217
  "Something or Nothing" on, 208, 220
"Incommensurable Quantities" (Ellis), 217
"Infinitesimal or Zero?" (Miller *et al.*), 205–6,
  207, 213–18
Infinitesimals, 18–21, 22n29, 34, 214, 216, 219
Infinities, 19, 206
Integral calculus. *See* Calculus
*Introduction to Boole's "The Laws of Thought,"
  An* (Hughlings), 201
Inversions, 202–3
Irrational numbers, 206, 207

Jabberwocky (journal), 350, 372
Jacobi, Carl G. J., 164, 170
Johnston, R. H., 164n9
*Journal für die Reine und Angewandte
  Mathematik,* 170

Kahn, David, 335n13
"Kalender-Formeln" (Zeller), 230n2
Kasiski, Friedrich, 327, 334
Katz, Victor, 118n3
Kerckhoffs, Auguste, 327
Key-Vowel Cipher (item 37), 332–34, 336–37
Klein, Felix, 14

Kline, Morris, 118n3
Knowledge (journal), 32–3n14, 229, 277
Knowles (journal correspondent), 213, 217–18
Knuth, Donald, 167, 168
Kolmogorov, A. N., 158, 205, 206
Kronecker-Capelli theorem, 158
Kruskal, William, 169n23

Lady Margaret Hall (Oxford), 8
Lambert, Johann, 118
Laplace's expansion theorem, 162, 163
Lawn tennis tournaments. *See* Tournaments
Lawn Tennis Tournaments, 168–69
Leibniz, Gottfried Wilhelm, 20–21, 162
Lennon, Florence Becker, 16n20
Leonardo of Pisa (Leonardo Pisano), 232
Lewis Carroll Circular, The, 288, 336, 339, 344
Lewis Carroll Handbook, The
  on "Abridged Long Division," 312
  on *Algebraical Formulæ*, 181
  on *Algebraical Formulæ and Rules*, 190
  Arithmetic, II not included in, 269
  "Divisibility by Seven" not included in, 277
  on *Educational Times* contributions, 150
  on *The Enunciations of Euclid, Books I and II*, 42
  on *Formulæ in Algebra*, 184
  on "Note on Question 7695," 209
  on Question 9588, 221
  Response to "Infinitesimal or Zero?" not included in, 213
  on *Three Problems by Oedipus*, 117n1
Lewis Carroll Picture Book, The (Collingwood), 18n23, 32, 227, 273, 298, 312
Liber Abbaci (Leonardo of Pisa), 232
Library, contents of Dodgson's, 18, 201
Liddon, Henry Parry, Dr., 9
Lindemann, Ferdinand, 118
Linear algebra. *See* Algebra
Linear algebraical geometry. *See* Algebraical geometry
Lipson, Stanley, 325, 335n13, 339n1
Literary work, Dodgson's, xi–xii, 158, 164
Lobatschewsky, N. (Lobachevski, Nicholas), 17
Logic of Chance, The (Venn), 201
London Mathematical Society, 2, 15, 16, *23*
"Long Multiplication." *See* Algorithms
Lotkin, Mark, 164n9
Lovett, Charles, 140

MacColl, Hugh, 206, 213
Macdonald Conjecture, 166
Machin, John, series, 118, 137
MacMillan, R. H., 164n9
Madan, Falconer, 146n1, 277, 326
Many Papers and Solutions in Addition to Those Published in the "Educational Times." See *Educational Times* Questions
Mathematical Questions and Solutions, from the "Educational Times." See *Educational Times* Questions
Mathematical Recreations and Problems (W. W. R. Ball), 230, 291, 302, 303
Matrices, 159, 162–63, 165, 166, 170–80
Matrix Cipher (item 38), 328–30, 332, 334, 335
"Memoria Technica. for Numbers," 280n1
Méré, Antoine Gombaud, Chevalier de, 205
Method of Taking Votes on More Than Two Issues, A, 168
Miller, William John Clarke, 213
Mind (journal), 16
Minimum-comparison selection, 167
Mirror images, 165–66
Mnemonic systems, 280n1, 307–9
Moderations, 10, 11, 52, 88
Modular arithmetic, 329, 332
Monalphabetic cipher techniques, 326, 327. *See also* Ciphers
Monatliche Correspondenz, 302–3
"Monkey and Weight Problem, The" 227
Motion, Law of, 227
Muir, Thomas, 157–58, 164n12

Nash, Professor, 283, 284
Nature (journal), 16, 229, 233
  "Abridged Long Division," 227, 232, 312
  "Brief Method of Dividing a Given Number by 9 or 11," 227, 293, 298
  "To Find the Day of the Week for any Given Date," 227, 280, 291
Neumann, John von, 335
Newnham College (Cambridge), 8
Newton, Isaac, Sir, 21, 227
Non-Euclidean geometries. *See* Geometry
Notation, Dodgson's system of, 117–18, 131, 159, 162
"Note on Question 7695" (item 17), 203–5, 207, 209–12, 213, 221
Notes on the First Part of Algebra, 121, 157
Notes on the First Two Books of Euclid (item 1), 10, 27, 30–31, 35–41, 52, 121
"Number-Guessing" (item 31), 227, 285, 288–90
Number-guessing puzzles, 285, 288–90

"On Direct Probabilities" (De Morgan), 201
"On Inverse Probabilities" (De Morgan), 201

"On Systems of Linear Indeterminate Equations and Congruences" (Smith), 159
"On the Hypotheses Which Lie at the Bases of Geometry" (Riemann), 16
On the Origin of Species (Darwin), 2
O'Regan, J., 201, 203, 204, 209
Oxford University, 3, 8, 9, 10–11, 349

"Pairs of Lines Treated on Direction-Theory," 1n
Pall Mall Gazette, 2
Pappus of Alexandria, 32
Parallel axiom, alternative to, 34
Parallel lines, 14, 17, 18, 22
Pascal, Blaise, 204–5
Passmen, 11, 29–30, 53, 88
Peacock, George, 23, 157
Peirce, B., 150
Penny Cyclopaedia, 29
Permutation matrices, 166
Philosophical Transactions of the Royal Society, 14
π, 24, 117
   circle-squarers and, 118–19
   Question 11530 on, 119–20, 150–54
   Simple Facts about Circle-Squaring on, 146–47
"Pillow Problems." See *Curiosa Mathematica, Part II*
Pisano, Leonardo, 232
Plane geometry. See also Euclidean geometry; Algebraical geometry
   Curiosa Mathematica, Part II on, 32
   syllabus for, 357, 379
Plato, 12
Playfair's axiom, 17, 34n15
Polyalphabetic ciphers, 325–26, 327. See also Ciphers
"Practical Hints on Teaching" (item 26), 227, 233–34, 273–76
Preference behavior, 169, 203. See also Election procedures
Price, Bartholomew, 3, 4, 11, 20–21, 162
Probability, 6, 7, 22, 199–223
Proceedings of the Royal Society, 170
Proclus (5th cent.), 32
Proctor, Richard A., 32–33n14
Professionalization of science, 3
Projective geometry. See Geometry
Proof Sheets Accompanying the "Circular to Mathematical Friends" (item 42), 353–71
Proof Sheets: Propositions I, II (item 8), 148–49

Proportional representation, 23, 167, 169. See also Election procedures
Proportionality, 14, 19, 28–29
The Fifth Book of Euclid Treated Algebraically on, 30, 52–87
Public Schools Commission Report, 11
Pusey, Edward Bouverie, Dr., 9
Puzzles, xi–xii, 32, 168, 227, 285, 288, 302. See also Number-guessing puzzles
Pycior, Helena, 24

Questions (*Educational Times*). See *Educational Times* Questions

Radhakinshuan, Professor, 285–86
Random experiments, 203–4, 210–11
   Response to "Infinitesimal or Zero?" on, 205–6, 213–18
   Question 9588 on, 206–7, 221–23
   "Something or Nothing?" on, 219–20
Ranking processes, 166. See also Election procedures
Rational numbers, 206, 207
Ratios, 19, 28, 30, 52
Rectangular arrays. See Matrices
Religion, 2, 8, 13. See also Christianity
Response to "Infinitesimal or Zero?" (item 18), 213–18
Responsions, 10, 11
Richard L., 312–13
Richards, Joan L., 15n16
Richardson, G., 350
Rickard (mathematician), 277
Riemann, George Friedrich Bernhard, 16
Riemannian space, 30n9
Ripon Cathedral, 52
Rix, Edith, 8
Robbins, David P., 166
Row transposition, 163. See also Zero in matrices
Royal Society, 2–3, 162, 170
   clerical members of, 4
   Philosophical Transactions, 14
   Proceedings, 170
   Spottiswoode and, 15, 156
   Todhunter and, 16
Rudakov, A. N., 158
"Rule for Finding Easter-Day for Any Date till A.D. 2499" (item 35), 227, 302–11
Rumsey, Howard, Jr., 166n14

St. Clair, J. C., 198
St.-Cyr cryptographic system, 327
St. James's Gazette, 167
Salmon, George, 14

INDEX 419

Sampson, Edward Frank, 3, 288, 349
Sang, Alfred, 312
Saturday Review, 17
Schering, C., 153, 302 n1
Science, 2, 8
Scientific organizations, 15
Self-verification, Dodgson's need for, 6–7
Seneta, Eugene, 158–59, 165–66, 201, 202
Series summation, xviii, *1, 20–21*
Sestri, Giovanni, 326
Shanks, W., 120, 152
Shenitzer, Abe, 158 n3
Simmons, Thomas Charles on
    Question 7695: 204, 209, 211–12, 213
    Question 8200: 205, 213, 214–16, 218
    Question 9588: 207, 221, 223
Simple Facts about Circle-Squaring (item 7), 144–47
Simson, Robert, 27–28
    The Enunciations of Euclid I–VI and, 32
    The Fifth Book of Euclid Treated Algebraically and, 29, 30, 42
    Notes on the First Two Books of Euclid and, 31
    Todhunter and, 30
Singmaster, David, 32–33 n14
"Sixth Memoir upon Quantics" (Cayley), 14
Smith, Henry John Stephen, xviii, 3, *5,* 16–17
    on Euclid's Book V, 27
    on geometry, 13
    non-Euclidean geometries and, 18
    "On Systems of Linear Indeterminate Equations and Congruences," 159
    on series summation problem, *1*
Solid geometry, 361, 383. *See also* Algebraical geometry
Somerville Hall (Oxford), 8
"Something or Nothing" (item 19), 206, 207, 208, 219–20, 222
Sonne, Otto, 298, 313
Spottiswoode, William
    Condensation of Determinants and, 162, 170
    on determinants, *160–61*
    projective geometry and, 14
    as Royal Society president, 15, *156,* 162
    Sylvester and, 3
Squaring the circle, 118–19, 144–47, 217
Steen, Adolph, 298, 313
Stenarithmie (Richard), 312–13
Stern, Jeffrey, 16 n18, 201 n1
Stored program, 335
Suggestions as to the Best Method of Taking Votes, When More Than Two Issues Are to Be Voted On, 169 n21
Summation of series. *See* Series summation

Supplement to Euclid and His Modern Rivals, 17, 27
Syllabi, 347–404
Syllabus of Plane Algebraical Geometry, A, 10, 27, 117, 121, 122, 349
Sylvester, James Joseph, 2, 3, 14, 150
Symbolical algebra. *See* Algebra

Tangents, theorem of, *116*
Tanner, Professor, 206, 207, 221, 223
Telegraph, 325. *See also* Victorian era
Telegraph Cipher (item 40), *324,* 326–27, 328, 335, 344–45
Tennis tournaments. *See* Tournaments
Theater, Dodgson's love of the, xi, 8–10
Theorie und Anwendung der Determinanten (Baltzer), 16, 164 n12
Theory of Determinants in the Historical Order of Development, The (Muir), 157–58, 164 n12
Three Problems by Oedipus (anonymous), 117 n1
"To Find the Day of the Week for any Given Date" (item 28), 227, 229–30, 280–82, 291
Todhunter, Isaac, *15,* 30
    Condensation of Determinants and, 157
    on Euclid I, 11, 31
    on geometry, 16
    History of Probability, 201
    on Simson, 27 n2
Tournaments, 167–69
Traicté des Chiffres (Vigenère), 325
Traité Élémentaire de Cryptographie (Delastelle), 327
Transitive preferences, 169. *See also* Tournaments
Treatise on Algebra (Peacock), 157
Trigonometry, 115–54, 360, 380–81
    symbolical, 117–18, 122–26, 131
Triple-elimination tournaments. *See* Tournaments
Tucker, Robert, 15, 16
Turner, Frank M., 4 n3

Universities, Euclid read at British, 10
University College London, 14
Unknown Lewis Carroll, The (Collingwood). *See The Lewis Carroll Picture Book*

Variant Beaufort cipher, 329, 330, 331–35
Venn, John, 201
Victorian era, 2–3, 15 n16, 18, 24, 202, 325, 334
Vigenère, Blaise de, 325
Vigenère cipher, 325–26, 327, 329, 331, 333

Von Neumann, John, 335
Voting procedures. *See* Election procedures

Wakeling, Edward, 372 n1
Wallace, William, 120, 151
Weaver, Warren, 16, 17, 22 n29, 32 n13, 293 n1
Weekdays. *See* Days of the week
Whitworth, W. A., 201
Wiener, L., 198
Wilberforce, Samuel, Bishop, 8
Wilcox, W. M., 8–9
Wilson, J. M., 34 n15
Wilson, John Cook, 3
Winkfield, Trevor, 336, 339, 344
Woodall, H. J., 120, 151–54

Yushkevich, A. P., 158

Zach, Franx Xaver, Freiherr von, 302–3
Zeller, Christopher, 230, 291
Zermelo, Ernst, 168
Zero
  and infinitesimals, 214, 216, 219
  in matrices, 163–64, 172, 173–75, 179
Zero probability, 204, 207
  Queston 8200 on, 205–6, 215, 216, 218
  Question 9588 on, 222–23
  "Something or Nothing?" on, 219–20
Zero-sum games, 167
Zerr, G. B. M., 283, 284

LIBRARY